Transition Metal Clusters

Dr. B. F. G. Johnson and R. E. Benfield,
University Chemical Laboratory,
Lensfield Road,
Cambridge CB2 1EW, U.K.

Dr. P. R. Raithby,
University Chemical Laboratory,
Lensfield Road,
Cambridge CB2 1EW, U.K.

Dr. K. Wade,
Department of Chemistry,
University of Durham,
Science Laboratories,
South Road,
Durham DH1 3LE, U.K.

Dr. R. Whyman,
ICI Corporate Laboratory,
P.O. Box 11,
The Heath,
Runcorn,
Cheshire WA7 4QE, U.K.

Dr. R. G. Woolley,
Department of Physics,
Cavendish Laboratory,
University of Cambridge,
Madingley Road,
Cambridge CB3 OHE, U.K.

List of Contributors

Dr. J. A. Connor,
    Department of Chemistry,
    The University of Manchester,
    Manchester M13 9PL, U.K.

Dr. A. J. Deeming,
    Department of Chemistry,
    University College London,
    20 Gordon Street,
    London WC1H OAJ, U.K.

Dr. C. D. Garner,
    Department of Chemistry,
    The University of Manchester,
    Manchester M13 9PL, U.K.

Dr. B. F. G. Johnson,
    University Chemical Laboratory,
    Lensfield Road,
    Cambridge CB2 1EW, U.K.

Acknowledgements

I am grateful to a number of my colleagues for reading the Chapters of this book and for their suggestions and corrections.

We are especially indebted to Miss Sally Stephenson for typing the final manuscript for the camera ready printing process and to Mrs. Cheryl Cook for all drawings and lettering.

British Library Cataloguing in Publication Data:

Transition metal clusters.
    1. Transition metal compounds
    I. Johnson, Brian Frederick Gilbert
    546'.3   QD172.T6   80-40496

    ISBN 0 471 27817 3

Printed in Great Britain

Transition Metal Clusters

Editor: **B.F.G. Johnson,**
University Chemical Laboratory,
Lensfield Road,
Cambridge,
England

A Wiley–Interscience Publication

JOHN WILEY & SONS
Chichester · New York · Brisbane · Toronto

Contents

Preface

The last ten years has seen a dramatic increase in interest in the chemistry of cluster compounds, especially those of the transition metals containing the carbonyl, hydrido- and halogeno-ligands. Principles underlying the synthesis, structure, bonding stereochemistry and chemical reactivity of such compounds have been established. Studies of their chemical reactivity have helped considerably in our understanding of surface phenomena and the ability of higher rhodium carbonyl clusters to function as active catalysts for the specific reduction of carbon monoxide to ethylene glycol has excited the attention of the chemical industry. It would appear, therefore, that a text devoted to the subject of Transition Metal Clusters is justified.

The purpose of this book, the first devoted to the subject, is twofold. First, as a source of reference to workers in the field and, secondly, as a text for honours undergraduates or graduate students specialising in the subject.

The book is divided into nine main chapters. Chapter 1, which serves as a general introduction to the field and as a guide to the remaining chapters, is devoted to simple concepts, definitions and principles which underpin the subject. Chapter 2, a review chapter, is concerned with the structures of the transition metal clusters with emphasis on the geometries adopted by simple polymetal aggregates.

We have chosen in Chapter 3 to review the chemistry of the so-called cubane compounds. Strictly, many of these compounds are not clusters according to the definitions offered by Cotton and others but represent an area of considerable importance and may be viewed from many points of view as members of the cluster class.

In Chapter 4, principles underlying the structure and bonding of cluster compounds are described and in Chapter 5, bond enthalpies, their determination and utilisation, are discussed. An area of considerable research activity, stereo-dynamic or fluxional behaviour of clusters, is classified and reviewed in Chapter 6. The chemical reactivity of cluster compounds, a subject still in its infancy, is reviewed in Chapter 7, and in Chapter 8 their application in industrial processes assessed. Finally, in Chapter 9, "Electrons in Transition Metal Cluster Carbonyls", the views of a non-chemist are presented.

We hope that this book will present a balanced view of transition metal cluster chemistry. We expect and look forward to constructive criticism.

Brian F. G. Johnson
20th September, 1979

Transition Metal Clusters
Edited by Brian F.G. Johnson
© 1980 John Wiley & Sons Ltd.

CHAPTER 1

Introduction

B. F. G. Johnson
University Chemical Laboratory,
Cambridge, U.K.

Much of the progress in transition-metal cluster chemistry has been made since the mid-1960's. Many of the advances could not have been made without the development of spectroscopic methods and more effective means of structure determination. However, the preparation of clusters remains almost entirely an accidental affair and few systematic routes have been devised.

A number of classes of binary clusters have been established. The binary carbonyls form the largest and most well-defined group of compounds. Within this class cluster sizes ranging from three to thirty and a wide range of cluster polyhedra have been established. To a large extent these polyhedra are deltahedra and as such may be compared to the clusters formed by the main group elements. Others contain square faces and several square planar, square based pyramidal and trigonal prismatic clusters are known. An especially attractive cluster, $Rh_{13}(CO)_{24}H_3{}^{2-}$, contains an Rh_{13} unit in which the thirteen rhodium atoms form a body-centred cuboctahedron which is the smallest fragment of the hexagonal close-packed metallic structure. The majority of smaller clusters represent smaller, sub-fragments, of this arrangement. Larger rhodium clusters

such as $Rh_{14}(CO)_{25}{}^{4-}$ apparently correspond to fragments
of body-centred cubic packing and the capped metal
polyhedra observed for the series of compounds $Os_7(CO)_{21}$,
$Os_8(CO)_{23}$ and $Os_{10}C(CO)_{24}{}^{2-}$ represent a steady progression
towards this same packing arrangement. In contrast, the
platinum clusters $[Pt_3(CO)_6]_n{}^{2-}$ (n = 2, 3, 4, 5, 6 and
10) bear no resemblance to the bulk metal form. These
anions form helical chains which are constructed from
stacked trigonal prisms. No less remarkable is the
platinum cluster $Pt_{19}(CO)_{12}{}^{2-}$ which possesses a five-
fold, rather than the six-fold symmetry commonly found
for metallic structures.

Another class of cluster is that formed by the
halides of the second and third-row transition-metals.
These span much of the transition-block but the most
extensively studied are those containing triangular
and octahedral metal units. Related to these are the
cluster compounds found in oxide and alkoxide systems.

Several definitions of a cluster compound have
been offered. For the purposes of this book we define
a cluster compound as a discrete unit containing at
least three metal atoms in which metal-metal bonding
is present. However, such a definition does not
include a range of compounds commonly known as cubanes
but which are included in this book. All have in common
a central cube M_4X_4 in which there may be strong, weak
or no metal-metal bonding interaction. They are of
considerable importance particularly with regard to
their relevance to the redox behaviour of biological
systems.

Theories of bonding within metal cluster compounds
remain primitive. Attempts to rationalise the bonding
interactions within discrete groups of transition-
metal ions are necessarily of a qualitative nature
because of the complexity of the problem. A number of

empirical rules have been applied with varying degrees of success. One of the more successful treatments has been the Skeletal Electron Pair Theory developed by Wade. A number of semi-empirical LCAO-MO treatments have also been applied to specific compounds by Mingos and Fenske. A general M.O. treatment of clusters has been made by Lauher and the interactions of cluster fragments examined by Hoffman. Bonding interactions in basic clusters have also been examined since these may serve as models for surface and bulk-metal phenomena and within this group of studies are numerous LCAO-MO calculations and the more sophisicated Xα treatments. Within this book we have, in general, restricted our approach to one of systematics and generalisation.

Although a large number of cluster compounds are now available, studies of their chemical reactivity have been limited and largely restricted to a few simple compounds. There is a view that molecular metal clusters may be reasonable models of metal surfaces and that studies of the reactions of small substrates may shed light of the interactions which occur on the metallic surface. Information exists which tends to support this view. The bonding modes which CO, H_2 and small organic molecules adopt on reaction with small clusters may be identified with certainty and their spectroscopic properties often identified with those of chemisorped species. However, the extent of this information is limited and restricted largely to observations made on clusters containing few metals (three or four). It is apparent that the reactions involving two or more metal centres are, in general, very different to those observed with the more conventional monometal systems. The idea that catalysis in the heterogeneous systems may be understood in terms of mechanisms established for monometal homogeneous systems is no longer valid. The formation of carbide-species, for example, is a moderately common feature in the chemistry of cluster carbonyls but is rarely observed for monometal systems.

This approach has also led to the view that clusters themselves may function as catalysts for a wide range of chemical processes. Certainly, many reactions are catalysed on addition of cluster compounds but whether or not the cluster retains its integrity throughout the reaction is a debatable point. Commercial interest in clusters remains high and is discussed in Chapter 8.

Our understanding of cluster chemistry is to some extent limited by the scarcity of good thermochemical data. Although a few trends have been established the values placed on such important terms such as metal-metal bond energies are, to a large extent, crude approximations. Yet, despite the value of such measurements, few laboratories are actively pursuing this line of enquiry.

In this book the various facets of cluster chemistry and other related topics such as stereochemical non-rigidity are considered. In part, we have chosen to be comprehensive and to set out _albeit_ in an abbreviated form much of the information at present available. Elsewhere, emphasis has been placed rather more on views and opinions. This is particularly the case in areas where the subject is less well understood or where information is sparse. We hope to have achieved a balanced view and at least to have considered most of the important areas of investigation.

Transition Metal Clusters
Edited by Brian F.G. Johnson
© 1980 John Wiley & Sons Ltd.

CHAPTER 2

The Structure of Metal Cluster Compounds

P. R. Raithby

*University Chemical Laboratory,
Cambridge, U.K.*

1. INTRODUCTION

At the time of writing there are nearly one
thousand cluster structures in the literature, and most
of these contain transition metal atoms. They can be
found principally among two classes of compounds, low
oxidation state metal halides and metal carbonyls.
Clusters are common in the lower halide chemistry of
niobium, tantalum, molybdenum and tungsten, and in the
carbonyl chemistry of elements of the iron, cobalt and
nickel subgroups. There is also a small but significant
number of clusters of the elements at the extreme right
of the d block which are stabilised by phosphorus donor
ligands.

Numerous methods have been used to elucidate the
structures of these clusters both in the solid and
solution states. Although i.r. spectroscopy is often
used as a fast method for characterising or identifying
known complexes, mass spectroscopy, n.m.r. spectroscopy
and X-ray diffraction have proved the most useful
techniques in the determination of their molecular
structures. In the solid state, particularly in the
case of clusters containing between four and twenty
metal atoms, single-crystal X-ray diffraction has been
found to be the most rewarding technique. However, in
recent years the application of single-crystal neutron

8.

diffraction to hydrido-cluster compounds, and to those where the X-ray scattering is dominated by the heavy metal contribution, has yielded valuable information not readily obtainable from X-ray data alone.

In this chapter we shall discuss the molecular geometries adopted by metal cluster compounds in the solid state and make some generalisations on the cluster conformations.

2. STRUCTURAL SURVEY OF METAL CLUSTER COMPLEXES

In this section we shall survey some of the more interesting types of clusters, most of which have been characterised by crystallographic techniques. As a large proportion of clusters contain transition metals, we shall deal with the d block group by group and follow this by a discussion of mixed clusters, that is those containing different metallic elements.

Table 1 shows the d block elements. The hatched squares indicate those metals for which clusters have been structurally characterised; the dotted ones show where the elements have been found to be involved in mixed clusters.

TABLE 1

Sc	Ti	V	Cr	Mn	Fe	Co	Ni	Cu	Zn
Y	Zr	Nb	Mo	Tc	Ru	Rh	Pd	Ag	Cd
La	Hf	Ta	W	Re	Os	Ir	Pt	Au	Hg

2.1. Titanium, Zirconium and Hafnium

The elements on the extreme left of the d block
do not, as a general rule, form cluster complexes.
There is infrared evidence for the existence of a Ti_3
triangle in the Ti_3Cl_{12} species, and it is probable
that triangular moieties exist in the compound
$[M_3(C_6Me_6)_3Cl_6]Cl$ (M = Ti, Zr) (1). The only
structurally characterised example for titanium is the
hexanuclear complex, $(\eta^5-C_5H_5)_6Ti_6O_8$ (2). The six Ti
atoms define a regular octahedron with an average
Ti-Ti distance of 2.891 Å. This is significantly
shorter than the metal-metal distance of 2.951 Å in
α-Ti, and is indicative of direct Ti-Ti interaction,
although there is delocalised Ti-O-Ti bonding involving
the eight oxygen atoms which cap the faces of the
octahedron. Each of the six cyclopentadienyl groups
coordinates to one metal atom as shown in Figure 1.

2.2. Vanadium, Niobium and Tantalum

Currently, there are no known clusters of vanadium
which have been structurally characterised.
By far the most common geometry observed in the
cluster chemistry of niobium and tantalum is the
octahedron. This is found in a number of halide cluster
units with the general formula $[M_6X_{12}]^{n+}$, where the
twelve halogen atoms, X, act as μ_2-bridging groups and
are located over the centres of the twelve edges of
the metal octahedron (Figure 2). These species are

generally the core of a larger unit with the formula
$[(M_6X_{12})X_6]^{n-}$. The additional six halogen atoms
coordinate terminally to the metals, as is illustrated
in Figure 3 for the $[(Nb_6Cl_{12})Cl_6]^{2-}$ cluster (5).
Table 2 summarises the structural data for these
systems.

In these complexes there is usually little
distortion of the octahedron. Only in the cases of
$(M_6Cl_{12})Cl_{6/2}(M = Nb, Ta)$ has a slight tetragonal
flattening been observed. From the three determinations
of the $[(Nb_6Cl_{12})Cl_6]^{2-}$ anion it can be seen that the
Nb-Nb distances are significantly longer, and the Nb-
Cl distances shorter, than those in $[(Nb_6Cl_{12})Cl_6]^{4-}$.
The dianion complexes have 14 electrons available for
cluster formation while the tetra-anion has 16 electrons.
The $(M_6Cl_{12})Cl_{6/2}$ (M = Nb, Ta) complexes, in which the
molecules are linked by the terminal halogens to give
infinite chains (the six terminal halogens each belong
to two neighbouring groups), have 15 electrons
available to form metal-metal bonds. The substitution
of the large bromine atoms for chlorines in the
$[(Nb_6X_{12})X_6]^{2-}$ systems result in the expected increase
in the metal-ligand bond distance, and also causes the
Nb-Nb bond to lengthen somewhat.

A comparison of analogous niobium and tantalum
complexes shows that the Ta-Ta distances are significantly shorter than the corresponding Nb-Nb bond
lengths. The Ta-Cl (terminal) distances are generally
slightly longer than the corresponding niobium distances.
This reflects the decrease in metal-terminal ligand
bonding as metal-metal interactions increase.

An exception to the $[M_6X_{12}]^{n+}$ geometry for niobium
and tantalum is $[Nb_6I_8]I_{6/2}$ (12), in which the eight
iodine atoms are μ_3-bonded and lie over the triangular
faces of the Nb_6 octahedron. This face-centred cubic

TABLE 2

Cluster Compound		M-M bond lengths (Å)	M-X (bridging)	M-X (terminal)
$(pyH)_2[(Nb_6Cl_{12})Cl_6]$	(6)	3.02	2.43	2.48
$(Me_4N)_2[(Nb_6Cl_{12})Cl_6]$	(4)	3.02	2.42	2.46
$(Ph_3AsOH)_2[(Nb_6Cl_{12})Cl_6]$	(5)	3.02	2.42	2.48
$(pyH)_2[(Nb_6Br_{12})Cl_6]$	(6)	3.07	2.56	2.46
$(pyH)_2[(Nb_6Br_6Cl_6)Br_6]$	(6)	3.03	2.57, 2.52	2.62
$(Nb_6Cl_{12})Cl_{6/2}$	(7)	2.92	2.41	2.58
$K_4[(Nb_6Cl_{12})Cl_6]$	(8)	2.91	2.49	2.60
$(Ta_6Cl_{12})Cl_{6/2}$	(9)	2.92	2.43	2.56
$H_2[Ta_6Cl_{12})Cl_6].6H_2O$	(10)	2.96	2.41	2.51
$(Ta_6Cl_{12})Cl_2.7H_2O$	(11)	2.96	2.28	2.35

arrangement is circumscribed by an octahedron of six iodine atoms which extend radially outwards from the apices of the Nb cluster. Each of these latter iodines symmetrically bridges two $[Nb_6I_8]^{3+}$ groups to form infinite chains. This general configuration $[M_6X_8]^{n+}$ is usually found in the lower-valency halide clusters of molybdenum and tungsten. The metals in these latter complexes possess a formal oxidation state of +2 and a closed-shell electronic configuration. The $[Nb_6I_8]^{2+}$ ion, on the other hand, has a non-integral Nb oxidation state, and the metals have an open shell configuration with 19 valence electrons.

A powder neutron study on the related compound $[HNb_6I_8]I_3$ (13) indicates the presence of the hydrogen at the centre of the Nb_6 octahedron.

There are reports of several trinuclear niobium complexes, notably the chloro (14) and the di-(7,7,8,8-tetracyano-p-quixodimethane){tcnq} (15) salts of the $[(Me_6C_6)_3Nb_3Cl_6]^{n+}$ cation. The complex with the chloro counterion is mono-cationic; the two tcnq ions, however, stabilise a paramagnetic di-cation. The molecular geometries of the two cations are similar: a triangle of Nb atoms, each edge of which is symmetrically bridged by two μ_2-Cl atoms. The three Me_6C_6 groups are approximately perpendicular to the Nb_3 plane and are η^6-bonded to the metals (Figure 4). The di-cation has crystallographic mirror symmetry, and the two independent Nb–Nb distances are 3.327(2) and 3.344(3) Å, with a mean of 3.335(9) Å. This is in close agreement with the value of 3.334(6) Å in the mono-cation. The difference between the two independent distances in the former is greater than 5σ indicating a small distortion from an equilateral triangle. This may be the result of the one-electron oxidation.

Nb_3 triangles have also been found in Nb_3Cl_8 which

14.

has a three-dimensional crosslinked structure (16).

2.3. Chromium, Molybdenum and Tungsten

Chromium atoms do not generally bond together to form cluster units. There is, however, infrared evidence which suggests that the species $[Cr_3(CO)_{14}]^{2-}$ exists, and it is probable that there is metal-metal interaction present. There are structural studies showing that chromium does bond to other metal atoms to give cluster fragments (see Section 2.10).

As for niobium and tantalum the structural chemistry of molybdenum and tungsten is dominated by low valence halide clusters, the metals of which have a formal oxidation state +2. The basic cluster unit is $[M_6X_8]^{n+}$ where X is a halogen (Figure 5). The six metal atoms define an octahedron, and the eight halogens act as μ_3-ligands symmetrically capping the triangular faces of the metal skeleton. Frequently, the six metal atoms are also coordinated to six terminal ligands which extend radially outwards from the apices of the metal cluster. The structural data for a number of Mo and W hexanuclear clusters is summarised in Table 3.

The metal atom octahedra show little distortion from idealised $\underline{O}_h$ symmetry. The substitution of a μ_3-bridging chlorine atom by a bromine causes a slight lengthening of metal-metal bonds, as can be seen by comparing the Mo-Mo distances in $Cs_2[(Mo_6Cl_8)Br_6]$ and $[(Mo_6Br_8)Br_4(H_2O)_2]$, although both are close to the value for a single Mo-Mo bond. In the latter complex the two water molecules coordinate <u>trans</u> to each other on the cluster skeleton. The two Mo-O(water) distances

TABLE 3

Cluster Compound	M-M distance (Å)	M-X (bridging)	M-X (terminal)
$Cs_2[(Mo_6Cl_8)Br_6]$ [17]	2.62	2.48	2.58
$[(Mo_6Cl_8)Cl_2]Cl_{4/2}$ [18]	2.61	2.47	2.50
$[(Mo_6Br_8)Br_4(H_2O)_2]$ [19]	2.64	2.61	2.59, 2.19
$Hg[(Mo_6Cl)_8Cl_6]$ [20]	2.62^a	-	-
$Cs_2[(W_6Cl_8)Br_6]$ [17]	2.62	2.49	2.57
$[(W_6Br_8)Br_4][Br_4]$ [21]	2.64	2.58	2.58

^{a}X-ray powder data only

are quite short. This may be due to increased oxygen
participation with the extensive delocalized bonding
within the cluster, although the bonds are formally
single. $[(Mo_6Cl_8)Cl_2]Cl_{4/2}$ has a chain structure
linked by chlorine bridges.

Oxidation of the $[W_6Br_8]^{4+}$ species with bromine
leads (22) to $[W_6Br_8]^{6+}$ in which the mean W-W distance
(2.64 Å) is almost the same as the Mo-Mo distance in
$Hg[(Mo_6Cl_8)Cl_6]$ and $[(Mo_6Br_8)Br_4(H_2O)_2]$. This is in
contrast to the case of $[Nb_6X_{12}]^{2+}$, where a two electron
oxidation increases the Nb-Nb and decreases the Nb-Cl
(axial) distances. The similarity in metal-metal
distances of $[Mo_6X_8]^{4+}$ and $[W_6X_8]^{6+}$ is because the
electrons involved in the oxidation process lie in non-
bonding orbitals and do not effect the M-M distances:
both of the $[M_6X_8]^{4+}$ species have 24 electrons available
for metal-metal bond formation.

An exception to the formation of $[M_6X_8]^{n+}$ species
for molybdenum and tungsten is $(W_6Cl_{12})Cl_6$ (23), which
has the $[M_6X_{12}]^{n+}$ structure normally found for the
hexanuclear cluster halides of niobium and tantalum.

There are several examples of molybdenum and tungsten
clusters containing fewer than six metal atoms. In the
salt $[NBu^n_4]_2[Mo_5Cl_{13}]$ (24), the five Mo atoms in the
anion exhibit a square-based pyramidal geometry.
Eight of the chlorine atoms bridge either faces or
edges to form a cube, and the remaining five each
bound terminally to one Mo (Figure 6). This structure
corresponds to $[(Mo_6Cl_8)Cl_6]^{2-}$ with one MoCl unit
removed. The Mo-Mo distances are in the range 2.563(3)-
2.602(3) Å. These are slightly shorter than those
found in the Mo_6 complex.

The oxide $Zn_2Mo_3O_8$ contains triangular Mo_3 units
with a Mo-Mo distance of 2.53(1) Å (25).

The cation $[(\eta^5C_5H_5)_3Mo_3S_4]^+$ (Figure 7) displays

$\underline{C}_{3v}$ symmetry. The three Mo atoms define an equilateral triangle with an edge length of 2.812 Å which is short enough to indicate that there is direct Mo-Mo interaction. A μ_3-sulphur atom sits above this triangle, and three μ_2-S atoms bridge the triangular edges. The average μ_2-S distance of 2.293 Å is shorter than might be expected for a Mo-S single bond. It is probable that there is considerable π-bonding in the Mo_3S_4 unit. Each of the three cyclopentadienyl ligands coordinates to one Mo and donates five electrons to the cluster. Even so, each Mo atom is two electrons short of the number required for a closed-shell electronic configuration (26).

The anion in $(NH_4)_5[W_3O_4F_9]3H_2O$ has $\underline{C}_{3v}$ symmetry (27). The tungsten-tungsten distance in the equilateral triangle is 2.515(1) Å (Figure 8). The W-F bonds have lengths in the range 1.96-2.04(2) Å expect those trans to the μ_3-oxygen; these average 2.07(2) Å. The average W-(μ_3-O) distance is also 2.07(2) Å, while the lengths of the bonds between W and the μ_2-O atoms are in the range 1.91-1.99(2) Å.

By comparison the tungsten-tungsten distances for the W_3 isosceles triangle in $[W_3(OCH_2CMe_3)O_3Cr_3(O_2CCMe_3)_{12}]$ are 2.608(1) (twice) and 2.614(1) Å (28). In this complex (Figure 9) the tungsten triangle is capped by the oxygen atom of a 'OCH$_2$CMe$_3$' group, and there are three μ_2-"oxo" oxygen atoms bridging the three W-W edges and lying on the opposite side of the plane. The chromium atoms are bound to the latter atoms to form planar W_2OCr units. Presumably the longer W-W bond lengths in this complex indicate a reduction in direct metal-metal bonding and an increase in interaction with the oxygen atoms.

2.4. Manganese, Technetium and Rhenium

There is an extensive cluster chemistry for manganese and rhenium, and both halide and carbonyl complexes are known. Complexes of technetium have not been studied widely, and there is no solid state structural data on clusters of this element. There is, however, some evidence for the existence of $H_3Tc_3(CO)_{12}$.

Halide complexes are confined to rhenium in the +3 oxidation state, and all the structures reported contain the Re_3 triangular unit. The $[Re_3Cl_{12}]^{3-}$ anion, crystallised as the cesium salt, has the structure shown in Figure 10 (29). The three Re atoms define an equilateral triangle with a short Re-Re distance of 2.47 Å. There are three μ_2-chlorine atoms bridging the triangular edges and lying in the plane of the metals. The remaining chlorines are bonded terminally to the rheniums - three to each metal, two in axial and one in equatorial sites. The short metal-metal bond is indicative of strong Re-Re bonding, and if the 18-electron rare gas electronic configuration is to be achieved by each metal, they should be considered as double bonds. A similar Re_3 triangle (Re-Re 2.49 Å) has been found in $Re_3Cl_9(PEt_2Ph)_3$, where the only difference between this structure and that reported for $[Re_3Cl_{12}]^{3-}$ is that the three equatorial non-bridging positions are filled by PEt_2Ph groups instead of Cl^- ions (30).

The Re-Re bond lengths in Re_3Cl_9 also average 2.49 Å. The molecules have rigorous $\underline{C}_{3v}$ symmetry and are held together by loose chloride bridges to give a two-dimensional polymeric sheet structure with the repeating unit as shown in Figure 11 (31). Replacement of the terminal chlorine ligands by bromines to give the complex $Re_3Br_6Cl_3$ reduced the metal-metal distances further to an average value of 2.46(1) Å (32). The related iodide complex with the repeat unit 'Re_3I_9' has a fibrous one-dimensional polymeric structure (33).

The complex with stoichiometry $[C_9H_8N]_2[Re_4Br_{15}]$ is actually an adduct on the type $[C_9H_8N]_2ReBr_6 \cdot Re_3Br_9$, containing both Re(IV) and Re(III). Here the Re-Re distances in the Re triangle of the Re_3Br_9 species average 2.465 Å (34).

The X-ray structures of the anions $[Re_3X_{11}]^{2-}$ (X = Cl or Br) show that both have an isosceles Re triangle with two 2.43 (for Cl) to 2.44 Å (for Br) sides and one 2.48 (for Cl) to 2.49 Å (for Br) side (Figure 12) (35,36).

In $Cs[Re_3Cl_3Br_7(H_2O)_2]$ (Figure 13) the asymmetry of the anion due to the presence of three different types of ligands has little effect on the geometry of the Re_3 triangle. It is essentially equilateral with sides of 2.45 Å (37).

It seems that rhenium triangles of the type $[Re_3X_{12}]^{3-}$ are equilateral even when all the X ligands are not equivalent, whereas Re_3 triangles of the type $[Re_3X_{11}]^{2-}$ are isosceles even when all the X ligands are equivalent. It is the number, rather than the type, of ligands bonded to the triangle that has the larger effect on the geometry of the triangle.

In the neutral complex $Re_3Cl_3(CH_2SiMe_3)_6$ (Figure 14), stabilised by the presence of the bulky trimethylsilyl-methyl groups, the three Re atoms define an equilateral triangle with a very short edge length of 2.387 Å. The three chlorine ligands act as equatorial μ_2-bridging

groups. The two CH_2SiMe_3 ligands terminally bound to
each metal are in 'pseudo-axial' positions and their
bulk prevents the coordination of any other ligands in
equatorial plane (38).

There are no high nuclearity manganese carbonyl
clusters, but mixed clusters containing manganese,
with up to five metal atoms, are known. The basic
trinuclear hydrido-carbonyl complex, $H_3Mn_3(CO)_{12}$, is
shown in Figure 15. The structure consists of an
equilateral triangle of Mn atoms (Mn-Mn 3.111(2) Å),
each with two axial and two equatorial carbonyl
ligands. The hydrides were located in the X-ray study
by a technique in which only low angle data are used
to calculate the electron density difference maps.
The three hydrides were found to be approximately
coplanar with the Mn_3 triangle and to symmetrically
bridge the three Mn-Mn edges. The average Mn-H
distance was found to be 1.72(3) Å, and the Mn-H-Mn
and H-Mn-H angles were 131(7)$^\circ$ and 108(6)$^\circ$, respectively
(39). The Mn-Mn distance is longer than would be
expected for a single bond, but as we shall see
elsewhere, the presence of a μ_2-H ligand along an edge,
when no other bridging ligands are present, always
causes a lengthening of the metal-metal bond. The
difference between the average axial (1.854(4) Å) and
average equatorial (1.789(1) Å) Mn-C (carbonyl) bond
lengths is in keeping with the general observation
that M-L distances opposite π-bonding ligands, such as
CO, are longer than M-L distances opposite σ-bonding
ligands such as H. This effect can be attributed to a
greater competition for backbonding electrons, and a
concomitant weakening of the M-L bonds.

Treatment of $H_3Mn_3(CO)_{12}$ with alcoholic base
gives the $[Mn_3(CO)_{14}]^-$ anion. In this species the
metal triangular arrangement has been broken, and a

and a linear Mn_3 unit (Mn-Mn 2.895 Å) is observed (40).
The geometry of the molecule is essentially $\underline{D}_{4h}$
(Figure 16). The equatorial carbonyl groups on the
terminal Mn atoms are staggered with respect to those
on the central metal and are bent away from the axial
carbonyl ligands. It is probable that there are
interactions between a metal atom and the carbonyl
groups on a neighbouring metal which cause these
geometrical features.

We have seen examples of triangular and linear
Mn_3 systems, but the intermediate bent configuration
also exists (Figure 17). In $CH_3N_2[Mn(CO)_4]_3$ the
terminal diazo nitrogen, which forms a μ_2-bridge with
two of the metals, acts as a one electron donor to one
Mn and a two electron donor to the other. Thus the
two Mn-N distances differ (1.964(8) and 1.879(8) Å).
The other nitrogen, which acts as a two electron donor,
coordinates to only one tetracarbonyl fragment. The
N-N distance of 1.224 Å corresponds to a typical double-
bond. The metal coordination at the central nitrogen
seems to have no effect on this bond length, but it
does cause a considerable reduction of the N-N-C angle
from the idealised value of 120° to 112.3°. The Mn-Mn
distances of 2.826(3) and 2.807(4) Å are typical for
single bonds. The Mn-Mn-Mn angle is $107.4(7)^{\circ}$ (41).

An example of a trinuclear manganese cluster
which does not contain carbonyl ligands is $(\eta^5\text{-}C_5H_5)_3$
$Mn_3(NO)_4$ (42). It is isostructural with
$[(C_5H_5)_3Mo_3S_4]^+$. One nitrosyl ligand sits above the
Mn_3 equilateral triangle and coordinates to all three
metals in a μ_3-fashion. The other three nitrosyls,
acting as μ_2-ligands, bridge the three Mn-Mn edges and
lie on the opposite side of the metal plane to the
μ_3-NO group. Each manganese is also π-bound to a
cyclopentadienyl ring (Figure 18). The nitrosyl

ligands each contribute three electrons to the cluster, while the cyclopentadienyl rings act as five electron donors. The average Mn-Mn distance of 2.506(3) $\overset{o}{A}$ is indicative of direct metal-metal bonding, which is required for the complex to achieve a 'closed shell' electronic configuration.

A large number of rhenium clusters containing three or four metal atoms have been investigated crystallographically. Nearly all require the presence of hydride ligands to stabilise them, and most are isolated as anionic species. In all of them the carbonyl ligands are linear and coordinate as terminal groups. Table 4 summarises the data on the known hydrido-carbonyl clusters containing three or four Re atoms. The molecular structure of $H_3Re_3(CO)_{12}$ has not been determined because of disorder problems, but it is expected to be isostructural with the manganese derivative.

As there have been no neutron studies on rhenium cluster complexes, the information given in the table on the hydride positions has been deduced from observations of Re-Re distances and cis Re-Re-C (carbonyl) angles.

As we have already mentioned, hydride ligands, in the absence of other bridging groups, lengthen metal-metal bonds. The carbonyls tend to bend away from the hydrides to reduce steric crowding, thus increasing the cis M-M-C(carbonyl) angle. In the case of a Re_3 triangle, a typical cis Re-Re-C(carbonyl) angle for a carbonyl group adjacent to a bridged edge is 118°, while that for one adjacent to an unbridged edge is 100°.

We have already seen that in the case of the halide complexes there is a tendency for the formation of Re-Re double bonds. Those Re_3 complexes with less

TABLE 4

Compound	Metal Arrangement	Re-Re distances (Å)	Type of bonds	No. of Valence Electrons
$[H_2Re_3(CO)_{12}]^-$ (43)	Triangular	3.173(7), 3.181(7) 3.035	Single, μ_2-H Single unbridged	48
$[HRe_3(CO)_{12}]^{2-}$ (44)	Triangular	3.125(3) 3.014(3), 3.018(3)	Single, μ_2-H Single, unbridged	48
$[H_3Re_3O(CO)_9]^{2-}$ (45)	Triangular	2.963, 2.973(1)	Single, μ_2-H, μ_3-O	48
$[H_4Re_3(CO)_{10}]^-$ (46)	Triangular	3.173(7), 3.194(7) 2.821(7)	Single, μ_2-H Double, bis (μ_2-H)	46
$[H_3Re_3(CO)_{10}]^{2-}$ (47)	Triangular	3.031(5) 2.797(4)	Single, one μ_2-H disordered on two bonds Double, bis (μ_2-H)	46
$HRe_3(CO)_{14}$ (48)	Bent Re-Re-Re $\approx 90°$	3.295(2)	Single, μ_2-H	50
$[H_6Re_4(CO)_{12}]^{2-}$ (49)	Tetrahedral	3.142 - 3.172(3)	Single, μ_2-H	60
$H_4Re_4(CO)_{12}$ (40)	Tetrahedral	2.896 - 2.945(3)	Partially double, four μ_3-H	56
$[H_4Re_4(CO)_{13}]^{2-}$ (51)	Tetrahedral	3.026(5) - 3.161(4)	Single, four μ_2-H, unclear assignment based on distances	60
$[H_4Re_4(CO)_{15}]^{2-}$ (52) (three crystal forms)	Triangular with one terminal Re	3.026 (av. on 3 values) 3.195 (av. on 6 values) 3.287 (av. on 3 values)	Single, unbridged Single, μ_2-H Single, μ_2-H	64
$[Re_4(CO)_{16}]^{2-}$ (54)	Two fused coplanar triangles	2.956 - 3.024(7)	Single, unbridged	62

than 48 valence electrons, and Re_4 complexes with less than 60 electrons are formally unsaturated. Of particular interest is the $[H_4Re_3(CO)_{10}]^-$ anion (Figure 19) in which two hydride ligands μ_2-bridge the short, unsaturated Re-Re edge. $H_4Re_4(CO)_{12}$ formally has two double bonds, but a number of resonance hybrids can be drawn so that there are no long and short bonds as such. The carbonyl configuration in this molecule is different to that found in other tetranuclear hydrido-carbonyl clusters (e.g. $H_4Ru_4(CO)_{12}$, $H_4Os_4(CO)_{12}$ and $Ir_4(CO)_{12}$). The carbonyls are eclipsed with respect to the metal-metal bonds (Figure 20) rather than staggered. This arrangement of the carbonyls requires the metal orbitals <u>trans</u> to the coordinated CO to point towards the centre of the Re_3 triangular face, rather than along Re-Re edges. The hydride is attracted to the centre of the face because of the build up of electronic charge (55). Hence an eclipsed carbonyl configuration requires the hydrides to be face capping. The carbonyl configurations in $[H_6Re_4(CO)_{12}]^{2-}$ (Figure 21) and $[H_4Re_4(CO)_{13}]^{2-}$ (Figure 22) are staggered.

The geometry adopted by the $[Re_4(CO)_{16}]^{2-}$ anion (Figure 23) is relatively uncommon in that the four metal atoms define a parallelogram with a Re-Re bond across the shorter diagonal. Each metal is associated with two axial and two equatorial terminal carbonyls. The structure of $[H_4Re_4(CO)_{15}]^{2-}$ (Figure 24) may be derived from $[Re_4(CO)_{16}]^{2-}$ if the loss of a carbonyl and the addition of four hydrides is considered as the addition of two electrons. This results in the breaking of a Re-Re edge bond to give the observed terminally coordinated metal. Two of the hydrides bridge triangular edges, a third bridges the terminal Re-Re bond, and the fourth is probably associated only with the terminal metal.

Two isostructural Re_4 clusters are worthy of note, $(Ph_3P)_4[Re_4X_4(CN)_{12}]3H_2O$, where X = S or Se (56) (Figure 25). The anions $[Re_4(\mu_3-X)_4(CN)_{12}]^{4-}$ each contain a tetrahedron of Re atoms held together by short Re-Re bonds and one triply bridging S are Se over each face of the tetrahedron. In the sulphur derivative the average Re-Re distance is 2.755(5) Å, while in the selenium complex it is 2.805(5) Å.

2.5.   Iron, Ruthenium and Osmium

There have been solid state structural studies on carbonyl complexes containing between three and seven metal atoms for the iron triad.

The trinuclear carbonyl complexes have the molecular formula $M_3(CO)_{12}$. The iron complex consists of a slightly distorted isosceles triangle of metal atoms. One iron atom is coordinated to two axial and two equatorial terminal carbonyls, the other two metals are each bonded to three terminal and to two μ_2-bridging carbonyls (Figure 26). These two groups both bridge the same edge of the triangle which is 0.12 Å shorter, at 2.558(1) Å, than the mean of the other two edges (2.660 Å). The bridging system is highly asymmetric with average long and short distances of 2.16(5) and 1.94(2) Å. The polyhedron defined by the carbonyl ligands is an icosahedron (57).

The ruthenium and osmium complexes have a different structure (Figure 27). All the carbonyl ligands are terminal, four being bonded to each metal atom, two in axial and two equatorial positions. The ligand poly-hedron is now best described as an anti-cubooctahedron.

Some structural data for the two compounds is
summarised in Table 5. The metal triangles are
essentially equilateral, small deviations have been
attributed to packing effects. All the carbonyl
groups are linear although there is a tendency for the
axial ligands to bend because of van der Waals
repulsions between axial oxygen atoms. As found in
other carbonyl complexes the M-C distances for the
axial carbonyls, those trans to another carbonyl
ligand, are longer than the equatorial ones, because
there is increased competition for back-donated
electron density causing a weakening of the individual
M-C bonds.

TABLE 5

	$Ru_3(CO)_{12}$ (58)	$Os_3(CO)_{12}$ (59)
Av. M-M ($\overset{o}{A}$)	2.854(4)	2.877(3)
M-C (axial) ($\overset{o}{A}$)	1.942(4)	1.946(6)
M-C (equatorial) ($\overset{o}{A}$)	1.921(5)	1.912(7)
C-O (axial) ($\overset{o}{A}$)	1.133(2)	1.134(8)
C-O (equatorial) ($\overset{o}{A}$)	1.127(2)	1.145(5)
M-C-O (axial) (^{o})	173.0(10)	175.3(10)
M-C-O (equatorial) (^{o})	178.9(10)	178.4(10)

The mono-anion, $[HM_3(CO)_{11}]^-$, has been structurally
characterised for iron (60) and ruthenium (61). They
have the same basic structure (Figure 28), an isosceles
triangle of metal atoms with a μ_2-CO ligand and a μ_2-
hydride symmetrically bridging the shortest metal-
metal edge. In the case of the ruthenium complex the
hydride has been located by X-ray diffraction. The
long and short metal-metal bond lengths are 2.691
(average) and 2.577 $\overset{o}{A}$ for Fe, and 2.844 (average) and

2.815 Å for Ru. It is interesting to note the change in structure between $Ru_3(CO)_{12}$ and $[HRu_3(CO)_{11}]^-$, as the latter has bridging groups and the former does not. The triangle in the anion is significantly smaller which would tend to favour the icosahedral arrangement of ligands.

There are many trinuclear complexes of Fe, Ru or Os which have their structures based on those of the neutral dodecacarbonyls or the carbonyl anions. In these one or more of the carbonyl groups have been replaced by other ligands. In $Fe_3(CO)_{11}(PPh_3)$ (62) the replacement of an equatorial CO by triphenyl-phosphine has little effect on the Fe-Fe distances and the asymmetry of the bridging carbonyls is retained. Three equatorial carbonyls are replaced by phosphines in $Fe_3(CO)_9(PMe_2Ph)_3$ (63). This has the effect of shortening the μ_2-CO bridged Fe-Fe edge by only 0.18 Å compared to the distance in $Fe_3(CO)_{12}$ and leaves the other two metal-metal bonds essentially unchanged at 2.668 Å. In $Os_3(CO)_{11}P(OMe)_3$ (64) the mean Os-Os distance of 2.897(10) Å is 0.02 Å longer than in $Os_3(CO)_{12}$. This may be due to the bulky phosphite ligand causing an expansion of the ligand packing; the Os_3 triangle also expands to maintain the optimum overall molecular geometry. In $Os_3(CO)_8(NO)_2P(OMe)_3$ four carbonyls on one Os atom are replaced by an equatorial $P(OMe)_3$ group and two terminal nitrosyls (65). The Os-Os bond lengths range from 2.790(5) to 2.860(6) Å.

$H_2Os_3(CO)_{11}$ is isomorphous with $Os_3(CO)_{12}$. One axial carbonyl group on one metal atom is replaced by a terminal hydride, and a μ_2-hydride bridges this Os atom and an adjacent one (Figure 29). The effects of this bridging hydride, which lies in the equatorial plane, can be assessed by comparing the geometry of the equatorial $Os_3(CO)_6$ fragments in $Os_3(CO)_{12}$ and $H_2Os_3(CO)_{11}$ since all structural dissimilarities between the two must arise from the presence or absence of the hydride. The bridged Os-Os bond length of 2.989(1) Å is significantly longer than the average of 2.884 Å for the non-bridged distances in the same molecule and is 0.112 Å longer than the average Os-Os distance in $Os_3(CO)_{12}$. The hydride causes adjacent equatorial carbonyl ligands to be pushed back relative to their position in $Os_3(CO)_{12}$; cis Os-Os-CO angles adjacent to the hydride average 112.9^O in $H_2Os_3(CO)_{11}$ (59). In the closely related structure of $H_2Os_3(CO)_{10}PPh_3$ (66) the triphenylphosphine ligand occupies an equatorial site on one Os atom and the terminal hydride occupies an axial site on another. These two metals are bridged by the μ_2-hydride. The effects of this hydride are similar to those found in $H_2Os_3(CO)_{11}$.

The locations of the hydride ligands in $H_2Os_3(CO)_{10}$ have been established from a neutron diffraction study (67). The osmium atoms define an isosceles triangle with two long (mean 2.814 Å) and one short (2.680(2) Å) edge. One metal is bonded to two axial and two equatorial carbonyls and the other two metals are each bonded to three terminal carbonyl groups. The two hydride ligands μ_2-bridge the short Os-Os edge, lying on opposite sides of the Os_3 plane (Figure 30). In terms of electron counting, with the hydride acting as a one electron donor, this Os-Os bond has formal double bond character and this may partially explain the reactivity of this complex. The Os-H-Os units are symmetrical, with a mean Os-H distance of 1.850(5) Å, a mean Os-H-Os angle of $92.9(2)^O$ and a H...H contact distance of 2.397(8) Å. All trans H-Os-C angles lie in the range $172.8 - 174.2^O$ which supports the

description of the Os-H-Os bridges as 'open' 2-electron
3-centre bent bonds.

$HOs_3(CO)_{10}(SEt)$ and $Os_3(CO)_{10}(OMe)_2$ have structures
similar to $H_2Os_3(CO)_{10}$. However, the thiolato ligand
is a three electron donor and the methoxy groups donate
three electrons each. Thus, in terms of electron
counting, the dibridged Os-Os bond in $HOs_3(CO)_{10}(SEt)$
has a bond order of 1 which is in agreement with the
observed bond length of 2.863 Å, while the bond order
for the equivalent bond in $Os_3(CO)_{10}(OMe)_2$ is zero,
and the observed distance is 3.078 Å (68).

There are many other neutral complexes with the
formula $(\mu_2-H)M_3(CO)_{10}(\mu_2-X)$ with essentially the same
structure as $HOs_3(CO)_{10}(SEt)$. In all of them X formally
acts as a three-electron donor. Since the influence
of the bridging hydride is approximately the same the
whole series the differences in metal-metal bond
lengths and in metal-CO bond distances trans to X must
reflect the donor and π-acceptor properties of the
various X ligands. Data for these compounds are given
in Table 6. In all the complexes except for
$HRu_3(CO)_{10}SCH_2COOH$ the bridged metal-metal bond is
shortened compared to the non-bridged bonds. The bond
shortening effect of these ligands more than compensates
for the lengthening effect of the bridging hydride.
The strength of the Ru-X bond appears to increase in
the order

$$SCH_2OOH \sim NO < CNMe_2 < COMe$$

since there is an increase in the competition with the
trans carbonyl for back-bonding from the filled
ruthenium orbitals which causes the trans Ru-C bond
distances to lengthen.

A compound closely related to this series is
$HOs_3(CO)_9(SMe)(C_2H_4)$ (Figure 31). The hydride and thiol

ligands both bridge the same Os-Os bond (2.842(1) $\overset{o}{A}$).
One of the equatorial carbonyls on the unique Os atom
has been replaced by a π-bonded ethylene the carbon
atoms of which in the plane of the Os_3 triangle (76).
The Os-C (ethylene) bond lengths are 2.23(4) $\overset{o}{A}$ and the
C-C distance is 1.42 $\overset{o}{A}$. A similar configuration for the
unsaturated fragments of the butadiene species is observed
in $Os_3(CO)_{10}$(S-<u>trans</u> C_4H_6). This ligand bridges one
Os-Os bond of the triangle and bonds to equatorial
sites of adjacent metal centres. The diene ligand in
the <u>cis</u> isomer is bonded to one metal atom only,
coordinating to one equatorial and one axial site (77).

In contrast to $HRu_3(CO)_{10}NO$ the dinitrosyl complex
$Ru_3(CO)_{10}(NO)_2$ has a very long Ru-Ru distance of 3.15 $\overset{o}{A}$
which is supported by both μ_2-nitrosyl ligands. The
two non-bridged Ru-Ru edges with a mean value of
2.866(5) $\overset{o}{A}$ are also longer than in the hydrido nitrosyl.
The $Ru-\mu_2N$ distances lie in the range 1.94 - 2.05 $\overset{o}{A}$
(78). Another complex with a Ru-Ru edge that is
bridged by two donor nitrogens is $(C_6H_{10}N_2)Ru_3(CO)_{11}$
(Figure 32) but here the bridged Ru-Ru edge is 2.841(1) $\overset{o}{A}$
(79). Both $HRu_3(CO)_{10}(NO)$ and $(C_6H_{10}N_2)Ru_3(CO)_{11}$ are
'electron precise' closed shell systems but $Ru_3(CO)_{10}(NO)_2$
with both nitrosyl ligands acting as three-electron
donors has two additional electrons. This causes the
breaking of a metal-metal bond; the bridged Ru-Ru
distance in $Ru_3(CO)_{10}(NO)_2$ is considered to be non-
bonding.

Other trinuclear clusters containing ten carbonyl
groups include a number of clusters in which organic
groups lie over the metal triangle and coordinate to
two or three of the metal atoms. In $(C_4H_4N_2)Ru_3(CO)_{10}$
two <u>cis</u> axial carbonyl ligands are replaced by the
bidentate 1,2-diazene ligand. The short Ru-Ru edge
(2.744(1) $\overset{o}{A}$) is bridged by both a carbonyl and the

TABLE 6

Effects of Different Bridging Groups in $HM_3(CO)_{10}X$ Systems

Compound	X	M-M bond lengths (Å)	M-X (Å)	M-trans (CO)	Ref.
$HFe_3(CO)_{10}COMe_2$	$COMe_2$	-	-	-	69
$HRu_3(CO)_{10}COMe_2$	$COMe_2$	2.803(2) bridged, 2.810(2), 2.821(2)	1.976(6), 1.978(7)	1.989(7), 1.987(8)	70
$HRu_3(CO)_{10}CNMe_3$	$CNMe_3$	2.801(6) bridged, 2.831(6), 2.825(6)	2.029(4), 2.036(4)	1.968(6), 1.966(6)	71
$HRu_3(CO)_{10}SCH_2COOH$	SCH_2COOH	2.839(4) bridged, 2.839(4), 2.826(5)	2.387(6), 2.388(3)	1.90(1), 1.89(1)	72
$HRu_3(CO)_{10}NO$	NO	2.816(2) bridged, 2.843(2), 2.856(2)	1.989(8), 1.973(5)	1.900(12), 1.929(9)	73
$HOs_3(CO)_{10}(CHCHNEt)$	$(CHCHNEt)$	2.785(2) bridged, 2.870(2), 2.866(2)	2.15(3), 2.16(3)	-, -	74
$HOs_3(CO)_{10}(CHCH_2PMe_2Ph)$	$(CHCH_2PMe_2Ph)$	2.800(1) bridged, 2.873(6), 2.869(6)	2.148(9), 2.173(8)	1.917(11), 1.920(11)	75

diazene group. The other two Ru-Ru edges with lengths
of 2.861(4) and 2.859(1) Å are bridged by carbonyl
groups. The unique Ru atom is also coordinated to
three terminal carbonyls, and the other two metals to
two each (80).

In $Os_3(CO)_{10}(PhC_2Ph)$ the diphenylacetylene ligand
bridges one Os-Os bond by σ-bond formation and π-bonds
to the third Os atom. The metals which form σ-bonds
to the acetylene differ in the number of coordinated
carbonyl ligands, one with three and the other with
four (Figure 33). The Os-Os bond between these metals
may be considered as a donor-acceptor bond between
electron rich and deficient metal centres. The <u>trans</u>
carbonyl ligands on the electron rich metal appear to
be of the semi-bridging type interacting with the
adjacent Os atoms. The Os-Os bond lengths range from
2.711(1) to 2.888(1) Å (81). In $Os_3(CO)_{10}(HC_2Ph)_2$ the
two phenylacetylene groups are linked through a carbon
and an oxygen atom derived from the opening of a
carbonyl C-O bond. One Os atom is π-bonded to the
allylic carbons and the other two metals are involved
in σ-bonding (Figure 34). Eight terminal carbonyl
groups occupy equatorial and axial sites on the three
metal atoms. One carbonyl asymmetrically bridges the
longest (2.880(2) Å) bond. The other two metal-metal
distances are 2.794(2) and 2.857(2) Å (82).

A neutron diffraction study $HOs_3(CO)_{10}(HCCH_2)$
shows that the μ_2-hydride and the vinyl group bridge
the same edge of the Os_3 triangle (Figure 35), and
that all the carbonyl ligands are terminal. The vinyl
group is formally σ-bound to one Os atom and π-bound
to another. If the vinyl group is considered to
donate two electrons in forming the π-bond the OsHOs
unit forms a two-electron three-centre bond. Although
the bridged Os-Os distance is quite short it may be due

to the geometry of the bridging groups and does not
necessarily imply direct Os-Os bonding. The vinyl
group and σ-bonded Os atom are coplanar to within
0.15 Å, the maximum deviations being shown by the two
vinylic hydrogens which bond away from the adjacent
Os atom. The C-C distance of 1.396(3) Å is longer
than expected for a normal C-C double bond. A
relatively weak π-interaction would be consistent with
the asymmetry of the hydride bridge; $Os(\mu_2-H)$ 1.813(4)
and 1.857(4) Å. The shorter Os-H bond is made by the
vinyl π-bonded Os atom (83). The C-C bond length in
the μ-but-1-enyl analogue of 1.40(3) Å is very similar
to the value obtained from the neutron study. In fact
the overall molecular structure of $HOs_3(CO)_{10}(HCCHEt)$
(84) closely resembles that of the vinyl complex.

The vinyl ligand in $HOs_3(CO)_{10}(CF_3CCHCF_3)$ adopts
a different mode of coordination to that found in the
two previous clusters. Here one vinylic carbon forms
σ-bonds with two metal atoms bridging one edge of the
Os_3 triangle while the other carbon is σ-bonded to the
third Os atom so that the ligand lies over one side of
the triangle. Thus one Os atom is bonded to four
terminal carbonyls and the vinyl ligand (Figure 36).
The hydride is assumed to bridge one of the longer
Os-Os edges (2.90 and 2.91 Å) (85).

In $(\mu-H)Os_3(CO)_{10}(\mu-\eta^2-C_6H_5CNCH_3)$ the Os atoms
define a triangle. Two metals each have three terminal
carbonyls while the other has four. The $\eta^2-C_6H_5CNCH_3$
ligand bridges the longest edge (2.918(1) Å); the
nitrogen atom bonds to one metal (Os-N 2.111(6) Å) and
the carbon to the other (Os-C 2.090(8) Å) (Figure 37).
The N-C distance of 1.278(10) Å is indicative of
multiple bond character (86).

Another complex with ten carbonyls is
$Fe_3(CO)_{10}(NSiMe_3)$. In this case nine of the carbonyls

are terminally bonded to the three Fe atoms in the triangle but the tenth caps the Fe_3 plane. The nitrogen of the $NSiMe_3$ group caps the other side of the metal triangle (87). The average Fe-Fe distance is 2.535(2) Å.

There are a large number of iron, ruthenium and osmium trinuclear clusters containing nine carbonyl ligands, and many of these have the triangle capped by an atom or group. The complex $Fe_3(CO)_9(PhCCPh)$ consists of an isosceles Fe_3 triangle with three terminal carbonyl ligands bonded to each. The diphenylacetylene lies above the plane of the metal atoms such that one acetylenic carbon is coordinated to all three Fe atoms while the other carbon is bonded to two metal atoms. The Fe-Fe distances lie in the range 2.480(10) – 2.579(11) Å and the acetylenic C-C bond is 1.409(22) Å long (88).

In $Fe_3(CO)_9S_2$ the iron atoms adopt a bent arrangement with the sulphur atoms capping the metals to give an overall trigonal bipyramidal geometry (Figure 38). The two Fe-Fe distances are 2.58 and 2.61 Å, and the average Fe-S distance is 2.23 Å (89). A different structure is observed in $Fe_3(CO)_9(RNS)S$ where the three Fe atoms, the two sulphurs and the nitrogen define a distorted trigonal prism (Figure 39). The bonding Fe-Fe distance is 2.161 Å, the two Fe-N distances are 2.08 and 2.09 Å, and the Fe-S bond lengths range from 2.228(5) to 2.314(6) Å (90).

In $HFe_3(CO)_9(SC_3H_7)$ the iron atoms occupy the vertices of a near-equilateral triangle. The sulphido group bridges all three metal atoms. The hydride is located on the opposite side of the Fe_3 plane to the S atom and symmetrically bridges two Fe atoms so that it lies essentially on the plane defined by two Fe and S atoms. The bonding influence of the capping SC_3H_7 group causes the equilateral symmetry of the Fe triangle

to be virtually unaltered by the presence of the
Fe-H-Fe bridge. The Fe-Fe distances range from 2.640(2)
to 2.678(2) Å and the Fe-S distances are in the range
2.136(2) - 2.154(2) Å; the sulphur atom lies 1.50 Å
above the Fe_3 plane (91). In this complex the sulphur
ligand acts as a five electron donor. In the two
related compounds $[HOs_3(CO)_9S]^-$ and $H_2Os_3(CO)_9S$ (93)
the μ_3-sulphur atom acts as a four electron donor. In
both compounds the metal atoms define isosceles
triangles; the anion has one long (2.899(2) Å) and two
short (mean 2.767(3) Å) Os-Os bonds, and the neutral
complex has two long (mean 2.915(2) Å) and one short
(2.764(1) Å) Os-Os bond. The average metal to capping
sulphur atom distance in the two complexes are very
similar, at 2.389 Å for the anion and 2.390 Å for the
neutral species, but in the neutral species the sulphur
atom is closer to the Os_3 plane, at 1.72 Å. The nine
carbonyls in each compounds are all terminal, with
three bonded to each metal. The distribution of these
carbonyl ligands suggests that the hydrides bridge the
long Os-Os bonds. This has been confirmed in the case
of $H_2Os_3(CO)_9S$ by a neutron diffraction study (Figure
40). The two hydridic hydrogens lie on the opposite
side of the metal triangle to the sulphur atom. The
Os-H-Os bridges are significantly asymmetric, with the
hydrides about 0.025 Å closer to the unique Os atom;
the Os-H distances lie in the range 1.804(3) - 1.837(3) Å.
These units are best described as 'open' three-centre
two-electrons bent bonds.

The complex $H_3Ru_3(CO)_9CMe$ (94) has a symmetrically
capping CMe group with a Ru-C distance of 2.083(11) Å.
The hydrides were located on the other side of the Ru_3
triangle bridging the three Ru-Ru edges. The Ru-Ru
distances are 2.842(6) Å and the Ru-H distances are
1.72(7) Å. The molecule approximates closely to $\underline{C}_{3v}$

symmetry (Figure 41).

A neutron diffraction study of $HRu_3(CO)_9(C\equiv CBu^t)$ (95) shows that the hydride symmetrically bridges an Ru-Ru edge of the equilateral triangle (mean Ru-Ru 2.796 Å) to form a bent Ru-H-Ru tricentric bond. The Ru-H average distance is 1.793 Å. The acetylenic ligand bridges the same edge as the hydride by π-bonding to both the metals. It is also σ-bonded to the third metal atom. The carbonyl ligands, three terminally bound to each metal, are arranged to be in keeping with the steric requirements of the hydride and acetylenic ligands. In $HRu_3(CO)_9(EtCCCHMe)$ (96) the allene ligand bonds to all three Ru atoms through two π and one σ-bond (Figure 42). The Ru atoms define an isosceles triangle with one long (2.994(1) Å) and two short (mean 2.754 Å) metal-metal bonds. Low angle X-ray data was used to locate the hydride ligand which lies along the longest edge; the Ru-H distances were found to be 1.61(7) and 1.77(7) Å. The monocyclic ligand in $HRu_3(CO)_9(C_{12}H_{15})$ (Figure 43) is attached to the cluster by an allylic bonding arrangement involving only three carbon atoms. The three Ru atoms define an isosceles triangle with one long (2.921(3) Å) and two short (mean 2.776 Å) metal-metal bonds. Three terminal carbonyl ligands are bonded to each metal. Three of the ring carbon atoms are equidistant from on Ru atom (2.15 Å) and two are each equidistant from one of the other metal atoms (2.04 Å). This is consistent with one tricentric π-bond and two Ru-C σ-bonds (97). An analogous structure has been observed in $HRu_3(CO)_9(C_6H_9)$ (98).

One of the complexes formed when $H_2Os_3(CO)_{10}$ reacts with cyclohexa-1,3-diene is $HOs_3(CO)_9C_6H_7$ (99). The metal triangle is approximately isosceles and, as in other systems, the hydride is thought to bridge the

long Os-Os edge. In the organic species five of the
carbon atoms which form the pentadienyl fragment are
coplanar and the sixth lies 0.32 Å above the plane
(Figure 44). The pentadienyl plane slopes away from
the Os triangle so that one Os-C bond is significantly
shorter (2.17 Å) than the other four Os-C distances
(two at 2.28, and two at 2.45 Å). It is probable that
there is significant Os-C σ bonding in the shorter
bond.

The organic ligand in $H_2Os_3(CO)_9(\mu_3-OC_6H_3CH_2Ph)$
coordinates to one Os atom in the triangle through the
oxygen <u>via</u> a σ-bond (Os-O 2.09(1) Å), and to the other
two atoms through electron deficient Os-C bonds (2.21(1)
and 2.24(1) Å) from one carbon of the six-membered
ring (Figure 45). The hydrides are thought to bridge
the long Os-Os edge (2.949(1) Å) and one of the
shorter edges (2.786(1) Å). In the latter case the
metal-metal bond lengthening effect of the bridging
hydride is compensated by the shortening effect of the
organic group (100).

An uncommon mode of ligand coordination is
observed in $Ru_3(CO)_9[MeSi(PBu_2)_3]$. The metal atoms
define an equilateral triangle (average Ru-Ru 2.917 Å)
and each is bonded to two terminal carbonyl ligands.
The three other carbonyls form asymmetric bridges. The
$MeSi(PBu_2)_3$ group lies above the Ru_3 plane and one
phosphorus atom coordinates to each metal atom (Figure
46). This gives additional stability to the cluster,
the tridentate ligand holding the system together (101).
A more common bidentate ligand is $[Me_2AsC:C.AsMe_2.CF_2.
CF_2]$ which bridges an edge of the Ru_3 triangle in
$Ru_3(CO)_{10}(Me_2AsC:C.AsMe_2.CF_2.CF_2)$ (102).

Cluster compounds of iron, ruthenium and osmium
containing fewer than nine carbonyl ligands retain the
same metal geometry as those already described and

carbonyl groups are often replaced by π-bonded ligands.
In the structure of $(\pi\text{-}C_5H_5)Fe_3(C_2Ph)(CO)_7$ one of the
acetylenic carbon atoms bridges all three metals while
the other bridges two. There are two $Fe(CO)_3$ units
but the third Fe atom is now bonded to a cyclopentadienyl and a semibridging carbonyl as well as the
acetylenic species (103). In $Fe_3(CO)_7(HC_2Et)_4$ oligo-
merization of the alkyne and cleavage of an acetylenic
triple bond produces a 1,2,3-trimethylcyclopentadienyl
unit which η^5-bonds to one Fe atom and an ethylallyl
unit which is σ- and η-bonded to all three Fe atoms in
the triangle. One asymmetrical carbonyl bridges two
Fe atoms and another very unsymmetrically triple
bridges all three metals (104). The related compound
$Fe_3(CO)_7\{Ph_2PC_4(CF_3)_2\}(PPh_2)$ contains a dimerized
phosphinoacetylene in the form of a $\underline{\text{trans}}$-butadiene
coordinated simultaneously to two Fe atoms $\underline{\text{via}}$ a Fe-C
π bond (2.14(8) Å) and a σ-bond (2.02(3))Å, a bridging
diphenylphosphide group (Fe-P 2.170(6) and 2.288(3) Å)
formed by cleavage of a P-C(sp) bond of one phosphinoacetylene, and two Fe-Fe bonds (2.662(7) Å and 2.531(10)
Å). The $Ph_2PC_4(CF_3)_2$ ligand remains bonded to iron
$\underline{\text{via}}$ a Fe-P bond (2.238(3) Å) (105).

In $Fe_3(CO)_8(\text{tmhd})$ [tmhd = 1,3,6-trimethylhexa-1,3,
5-triene-1,5-diyl] the unsymmetrical cluster of Fe
atoms is coordinated by eight carbonyl groups and a σ-
and π-bonded organic ligand derived from three methyl-
acetylene molecules. This organic species chelates to
one Fe atom $\underline{\text{via}}$ two σ-bonds so that a nearly planar
six atom heterocycle is formed. One of the carbonyl
groups very asymmetrically bridges two Fe atoms (106).

The molecule $Fe_3(CEt)\{C_5H_2Me_2(C_2H_3)\}(CO)_8$ contains
an Fe_3C core very similar to the Co_3C core found in
$Co_3(CY)(CO)_9$ [Y = alkyl, aryl or other group].
Trimerization of the alkyne leads to the formation of

a cyclopentadienyl unit η^5-bonded to one Fe atom. There are six terminal carbonyls and the other two bridge the shortest sides (mean 2.520 Å) of the Fe_3 triangle (107).

Two pentalene derivatives of $Ru_3(CO)_{12}$ have been structurally characterised. The 'X' substituents are H(a) or $SiMe_3$(b) (Figure 47). The unbridged Ru-Ru distances average 2.81 Å while the bridged is extended to 2.94 Å. The Ru-C(allylic) bond lengths average 2.23 Å, with the bridgehead carbon 2.48 Å from ruthenium. The plane of the pentalene unit makes an angle of 50° with the Ru_3 triangle (108). Two structurally different isomers of $Ru_3(CO)_8[1,3,5-(Me_3Si)_3C_8H_3]$ have been observed. They differ in the orientations of the pentalene ligands to the Ru_3 triangle, in one the structure is similar to that of the disubstituted derivative (Figure 48) but in the other one ring η^5-bonds to one Ru atom and the other ring bridges the opposite Ru-Ru edge (Figure 49) (109).

In $Ru_3(CO)_7(C_{10}H_8)$ (Figure 50) each Ru atom is bonded to two terminal carbonyls and two of the metals are bridged by the remaining CO group. The bridged and unbridged distances are 2.740(4) and 2.944(5) Å respectively. The azulene ligand coordinates to one Ru atom through the five membered ring, and to the other two metals through the seven membered ring (110).

In $Ru_3(CO)_6(C_7H_7)(C_7H_9)$ the $[C_7H_9]^-$ unit acts as a six electron donor, bonding in a η^5-fashion to one metal. The $[C_7H_7]^+$ ligand bridges two Ru atoms by π-allyl bonding to each metal and also forms a RuCRu linkage with the seventh carbon atom. Two carbonyl groups are bridging and the others are terminally bound to the metals involved in the RuCRu linkage (111).

$(C_8H_8)_2Ru_3(CO)_4$ has each cyclooctatetraene bound to a pair of Ru atoms with Ru-C distances in the range

2.14 - 2.91 Å. The Ru_3 atoms define an isosceles
triangle with two long (mean 2.938 Å) and one short
(2.782 Å) edge. The longer edges are bridged by the
organic rings. Two of the Ru atoms each have two
terminal carbonyl ligands (112). $(C_8H_8)(C_7H_7)Ru_3(CO)_6$
has two terminal carbonyl ligands bonded to each metal
in the Ru_3 triangle (113). Each organic species is
coordinated to two metal atoms (Figure 51).

In $Ru_3(CO)_6(C_{13}H_{20}O)(C_{12}H_{20})$ the three Ru atoms
define an isosceles triangle with two terminal carbonyl
ligands on each. The two organic ligands each donate
six electrons and are bonded to the cluster on opposite
sides of the Ru_3 plane. Two molecules of 3,3-dimethyl-
butyne-1 are joined _via_ a C-O bridge derived from the
opening of a carbonyl group, and this species
coordinates to the cluster through three σ- and π-
allyl bond. The second organic group, derived from a
head-to-tail joining of two alkyne molecules with the
shift of an H atom, is linked _via_ one π-allyl, one σ-
and one π-bond (114).

The osmium cluster $Os_3(CO)_9(CPh)_4$ (Figure 52)
exists in two crystalline modifications but with the
same structure. The Os-Os distances range from 2.734
to 2.905 Å with the longest bond between the $Os(CO)_3$
and $Os(CO)_4$ units. The Os-C σ-bonds have an average
length of 2.10 Å, and the Os-C π bonds range from 2.04
to 2.31(3) Å (115). In $Os_3(CO)_8(CPh)_4$ (Figure 53) one
metal atom forms σ-bonds with the two extreme acetylenic
atoms, while another forms σ-bonds with two acetylenic
carbons and a π-bond with the other two. The third
metal has sixteen electrons unless there is ortho-
metallation from one of the phenyl rings where the Os-C
distance is 2.18(7) Å (116). The third compound in
this series is $Os_3(CO)_7(CPh)_4(PhC_3Ph)$ (Figure 54).
The three metal atoms define a nearly equilateral

triangle. Two of the Os atoms are bonded to two
carbonyl ligands and the third metal to three. While
the chelation between the Ph_4C_4 ligand and the cluster
is attained _via_ σ-bonds and the donation of π-electrons
of the osmacyclopentadiene ring. The PhC_2Ph unit
appears to be a four-electron donor to the whole
cluster. The shortest Os-Os bond (2.680(2) Å) faces
the $(CPh)_4$ ligand (117).

The pyrolysis of $Os_3(CO)_{12}$ with triphenylphosphine
yields six compounds based on the Os_3 unit which have
been characterised crystallographically (118). In
these compounds, which are illustrated in Figure 55,
C-H bonds on some of the phenyl rings undergo oxidative
addition which results in phenyl carbon atoms bridging
Os-Os bonds _via_ σ-bonds. Triphenylphosphine molecules
also loose phenyl rings to give μ_2-bridging PPh_2 groups.

The molecule $(\mu-H)Os_3(CO)_8[\overline{C(O-)C(CHMe-)CHCHCEt}]$
contains a triangular arrangement of Os atoms with the
three Os-Os distances of 2.819(1), 2.889(1) and
3.007(1) Å. The long bond is associated with a single
equatorial μ_2-bridging hydride ligand. Two metal atoms
are each linked to three terminal carbonyls, while the
third Os atom is associated with two such ligands.
This metal is π-bonded to a trisubstituted cyclopentadienyl ligand which is additionally bridged to
the other Os atoms _via_ C(ring)-O-Os and C(ring)-CHMe-Os
linkages. The μ-hydride ligand has been formed by
oxidative addition of an aliphatic C-H bond to the
Os_3 cluster (119).

The exception to the triangular arrangement in
triosmium clusters is $Os_3(CO)_{12}I_2$ where there is a
straight chain of Os atoms. The iodine atoms occupy
equatorial sites _trans_ to one another, on the terminal
Os atoms. Each metal has an octahedral environment,
and, apart from the two I atoms all the other sites

are occupied by CO groups. The equatorial ligands on the terminal atoms are staggered with respect to the four CO groups on the central osmium, while the two terminal CO groups are collinear with the molecular axis. The Os-Os distance is 2.93 $\overset{o}{A}$ and the Os-I distance 2.77 $\overset{o}{A}$. The equatorial carbonyls on the terminal metals lean inwards towards the central Os atom (120).

The tetranuclear hydrido-carbonyl clusters $H_4Ru_4(CO)_{12}$ (121) and $H_4Os_4(CO)_{12}$ are isostructural (Figure 56). Both suffer from packing disorder, but it is clear that the molecules consist of a distorted tetrahedral core ($\underline{D}_{2d}$ symmetry) with four long and two short metal-metal distances. Each metal is coordinated to three terminal carbonyls. The arrangement of these ligands indicates that the four hydrides edge bridge the four long metal-metal bonds. The CO groups are staggered with respect to the metal-metal bonds in contrast to $H_4Re_4(CO)_{12}$ where the hydrides cap the triangular faces of the Re_4 tetrahedron. In $H_4Ru_4(CO)_{12}$ the long Ru-Ru bonds have an average length of 2.950 $\overset{o}{A}$ and the average length of the short Ru-Ru bonds is 2.786 $\overset{o}{A}$. In $H_4Os_4(CO)_{12}$ the long and short Os-Os bonds have average lengths of 2.964 and 2.816 $\overset{o}{A}$, respectively. Both compounds are 'electron-precise' 60-electron systems.

In the anionic species $[H_3M_4(CO)_{12}]^-$ (M = Ru, Os) there are several possible isomeric forms. Two isomers have been crystallographically characterised for the ruthenium complex (122). It is only the position of the bridging hydrides which differs between the isomers and they have idealised $\underline{C}_2$ and $\underline{C}_{3v}$ symmetry. In the $\underline{C}_2$ species (Figure 57) three Ru-Ru bonds are significantly longer (mean 2.923(1) $\overset{o}{A}$) than the other three (mean 2.803(1) $\overset{o}{A}$), and the carbonyl ligands bend away from these edges. This is consistent with the three hydrogens bridging these longer edges. In the C_{3v} isomer one triangular face of the tetrahedron has significantly longer Ru-Ru distances (mean 2.937(1) Å) than the remainder (mean 2.787(1) Å) (Figure 58). The carbonyl ligands appear to be pushed back from these edges which suggests that they are bridged by the three hydrides. Only the C_2 isomer has been crystallographically characterised for the corresponding osmium complex (123). Here the average long and short Os-Os distances are 2.939 Å and 2.794 Å.

The structure of the dianion $[H_2Os_4(CO)_{12}]^{2-}$ (124) shows that the Os atoms define a distorted tetrahedron with approximate D_{2d} symmetry (Figure 59). In this case, however, there are two long (mean 2.934 Å) and four short (2.798 Å) Os-Os bonds. Each metal atom is coordinated to three terminal carbonyl ligands, and, where appropriate, these groups bend away from the two long edges. This indicates that the hydrides bridge the two long trans edges. Like $H_4Os_4(CO)_{12}$ this dianion suffers from packing disorder.

In $H_2Ru_4(CO)_{13}$ the four ruthenium atoms define a distorted tetrahedron with four short Ru-Ru bonds (2.78(2) Å) and two long Ru-Ru bonds of 2.93(1) Å. The two long bonds are adjacent to each other (125). Three of the Ru atoms are each coordinated to three terminal carbonyl ligands. The fourth metal atom has four carbonyl groups bound to it but two of these form very asymmetric carbonyl bridges (1.93 and 2.40 Å) with two other metal atoms (Figure 60). [1]H n.m.r. data shows that there are two isomers in solution, differing in the position of the hydrides. In the solid state the hydrides are thought to bridge the longest Ru-Ru edges.

The cluster anion $[HFe_4(CO)_{13}]^-$ contains a

butterfly arrangement of metal atoms with a dihedral
angle of 117°. All five Fe-Fe distances are
equivalent to within experimental error with an
average value of 2.627 Å (126). Twelve of the
carbonyl groups are terminal, three bound to each
metal, but the thirteenth CO ligand behaves as a four-
electron donor and interacts with all four Fe atoms
(Figure 61). The conformation of the carbonyls
suggests that the hydride edge bridges the Fe-Fe
'hinge' bond of the butterfly. In the corresponding
dianion $\left[Fe_4(CO)_{13}\right]^{2-}$ the metal atoms adopt a tetrahedral configuration. An apical $Fe(CO)_3$ group is
symmetrically coordinated by only Fe-Fe bonds, with an
average length of 2.58 Å, to a basal $Fe_3(CO)_9$ fragment
which contains three identical $Fe(CO)_3$ groups located
at the corners of an equilateral triangle (mean edge
length of 2.50 Å); three of these carbonyls form very
asymmetric bridges with adjacent metals. The
thirteenth carbonyl triply bridges the three basal Fe
atoms (Figure 62) (127).

In the substituted Ru_4 complexes $H_4Ru_4(CO)_{10}(PPh_3)_2$
(121) and $H_4Ru_4(CO)_{11}P(OMe)_3$ the tetrahedron of metal
atoms has idealised $\underline{D_{2d}}$ symmetry with four long and
two short Ru-Ru bonds. It is believed that as in
$H_4Ru_4(CO)_{12}$ the four hydride ligands bridge the four
long edges. The two short Ru-Ru edges are trans to
each other. In $H_4Ru_4(CO)_{10}(PPh_3)_2$ the long edges
(mean 2.966(2) Å) are slightly longer than the value
in $H_4Ru_4(CO)_{12}$ and the shorter edges (mean 2.772(2) Å)
are slightly shorter. The average Ru-P distance is
2.359(4) Å.

By way of a contrast, the metal tetrahedron in
$H_4Ru_4(CO)_{10}(Ph_2PCH_2CH_2PPh_2)$ (128) is distorted in a
manner similar to that in $H_2Ru_4(CO)_{13}$. There are now
four long (mean 2.970 Å) and two short (2.791 Å)

edges and the two short bonds are adjacent to each
other. The diphosphine ligand chelates to the metal
atom which is coordinated to three of the μ_2-hydrides
bridging the long edges.

There are three tetranuclear ruthenium clusters
which display the butterfly configuration. In
$Ru_4(CO)_{12}(C_2Ph_2)$ the four Ru atoms and the two
acetylenic carbons describe a distorted octahedron
such that the PhCCPh group caps the Ru_4 butterfly
(129). The hinge Ru-Ru bond has a length of 2.85(1) Å
while the average Ru(hinge)-Ru(wingtip) distance is
2.73 Å. The acetylenic C-C distance is 1.46(2) Å and
σ-bonds are formed with the Ru(hinge) atoms and π-bonds
with the Ru(wingtip) atoms. Each metal is coordinated
to three terminal carbonyl groups (Figure 63). In
$Ru_4(CO)_{11}(C_8H_{10})$ one ethylenic group of the organic
species caps the Ru_4 butterfly while another π-bonds
to a wingtip ruthenium atom, replacing a terminal
carbonyl ligand (130). In $Ru_4(CO)_{10}(C_{12}H_{10})$ (Figure
64) the allylic system is attached to the two 'hinge'
Ru atoms by σ-bonds (2.14(4) Å) and π-bonded on both
sides to the two 'wingtip' metals. The Ru-Ru 'hinge'
bond length is 2.850(6) Å and the other metal-metal
bond lengths lie in the range 2.775 - 2.811(7) Å (131).

There are examples of tetra-osmium clusters with
the 'butterfly' geometry. In $H_3Os_4(CO)_{12}I$ the iodine
atom, which acts as a three-electron donor, caps the
Os_4 butterfly bonding to the two wingtip metal atoms
(Figure 65). Each metal is coordinated to three
terminal carbonyl ligands. The molecule has precise
$\underline{C}_2$ symmetry such that there are two long (3.052(1) Å)
and two short (2.876(1) Å) Os(wingtip)-Os(hinge) bonds.
The 'hinge' Os-Os bond is of intermediate length at
2.927(3) Å. From the locations of the carbonyl ligands
the hydrides are considered to bridge the three

longer edges (132). A similar structure is observed
in the cationic species $[H_4Os_4(CO)_{12}OH]^+$. The
hydroxyl group replaces the bridging iodine atom in
$H_3Os_4(CO)_{12}I$, and the hydrides are thought to edge
bridge the four long (mean 3.05 Å) Os(wingtip)-Os
(hinge) bonds.

The structure of product of the reaction cyclohexene with $H_4Os_4(CO)_{12}$ contains a distorted tetrahedron of Os atoms. The metal-metal bond lengths
range from 2.793(1) to 2.984(1) Å. The cyclohexene
ligand replaces a terminal carbonyl and a hydride
ligand to give a molecule with the molecular formula
$H_3Os_4(CO)_{11}(C_6H_9)$ (Figure 66). The cyclohexene ring
is σ-bonded to one metal atom and π-bound to another.
The C-C distance is 1.394(13) Å (133).

In one of the few Fe_4 clusters $Fe_4(CO)_{11}(HC_2Et)_2$
(Figure 67) the iron atoms are in a tetrahedrally
distorted square arrangement. Eight of the eleven
carbonyls are terminal, one is symmetrically bridging
two iron atoms, and two are very asymmetrically bridging. Each of the two alkyne ligands is σ-bonded to
two Fe atoms on the opposite vertices of the cluster
and π-bonded to the other two. The substituted
cluster has an Fe_4C_4 core in which the metal and carbon
atoms occupy the vertices of a triangulated dodecahedron. The four Fe-Fe bonds lie in the range 2.515(5)
- 2.644(4) Å (134). A square planar Fe_4 arrangement
is observed in $Fe_4(CO)_{11}(NEt)(ONEt)$ (Figure 68). The
two organic groups face bridge the two sides of the
metal cluster. Each metal atom has two terminal
carbonyl ligands, two of the three remaining CO groups
bridge Fe-Fe edges asymmetrically and the third
symmetrically (135).

There have been many more structural investigations
of ruthenium and osmium clusters containing five or

more metal atoms than for iron. With the heavier
metal clusters absorption effects often cause serious
problems in the X-ray diffraction studies and the
light atom positions tend to be inaccurate.

The binary carbonyl containing five osmium atoms
is $Os_5(CO)_{16}$ (136). In this molecule one $Os(CO)_4$ unit
and four $Os(CO)_3$ units are joined through metal-metal
bonds in such a way that the Os atoms define a
distorted trigonal bipyramid (Figure 69). The molecule
has precise $\underline{C}_2$ symmetry and the unique $Os(CO)_4$ group
occupies an equatorial site of the trigonal bipyramid.
The Os-Os bonds involving this unique metal atom are
significantly longer (mean 2.878 Å) than for bonds
involving the other metals (mean 2.750 Å). All the
carbonyl ligands are formally terminal but there are
two short non-bonded Os...C contacts (2.66(7) and
2.89(4) Å), both associated with significant departures
from linear Os-C-O units. These structural features
may be explained by investigating the formal oxidation
state of each metal. The unique Os atom has an
oxidation state of +2, the other two metal atoms in
the equatorial plane of the trigonal bipyramid have an
oxidation state of O, and the two axial metals have an
oxidation state of -1. Thus there is a highly polar
distribution within the metal framework. The shorter
metal-metal distances occur between atoms having the
same or at least similar oxidation states. A terminal
carbonyl on a metal of higher oxidation state forms a
semi-bridge bond to a metal of lower oxidation state
to partially redress the electron imbalance.

The corresponding hydrido-carbonyl cluster
$H_2Os_5(CO)_{16}$ has a metal skeleton based on an edge-
bridged tetrahedron (137). Four of the Os atoms are
coplanar and the fifth metal atom lies 2.39 Å above
this plane. All the carbonyl ligands are terminal and

there are no Os...C non-bonded contacts less than
2.93 Å, so there are no incipient bridging carbonyls
as found in $Os_5(CO)_{16}$. Spectroscopic data indicates
that there are two equivalent edge-bridging hydrides.
The disposition of the carbonyls suggests that the
hydrides edge bridge the two long (mean 2.962 Å) basal-
apical edges of the tetrahedron (Figure 70). The two
Os-H-Os bridges are considered to be two-electron
three-centre bonds. The anion $[HOs_5(CO)_{15}]^-$ has a
trigonal bipyramidal arrangement of metal atoms (138).
The structure may be derived from that of $Os_5(CO)_{16}$ by
the replacement of a carbonyl ligand attached to the
unique Os atom by a presumably bridging hydride.
There is evidence for incipient carbonyl bridge
bonding between Os atoms in different formal oxidation
states. A similar structure is observed for
$[IOs_5(CO)_{15}]^-$ (139). Here the iodine replaces one of
the terminal carbonyls on the unique $Os(CO)_4$ unit.
Semi-bridging carbonyl groups again occur. The Os-I
distance is 2.740(2) Å, and it is possible that the
negative charge is localised on this terminal ligand
weakening the bond. Iodine substitution seems to
lower the electron imbalance because the relatively
long Os-Os bonds involving the unique Os atom in
$Os_5(CO)_{16}$ are reduced by about 0.05 Å in the iodide.

We have seen that the pyrolysis of $Os_3(CO)_{11}PPh_3$
yields a number of clusters based on the Os_3 unit. In
contrast, the four products of the pyrolysis of
$Os_3(CO)_{11}P(OMe)_3$ which have been fully characterised
are based on the Os_5 unit. Their structures are shown
in Figure 71, the carbonyl groups have been omitted
for the sake of clarity. In $HOs_5C(CO)_{14}[OP(OMe)_2]$ the
metal atoms define a distorted trigonal bipyramid and
a carbide atom lies in the equatorial plane coordinating to all five Os atoms (140). One Os(equatorial)-Os

(equatorial) and one Os(equatorial)-Os(axial) bond
are lengthened to 3.98 Å and 3.74 Å, respectively, so
there can be no direct metal-metal interaction between
these atoms. The phosphonate ligand which acts as a
three electron donor coordinates to two Os atoms to
form an Os-P-O-Os linkage. In $HOs_5C(CO)_{13}[OP(OMe)_2]$
$[P(OMe)_3]$ the metals again define a trigonal bipyramid
with two extended edges and a carbide lies at the
centre of the cluster. One phosphite forms the Os-P-
O-Os linkage while the other terminally bonds to the
Os atom which is coordinated to the phosphite oxygen,
replacing a carbonyl group. In the third compound in
the series $HOs_5C(CO)_{13}[OP(OMe)OP(OMe)_2]$ the metal
skeleton is similar to that in the first two complexes
and the carbide occupies a similar location. This
compound may be derived from the previous one by a
combination of the two phosphite ligands to form a
complex Os-P-O-P(Os)-O-Os linkage (141). In the last
compound in the series $Os_5(CO)_{15}POMe$ the metal atoms
adopt a square-pyramidal geometry. The apical Os atom
sits 1.981 Å above the basal plane. The POMe group,
which acts as a four electron donor, lies 1.185 Å
below the basal Os_4 plane and coordinates to all four
metals. Each Os atom is also bonded to three terminal
carbonyl groups so that all five metals are seven
coordinate (142).

The only Fe_5 cluster so far investigated
crystallographically is $Fe_5C(CO)_{15}$ (143). This structure
is closely related to that of $Os_5(CO)_{15}POMe$. The iron
atoms define a square-based pyramid and the carbide lies
just below the centre of the basal plane (Figure 72).
The Fe(base)-C(carbide) bond lengths are 1.89 Å but the
Fe(apex)-C(carbide) distance is 1.96 Å. The average
Fe-Fe distance is 2.64 Å.

The carbide atom in $[Fe_6C(CO)_{16}]^{2-}$ lies in the

centre of the octahedron of iron atoms (Figure 73).
Five of the metal atoms are bound to two terminal CO
groups, and the sixth to three, and there are also
three μ_2-CO groups (144). In the corresponding neutral
ruthenium complex $Ru_6C(CO)_{17}$ there is only one bridging
carbonyl ligand. The two Ru atoms bridged by this
group are also coordinated to two terminal carbonyls
and the other four metal atoms are each bonded to
three terminal CO groups (Figure 74). The Ru-Ru
distances in the octahedron range from 2.840(6) to
3.034(5) Å, and the mean Ru-C distance to the central
carbide is 2.05 Å (145). The structure of $Ru_6C(CO)_{14}$
$(C_6H_3Me_3)$ is derived from that of $Ru_6C(CO)_{17}$ by the
replacement of three terminal carbonyls on one metal
atom by a π-bound arene ligand (146). The mean Ru-C
(arene) distance is 2.24 Å.

The cluster anion $[HRu_6(CO)_{18}]^-$ crystallises in a
number of modifications (147). In all of them the
ruthenium atoms define a slightly distorted octahedron
with a mean edge length of 2.87 Å. Each metal atom is
coordinated to three terminal carbonyl ligands and
preliminary neutron diffraction data shows that the
hydride lies in the centre of the octahedron. The
carbonyl ligands are eclipsed with respect to the Ru-
Ru bonds (Figure 75). The corresponding osmium complex
$[HOs_6(CO)_{18}]^-$ also has a slightly distorted octahedral
metal skeleton with three terminal carbonyl groups
associated with each Os atom (148). In this case one
face of the octahedron has significantly longer edges
(2.973 Å) than the others (2.863 Å) and the carbonyl
groups appear to bend away from it. This suggests that
the hydride ligand caps this face. The dianion
$[Os_6(CO)_{18}]^{2-}$ has precise $\underline{D}_3$ symmetry. There are two
equivalent parallel faces that are exactly equilateral
(edge 2.876 Å). These are slightly twisted from $\underline{O}_h$

symmetry so that the remaining bonds are alternatively long (2.886 Å) and short (2.814 Å). Three terminal carbonyls are bonded to each metal.

The structures of the two complexes $H_2M_6(CO)_{18}$ (M = Ru and Os) are different. The metal atoms in the ruthenium complex define an octahedron with two enlarged trans faces (Ru-Ru 2.950 - 2.959 Å). The other metal-metal bond length lies in the range 2.858 - 2.867 Å. Each metal is associated with three terminal carbonyls, and as these bend away from the enlarged faces it is presumed that the two hydrides μ_3-bridge these two faces (149). In $H_2Os_6(CO)_{18}$ the metal atoms adopt a monocapped square-pyramidal geometry (Figure 76). The Os-Os distances range from 2.805 Å to 2.965 Å. The two longest edges are the two parallel Os-Os bonds one of which is bridged by the μ_3-Os group. An examination of the distribution of the carbonyl ligands, three of which are bonded to each metal, indicates that the hydrides μ_2-bridge these two long Os-Os bonds.

The parent carbonyl in the other series of Os_6 clusters is $Os_6(CO)_{18}$. The osmium atoms define a bicapped tetrahedron. Each metal atom has three terminally bound carbonyls. The individual Os-Os bond lengths show variations which reflect the formal oxidation state of each metal. The Os-Os bond of the central tetrahedron which is capped by the two 'wingtip' Os atoms is 2.757 Å long. Each metal is in the +1 oxidation state and have five short metal contacts. The bond between the other two Os atoms in the tetrahedron is 2.731 Å long. These atoms are in formal oxidation state 0 and has four metal contacts. The bond lengths between atoms in different oxidation states are longer (150).

The structure of the cluster $Os_6(CO)_{16}(CNBu^t)_2$ is

derived from that of $Os_6(CO)_{18}$ by the replacement of
one carbonyl group on each of the capping metals by a
terminal isocyanide ligand (151). This substitution
has little effect on the bond lengths in the central
tetrahedron but causes a slight lengthening of the Os-
Os bonds involving the capping metals. Although the
isocyanides are positioned so as to induce $\underline{C}_s$ molecular
symmetry the metal cluster shows a marked $\underline{C}_2$ distortion
which contradicts the expected trans effect of the
$CNBu^t$ groups. In another isocyanide complex $Os_6(CO)_{18}$
$(CNC_6H_4CH_3)_2$ one of these ligands is terminal but the
other, which acts as a four-electron donor, bridges
three metal atoms (Figure 77). The structure may be
derived from that of $Os_6(CO)_{18}$ by breaking three metal-
metal bonds involving the central tetrahedron and one
capping metal, and inserting a triply bridging
isocyanide group which coordinates to one Os atom
through carbon and edge bridges the terminal Os-Os bond
through nitrogen. The terminal isocyanide group is bonded
to the terminal metal (152). Although the carbonyl
groups in both these clusters are formally terminal
the metals are in varying oxidation states and there
is evidence for incipient carbonyl bridge bonding (153).

The action of ethylene on $Os_6(CO)_{18}$ yields two
products which have been fully characterised. In the
first, $Os_6(CO)_{16}(CMe)_2$, the metals define a monocapped
square-based pyramid similar to that found in
$H_2Os_6(CO)_{18}$. The structure may be considered as
derived from the dihydride by the replacement of the
hydrogen atoms and two carbonyl ligands, one from each
Os atom in the basal plane which is bonded to the face
capping metal, by two CMe groups. One of these ligands
caps the square basal plane and the other an adjacent
triangular face, although both must be three-electron
donors (Figure 78). The Os-Os bond lengths in this

complex are similar to those in $H_2Os_6(CO)_{18}$ and the
longest bond is between the apical Os atom of the
pyramid and the capping Os atom (2.960 Å). It is this
bond which is broken upon the formation of the second
complex $Os_6(CO)_{16}(MeCCMe)C$. The square based pyramid
of Os atoms is retained but the sixth metal atom lies
on the opposite side of the basal plane to the apical
Os and edge bridges one side of the base (Figure 79).
The but-2-yne group caps the 'butterfly' configuration
now obtained by the breaking of the metal-metal bond.
It is π-bound to one 'wingtip' Os atom and σ-bound to
the two 'hinge' atoms. A carbide carbon atom lies just
under the square basal plane and donates four electrons
to the cluster (154). The reaction of phenylacetylene
with $Os_6(CO)_{18}$ yields the cluster $Os_6(CO)_{16}(CPh)_2$
which has the same structure as $Os_6(CO)_{16}(CMe)_2$ with
the CMe groups replaced by CPh (153).

The only Os_7 cluster which has been characterised
crystallographically is $Os_7(CO)_{21}$. Here the metals
defind a monocapped octahedron. Within the octahedron
the mean Os-Os distance is 2.855 Å while the capping
atom forms metal-metal bonds of average length 2.817 Å
(156). The carbonyl ligands are all terminal and
three bond to each metal (Figure 80).

In the 'pseudo' Os_9 anion $[HOs_3(CO)_{10}O_2C.Os_6(CO)_{17}]^-$
Os_6 and Os_3 units are linked by a CO_2 bridge (Figure
81). The Os_6 cluster may be regarded as a trigonal
bipyramid of five $Os(CO)_3$ groups with one triangular
face capped by an Os atom. The CO_2 bridge is attached
to this atom and may be regarded as either a carbene or
carboxylato substituent. The Os_3 cluster has a triangular
metal skeleton and the hydride ligand is thought to
bridge one of these edges. The Os-O-C-O-Os ring is
virtually planar and the stability of the complex may
be enhanced by some electron delocalisation in this ring
(157).

2.6. Cobalt, Rhodium and Iridium

The structural cluster chemistries of cobalt and rhodium are similar, although there are some important differences. Rhodium shows a greater tendency to form clusters containing more than eight metal atoms than does cobalt. There are a large number of trinuclear cobalt clusters with general formula $YCo_3(CO)_9$, where Y acts as a μ_3-capping group, which have been structurally characterised. Few Rh_3 complexes have been investigated. The structural chemistry of iridium clusters has been neglected until recently.

Structural data on cobalt complexes of the type $(\mu_3-Y)Co_3(CO)_9$ are summarised in Table 7. Included in this list are compounds in which one or more of the carbonyl groups in the basal $Co_3(CO)_9$ species have been replaced by other ligands. In the structures of the unsubstituted compounds the Co_3 triangle is usually close to the idealised equilateral geometry and the three carbonyl groups on each metal are terminally bound. When the basal CO groups are substituted the metal triangle may distort because of the influence of the new ligands. Figure 82 is a diagram of $CH_3CCo_3(CO)_9$, which may be considered as the parent compound of the series.

The complex with $Y = \mu_3-CCCo_3(CO)_9$ (Figure 83) is best described as a dimer, $[CCo_3(CO)_9]_2$, where the groups are linked through a C-C bond of length 1.37(2) $\overset{\circ}{A}$. It is possible to take this a step further. In $Co_8(CO)_{24}C_6$ (169) two $CCo_3(CO)_9$ units having approximate $\underline{C}_{3v}$ symmetry are linked by a four-carbon

chain containing two triple bonds, one of which
bridges a $Co_2(CO)_6$ unit (Figure 87).

It is not only possible to cap one face of the
Co_3 triangle to form a tetrahedral arrangement, as in
$SCo_3(CO)_9$, but also to cap both faces to form a
trigonal bipyramid. For example, this occurs in the
three clusters with the general formula $[(\eta^5-C_5H_5)_3Co_3$
$(\mu_3-X)(\mu_3-Y)]^{n+}$ (Figure 85) where:

        a; X = CO, Y = S, n = 0
        b; X = Y = S, n = 0
        c; X = Y = S, n = 1

The average Co-Co distance in (a) is 2.452(2) and in
(b) it is 2.687(3) Å. In (c) there are two distances
of 2.649(1) Å and one of 2.474(2) Å (169). The
variations in the metal-metal bond length are thought
to reflect the number of electrons in antibonding
orbitals: there are none in (a), two in (b) and one in
(c).

The complex $(\eta^5-C_5H_5)_3Co_3(\mu_3-CO)(\mu_2-CO)_2$ (170) is
interesting because it is found that the geometry of
the carbonyls is very sensitive to subtle changes in
the environment. In the solid state structure
(Figure 86) one carbonyl caps the Co_3 triangle acting
as a μ_3-ligand, while the other two bridge two of the
edges. In solution the structure is different. Two
of the carbonyl ligands bridge the same edge, and the
third coordinates linearly to the Co not associated
with the bridging groups.

Observations on the triangular clusters which
have been fully characterised for rhodium suggest that
both carbonyl and cyclopentadienyl ligands are required
to give stable electronic configurations. The structures
of two isomers of $(\eta^5-C_5H_5)_3Rh_3(CO)_3$ have been
determined. The less abundant isomer has a structure

TABLE 7

Trinuclear Complexes of Cobalt with the General Formula $(\mu_3-Y)Co_3(CO)_{9-n}X_n$.

Y	X	Co-Co bond length (Å)	Co-Y bond length (Å)	Ref
μ_3-CMe	-	2.467 (av)	1.90 (av)	(158)
μ_3-COB(Cl)$_2$(NEt$_3$)	-	2.472 (av)	1.90 (av)	(159)
μ_3-COB(Br)$_2$(NEt$_3$)	-	2.484 (av)	1.93 (av)	(160)
μ_3-COB(H)$_2$(NMe$_3$)	-	2.495 (av)	1.92 (av)	(161)
μ_3-CCo$_3$(CO)$_9$	-	*	1.96 (av)	(162)
μ_3-S	-	2.637 (av)	2.139 (av)	(163)
μ_3-CSSCCo$_3$(CO)$_9$	-	2.478 (av)	1.89 (av)	(164)
μ_3-CMe	P(C$_6$H$_{11}$)$_3$	2.52, 2.53, 2.38		(165)
μ_3-CMe	As$_2$Me$_4$(CF$_2$)$_2$ replaces two CO groups	2.44, 2.48, 2.47	1.903 (av)	(166)
μ_3-CEt	$(\pi-C_7H_8)$ replaces two CO groups	2.483, 2.475, 2.470	1.86, 1.88, 1.90	(167)
μ_3-CEt	$(\eta^5-C_5H_5)_2$ replaces five CO groups, one CO now μ_2-bridging	2.368, 2.477, 2.480	1.85 (av)	(168)
μ_3-CPh	$(\eta^6-C_6H_3Me_3)$ replaces three CO groups	2.441, 2.477 (twice)	1.89 (av)	(162)
μ_3-CPh	(C_8H_8) replaces three axial CO groups	*	*	(162)

*data not available

similar to that proposed for $(\eta^5 C_5H_5)_3Co_3(CO)_3$ Figure
(86) in solution. The Rh_3 triangle has edge lengths
of 2.62, 2.66 and 2.71 Å (171). In the other isomer
the metals define an equilateral triangle with a side
of 2.62 Å and the carbonyls bridge the three edges
(172).

$(\eta^5-C_5H_5)_4Rh_3H$ has the structure shown in
Figure 87. The Rh_3 triangle is equilateral with an
edge length of 2.72 Å. The unique cyclopentadienyl
ring and the hydride are bonded to the top and bottom
of the triangle, respectively (173).

A 'semi-triple-bridging' carbonyl ligand is
observed in the anion $[(\eta^5 C_5H_5)_2Rh_3(CO)_4]^-$ (Figure 88).
One Rh-Rh edge is bridged by two carbonyl groups,
while the unique Rh atom in the triangle is bonded to
two terminal carbonyls and formally carries the
negative charge. The two bridging carbonyls bond up
towards the unique Rh and the resulting Rh...C
contacts of 2.40 Å are very short (174).

The nature of the bridging carbonyls are different
in the two closely related complexes $(\eta^5 C_5H_5)_3Rh_3(CO)-$
$(C_6H_5C_2C_6H_5)$ and $(\eta^5-C_5H_5)_3Rh_3(CO)(C_6F_5C_2C_6F_5)$ (175).
In the diphenylacetylene complex the carbonyl lies on
the opposite side of the triangle to the organic group
in the μ_3-mode. In the second complex it sits on the
same side of the triangle as the acetylenic species and
bridges a Rh-Rh edge (Figure 89). It is proposed from
orbital energetic arguments that the composite π-donor
and π^*-acceptor interaction of the RC_2R ligand with the
unique Rh atom in competition with the relative π-
acceptor of the carbonyl as a triply bridging ligand
in its interaction with the metal is the prime driving
force which gives rise to one or other of the observed
structures.

Table 8 gives a list of the smaller binary carbonyl

clusters, hydrido-carbonyl clusters, and carbonyl
clusters containing carbido species for cobalt and
rhodium.

Both dodecacarbonyl clusters have idealised $\underline{C}_{3v}$
symmetry. In each the apical $M(CO)_3$ group is
coordinated by three metal-metal bonds to the basal
$M_3(CO)_9$ fragment which contains three chemically
identical $M(CO)_2$ groups linked to each other by metal-
metal bonds and μ_2-carbonyl groups (Figure 90). There
is little difference between the bridged and unbridged
metal-metal distances in each complex; and these
distances range from 2.457(3) to 2.527(3) Å for Co and
from 2.70(1) to 2.80(1) Å for Rh.

A number of substituted complexes based on the
tetrahedral $M_4(CO)_{12}$ unit have been studied. In
$Rh_4(CO)_8[P(OPh)_3]_4$ (187) one terminal carbonyl on each
metal has been replaced by a phosphite ligand. The
$Rh_4(CO)_8P_4$ species has approximate $\underline{C}_s$ symmetry, and
the mean Rh-Rh distance is 2.72 Å. A similar structure
is observed when the phosphite is replaced by a
bidentate phosphine, as in $Rh_4(CO)_8(Ph_2PCH_2PPh_2)_2$,
where the Rh-Rh distances are in the range 2.671(1) -
2.740(1) Å (186). A distorted Co_4 tetrahedron is
retained in $Co_4(NO)_4(\mu_3-NCMe_3)_4$ (Figure 91). The four
nitrosyl ligands are terminal, one coordinated to each
metal, and the isocyanide ligands cap the four faces
of the tetrahedron (188). Cyclopentadienyl ligands
may also replace carbonyl groups in Co_4 clusters. In
$H_4(\eta^5-C_5H_5)_4Co_4$ (189) where each Co is π-bonded to one
cyclopentadienyl ring, the average Co-Co distance is
2.467 Å. The four hydrides cap the faces of the
tetrahedron (Figure 92). The position of the hydrides
over the triangular faces may be explained in terms of
the directions of the orbitals, using the same argument
as in the case of $H_4Re_4(CO)_{12}$.

TABLE 8

Cobalt Clusters	Rhodium Clusters	Metal Polyhedron
$Co_4(\mu_2\text{-}CO)_3(CO)_9$ (176)	$Rh_4(\mu_2\text{-}CO)_3(CO)_9$ (177)	tetrahedron
$[Co_6(\mu_3\text{-}CO)_8(CO)_6]^{4-}$ (178)		octahedron
$[Co_6(\mu_3\text{-}CO)_3(\mu_2\text{-}CO)_3(CO)_9]^{2-}$ (179)		octahedron
	$[Rh_6(\mu_2\text{-}CO)_9(CO)_6C]^{2-}$ (180)	trigonal prism
	$[Rh_6(\mu_3\text{-}CO)_4(CO)_{11}I]^{-}$ (181)	octahedron
$Co_6(\mu_3\text{-}CO)_4(CO)_{12}$ (179)	$Rh_6(\mu_3\text{-}CO)_4(CO)_{12}$ (182)	octahedron
	$[Rh_7(\mu_3\text{-}CO)_3(\mu_2\text{-}CO)_6(CO)_7]^{3-}$ (183)	monocapped octahedron
	$[Rh_7(\mu_3\text{-}CO)_4(\mu_2\text{-}CO)_2(CO)_{10}(\mu_2\text{-}I)]^{2-}$ (184)	monocapped octahedron
$[Co_8(\mu_2\text{-}CO)_{10}(CO)_8C]^{2-}$ (185)		tetragonal antiprism
	$Rh_8(\mu_3\text{-}CO)_2(\mu_2\text{-}CO)_6(CO)_{11}C$ (186)	monocapped trigonal prism with one edge bridged

Larger distortions, involving the breaking of a metal-metal bond, occur when ethylenic or acetylenic ligands are allowed to interact with Co_4 species. In $Co_4(CO)_{10}(EtC_2Et)$ (190) the four metal atoms adopt a 'butterfly' configuration with a dihedral angle of 118° (Figure 93) and a non-bonding Co...Co distance of 3.55 $\overset{o}{A}$. The organic species sits above the 'hinge' bond and donates four electrons to the cluster through two σ-bonds to the hinge cobalt atoms and through a 'bent' μ_2-type bridge bond to the 'wingtip' atoms. The Co-Co hinge distance of 2.55 $\overset{o}{A}$ is about 0.1 $\overset{o}{A}$ longer than the mean bond length of the other four equivalent Co-Co bonds. The acetylenic C-C distance is 1.44 $\overset{o}{A}$ which is consistent with considerable electron donation to the Co_4 system.

The addition of two more electron pairs to the closed shell configuration in $Co_4(CO)_{12}$ causes two metal-metal bonds to break, leaving the rectangular Co_4 configuration observed in $Co_4(\mu_2\text{-CO})_2(CO)_8(\mu_4\text{-PPh})_2$ (Figure 94). The longer edges of the rectangle (2.697(2) $\overset{o}{A}$) are unbridged while the shorter ones (2.519(2) $\overset{o}{A}$) have μ_2-CO groups associated with them. The two phosphorus-containing ligands each donate four electrons to the cluster, giving a 64 electron system, and cap the two rectangular faces to form an octa-hedral-like Co_4P_2 core (191). The P...P distance of 2.544(3) $\overset{o}{A}$ indicates bonding interaction between these atoms. This interaction is probably stronger than the S...S interaction in $Co_4(\mu_2\text{-CO})_2(CO)_8(\mu_4\text{-S})_2$ (192).

In the Co_5 cluster, $Co_5(CO)_{11}(PMe_2)_3$ (193), the metal skeleton is based on a trigonal bipyramid with one Co(axial)-Co(equatorial) bond ruptured (Figure 95). In the central $Co_3(CO)_6$ unit the metal-metal distances lie in the range 2.41 - 2.47 $\overset{o}{A}$. The axial-equatorial

Co-Co bonds bridged by the PMe_2 groups are shorter (range 2.76 - 2.81 Å) than the unbridged distances (2.89 and 2.92 Å). The non-bonded Co...Co distance is 3.21 Å.

Both the anionic Co_6 clusters $[Co_6(\mu_3-CO)_8(CO)_6]^{4-}$ (Figure 96) and $[Co_6(\mu_3-CO)_3(\mu_2-CO)_3(CO)_9]^{2-}$ (Figure 97) have distorted octahedral metal skeletons. In the former, eight μ_3-CO groups cap the faces of the octahedron, and the remaining six carbonyls terminally bond to one Co atom such that the molecule has overall S_6 symmetry. The mean Co-Co distance is 2.50 Å. In the dianion the average Co-Co distance is 2.49 Å. Nine of the carbonyls are terminally coordinated, three bridge edges, and three cap faces. In a third Co_6 cluster, $Co_6(CO)_{12}S_2C$ (164), the metal skeleton is a trigonal prism, the two ends of which are capped by μ_3-S atoms. A carbide atom lies at the centre of the prism (Figure 98). Each metal atom is coordinated to two terminal carbonyl groups. Within the triangular faces the mean Co-Co distance is 2.437(4) Å, and in the rectangular ones it is 2.669(5) Å. The mean Co-S and Co-C (carbide) distances are 2.194(6) and 1.94(2) Å, respectively.

Preliminary crystallographic measurements have shown that $Co_6(\mu_3-CO)_4(CO)_{12}$ is isostructural with $Rh_6(\mu_3-CO)_4(CO)_{12}$ (Figure 99). The Rh-Rh distances in the Rh_6 octahedron average 2.776(1) Å. The four μ_3-CO groups are located on three-fold axes above four of the octahedral faces; the mean Rh-C distance is 2.168(12) Å. The remaining twelve carbonyls, two coordinated to each metal, are arranged such that crystallographic C_2 symmetry is retained. The average terminal Rh-C distance is 1.864(15) Å. The structure of $Rh_6(CO)_{12}[P(OPh)_3]_4$ (187) is derived from that of the parent carbonyl by the replacement of four terminal carbonyls by phosphites,

alternatively positioned two above and two below the
same Rh_4 plane. The mean Rh-Rh bond length, at 2.79 Å,
is slightly longer than that in $Rh_6(CO)_{16}$. The
anionic species, $[Rh_6(\mu_3-CO)_4(CO)_{11}I]^-$, is derived
from the hexadecacarbonyl by the replacement of a
terminal carbonyl by an iodine atom. This has little
effect on the average Rh-Rh distance (2.746 Å). The
Rh-I bond length is 2.709(6) Å. The presence of the
carbide at the centre of the $[Rh_6(CO)_{15}C]^{2-}$ anion
(Figure 100) may cause the metals to adopt the
observed trigonal prismatic arrangement. Carbonyl
groups symmetrically bridge the edges of the triangular
and rectangular faces, and one additional terminal
carbonyl is bonded to each of the six rhodium atoms.
Within the triangular faces the average Rh-Rh distance
is 2.776(3) Å and for the rectangular faces the average
distance is 2.817(2) Å. The mean Rh-C (carbide)
distance is 2.134(6) Å.

The two Rh_7 clusters, $[Rh_7(\mu_3-CO)_3(\mu_2-CO)_6(CO)_7]^{2-}$
(Figure 101) and $[Rh_7(\mu_3-CO)_4(\mu_2-CO)_2(CO)_{10}(\mu_2-I)]^{2-}$
(Figure 102), have monocapped octahedral arrangements
of metal atoms. In the first anion the Rh-Rh distances
all lie in the range 2.72 - 2.81 Å, while in the
second the range is 2.74 - 3.00 Å. The longer
distances are associated with the capping Rh atom, and
the bond destabilisation may be ascribed to the
presence of an electron pair in a low-lying antibonding orbital delocalised in the tetrahedral region
of the cluster. In the second cluster the source of
the electrons is the bridging iodine atom which acts
as a three electron donor. The average Rh-I distance
of 2.73 Å is similar to the Rh-I (terminal) distance in
$[Rh_6(CO)_{15}I]^-$.

The difference in the geometries of the metal
skeletons of $[Co_8(\mu_2-CO)_{10}(CO)_8C]^{2-}$ (Figure 103) and

$Rh_8(\mu_3-CO)_2(\mu_2-CO)_6(CO)_{11}C$ (Figure 104) may be due to the spatial requirements of the carbide atoms and to the steric hinderance caused by the additional carbonyl group in the rhodium cluster. In the tetragonal anti-prismatic geometry adopted by the Co_8 species all metal atoms interact with the carbide. Four of the Rh-C bonds average 1.99 Å in length while the remaining four average 2.15 Å. In the edge-bridged, mono-capped trigonal prismatic geometry of the rhodium cluster only six metal atoms are coordinated to the carbide at a mean distance of 2.127 Å. It may be that the space inside a trigonal prism of Co atoms is not large enough to accommodate a carbide. The average Co-Co distance is 2.52 Å. Each Co has one terminal carbonyl, and the remaining ten groups range from partially bent to symmetrically edge bridging. In the Rh complex the metal-metal distances range from 2.699 to 2.913(3) Å, with a mean of 2.81 Å.

At the time of writing seven rhodium clusters containing more than eight metal atoms have been characterised by X-ray diffraction. They are listed in Table 9, and diagrams of their structures are shown in Figure 105. In each of these compounds the arrangement of metal atoms is similar to that found in the bulk metal. The analogy between the situation in the clusters and that in the metal becomes closer as the number of metal atoms in the cluster increases.

TABLE 9

Higher Rhodium Clusters

$[Rh_{12}(CO)_{30}]^{2-}$ (194) $[Rh_{14}(CO)_{25}]^{4-}$ (197)

$[Rh_{12}(CO)_{25}C_2]^{2-}$ (195) $[Rh_{15}(CO)_{27}]^{3-}$ (197)

$[H_3Rh_{13}(CO)_{24}]^{2-}$ (196) $[Rh_{15}(CO)_{28}C_2]^{-}$ (198)

$[Rh_{17}(CO)_{32}(S)_2]^{3-}$ (199)

$[Rh_{12}(CO)_{30}]^{2-}$ is a centrosymmetric dimer of the $[Rh_6(CO)_{15}]^-$ unit. The catenation between the two sub-units is achieved by a Rh-Rh bond and two μ_2-CO groups. The cluster exhibits a range of metal-metal distances from 2.68 Å to 2.85 Å. In each subunit there are 10 terminal carbonyls, four μ_3-CO's asymmetrically bridging alternating octahedral faces, and one μ_2-CO which is involved in bridging the connecting Rh-Rh bond between units.

The structure of $Rh_{12}(CO)_{25}C_2$ cannot easily be described in terms of the fusion of smaller polyhedra. The metal atoms define an irregular, closed polyhedron consisting of larger packing of atoms. The mean Rh-Rh distance is 2.79 Å. The carbide atoms occupy an irregular cavity as a C_2 unit with a C-C distance of 1.48(2) Å. There are 14 Rh-C (carbide) contacts, nine short (mean 2.22 Å) and five long (mean 2.58 Å). Fourteen carbonyls are terminal, 10 are edge bridging, and one is face capping.

$[H_3Rh_{13}(CO)_{24}]^{2-}$ is comprised of three nearly parallel layers of metal atoms (mean Rh-Rh of 2.81 Å) to give the cluster approximate $\underline{D}_{3h}$ symmetry. This represents a fragment of hexagonal close packing. The central metal atom is twelve-coordinate, and the surface atoms are all seven coordinate, bonding to four other metals and three carbonyl groups. Twelve of the carbonyls are terminal and the rest act as μ_2-ligands bridging half the polyhedral edges.

The structures of $[Rh_{15}(\mu_2\text{-CO})_{14}(CO)_{13}]^{3-}$ and $[Rh_{14}(\mu_2\text{-CO})_{16}(CO)_9]^{4-}$ show a stepwise hexagonal close-packed/body centred cubic interconversion. $[Rh_{15}(CO)_{27}]^{3-}$ may formally be derived from the poly-hedron of $\underline{D}_{3h}$ symmetry which contains that part of the hexagonal close-packed array found in $[H_3Rh_{13}(CO)_{24}]^{2-}$ by capping two of the square faces. However, while one part of the polyhedron follows this close packing, the other part is distorted and the metals define one half of a cube with a metal atom at the centre. The metal-metal distances are rather scattered: there are 36 normal contacts ranging from 2.60 to 2.86 Å, 7 intermediate contacts from 2.90 to 3.00 Å, and 5 long interactions from 3.15 to 3.91 Å. The metal skeleton in $[Rh_{14}(CO)_{25}]^{4-}$ corresponds to an incomplete rhombic dodecahedron elongated about the crystallographic quaternary axis. This is essentially part of a body-centred cubic lattice. There are eight normal Rh-Rh bonds ranging from 2.63 to 2.79 Å, two intermediate ones at 3.00 and 3.08 Å, and two long contacts at 3.33 and 3.38 Å.

The $[Rh_{15}(CO)_{28}C_2]^-$ anion has precise $\underline{C}_2$ symmetry. The metal skeleton can be described either as a centred and tetracapped pentagonal prism in which the central atom is 12 coordinate, or as two fused octa-hedra sharing one vertex, plus four extra atoms forming tetrahedra fused with the octahedra and themselves. Rh-Rh distances range from 2.738 to 3.332(3) Å. The two carbide atoms occupy octahedral cavities, and the mean Rh-C distance is 2.04 Å. The carbonyl ligands are bound to the cluster surface; 14 linearly and 14 edge-bridging, such that each metal is bonded to three ligands.

In $[Rh_{17}(CO)_{32}(S)_2]^{3-}$ 16 of the Rh atoms are at the corners of four stacked parallel squares, and a S-Rh-S group is located inside the cluster. The Rh-Rh bonding distances range from 2.76 to 2.88 Å, and the Rh-S distances are of two types with mean values of 2.17 and 2.33 Å. 16 carbonyls are terminally bound to an equal number of Rh atoms. Eight of the remaining carbonyls bridge the edges of the basal planes, while the others are found along edges connecting the two internal Rh_4 planes.

The only structurally characterised trinuclear iridium complex is $[\{IrH_2(PCy_3)(C_5H_5N)\}_3(\mu_3-H)][PF_6]_2$. The Ir-Ir distances in the Ir_3 triangle range from 2.755(1) to 2.775(1) Å. Each metal atom is coordinated to a terminal phosphine (mean Ir-P 2.286 Å), to the nitrogen of a piperidine ligand (mean 2.139 Å) and to two terminal hydrides (deduced from the distribution of the other ligands). The other hydride caps the Ir_3 plane (200). The positions proposed for the hydrides are in agreement with the infrared and n.m.r. data.

The metal atoms in the parent tetranuclear iridium cluster $Ir_4(CO)_{12}$ define a tetrahedron with an Ir-Ir bond length of 2.693 Å. Each metal atom is coordinated to three terminal carbonyl groups. The $Ir(CO)_3$ units are arranged so that the carbonyls are staggered with respect to the Ir-Ir bonds; the polyhedron defined by the 12 ligands is a cubooctahedron (Figure 106). Although disorder problems make the determination of accurate atomic positions difficult it is clear that the geometry is close to the idealised $\underline{T}_d$ symmetry. The molecule is 'electron precise' in that it has the 60 electron closed shell configuration (201).

The reaction of halide and pseudo-halide ions with $Ir_4(CO)_{12}$ yields $[Ir_4(CO)_{11}X]^-$ anions (X = Cl, Br, I, CN, SCN). In the solid state structure of $[Ir_4(CO)_{11}Br]^-$ the carbonyl arrangement is different to that in $Ir_4(CO)_{12}$. It adopts the $Co_4(CO)_{12}$ structure except that one terminal CO in the basal $M_3(CO)_9$ species is replaced by the bromine atom and two of the bridging CO groups become markedly asymmetric (Figure 107). The Ir-Ir distances in the basal plane range from 2.695(2) to 2.745(2) Å, while the unbridged edges from the apical Ir to the three basal ones are shorter with an average distance of 2.696 Å. The Ir-Br bond length is 2.574(4) Å. The Ir-C distances in the asymmetric bridges are 1.97 and 2.24 Å (202). It is proposed that in this complex we are seeing a frozen-out intermediate between the terminal and bridging modes of coordination for these two CO groups.

The $Co_4(CO)_{12}$ type structure is also observed in the dianion $[H_2Ir_4(CO)_{10}]^{2-}$, but with two terminal CO groups in the basal plane replaced by terminal hydrides (Figure 108). The presence of terminal hydrides may be associated with the low steric requirements of the limited number of carbonyl ligands (203). In the structure three terminal carbonyls are coordinated to the apical metal, two to one of the basal Ir atoms, and one each to the other two basal iridiums. Two of the three μ_2-CO groups in the basal plane are asymmetric (average Ir-C of 1.90 and 2.12 Å) as was found in $[Ir_4(CO)_{11}Br]^-$.

The substitution of carbonyl groups by two or three triphenylphosphine ligands to give either $Ir_4(CO)_{10}(PPh_3)_2$ or $Ir_4(CO)_9(PPh_3)_3$ causes a change from the $Ir_4(CO)_{12}$ to the $Co_4(CO)_{12}$ arrangement of carbonyls. There seems to be little difference in Ir-Ir bond lengths for the bridged and unbridged edges (mean 2.73 Å) and the values are longer than found in $Ir_4(CO)_{12}$. The phosphines replace equatorial terminal carbonyl groups in the $M_3(CO)_9$ unit and bond to different metal atoms (204).

In the complex $Ir_4(CO)_5(C_8H_{12})_2(C_8H_{10})$ (205) the four iridium atoms adopt a 'butterfly' configuration (Figure 109). The 'hinge' bond is longer at 2.787(1) Å than the four Ir (wingtip)-Ir (hinge) distances which lie in the range 2.695(1) - 2.741(1) Å. Two carbons of the cycloocta-1-ene-5-yne ligand lie parallel to the Ir-Ir hinge bond and bridge the 'wingtip' metal atoms to give a pseudo octahedral arrangement as observed in $Co_4(CO)_{10}(C_2Et_2)$. Here the C-C distance is 1.49(2) Å, and the other unsaturated bond in the ligand is coordinated to one metal centre. Two carbonyl ligands unsymmetrically bridge opposite edges of the Ir_4 unit, with the shorter bonds associated with the 'wingtip' metals. Each of the two 1,5-cyclooctadiene ligands coordinates to one metal atom through its two unsaturated C-C bonds.

The addition of two electron pairs to the closed shell electronic configuration should cause the rupture of another metal-metal bond and result in a rectangular arrangement of the four metals. This is observed in the structure of $Ir_4(CO)_8[C_2(CO_2Me)_2]_4$ (Figure 110), a 64 electron system. The two short Ir-Ir edges (2.715(1) Å) are bridged by two of the substituted acetylenes. The other two acetylenic ligands lie 1.50 Å above and below the metal plane, capping the rectangular faces, with acetylenic C-C bonds approximately parallel with the longer Ir-Ir edges (2.810(1) Å) (206). Each Ir is also coordinated to two terminal carbonyl groups. The edge-bridging acetylenes donate two electrons each to the complex, and the face capping ones four.

The only structurally characterised iridium cluster containing more than four iridium atoms is $Ir_7(CO)_{12}(C_8H_{12})(C_8H_{11})(C_8H_{10})$ (207). The polyhedron adopted by the metals is a monocapped octahedron. Eleven of the carbonyls are terminally bound to the six metals of the octahedral fragment, two to each metal except for the one which has one terminal CO and shares the twelfth carbonyl with the capping metal. In addition to the bridging CO, the apical Ir coordinates a normal chelating 1,5-cyclooctadiene ligand. The bridging vinyl species is π bonded to one metal and σ bonded to another. Located beneath one

triangular face of the cluster is the acetylenic
portion of a cycloocta-1-en-5-yne ligand which
coordinates as a $2\sigma + \pi$ four electron donor (Figure
111). This cluster shows three stages of cycloocta-
diene coordination and dehydrogenation, while only two
were displayed by $Ir_4(CO)_5(C_8H_{12})_2(C_8H_{10})$. Both
complexes were isolated as products of the same
reaction.

2.7. Nickel, Palladium and Platinum

Triangular arrangements of metal atoms are common
in the structural chemistry of nickel and platinum,
and some of the larger clusters are built up by a
combination of these units. There are few examples of
palladium clusters, but those that have been characterised contain triangular or rectangular arrays of metal
atoms.

The nickel carbonyl dianion, which was isolated
as the bis (tetramethylammonium) salt, has the formula
$[Ni_3(CO)_3(\mu_2\text{-}CO)_3]_2^{2-}$ (208). The structure may be
discribed as a trigonal-antiprismatic array of metal
atoms from the dimerisation of two $Ni_3(CO)_3(\mu_2\text{-}CO)_3$
species through direct Ni-Ni interactions involving
the two additional anionic electrons. The symmetry-
related Ni-Ni bonds within both $Ni_3(CO)_6$ triangles are
2.38 Å in length, while the other six equivalent Ni-
Ni bonds between them are 2.77 Å. This elongated
distortion of the octahedron of nickel atoms along the
three-fold axis results in two smaller transoid,
equilateral triangular metal faces, whose edges are
symmetrically bridged by carbonyls, and six isosceles
triangular faces with two longer edges (112). The two

independent Ni-CO (bridging) bond lengths are both
1.90 Å which is 0.25 Å longer than the Ni-CO (terminal)
distance.

The transformation from a trigonal-antiprismatic
metal arrangement in the nickel complex to a trigonal-
prismatic one in the corresponding platinum analogue,
$[Pt_3(CO)_3(\mu_2-CO)_3]_2^{2-}$, represents an instance of metal-
metal interactions involving first-row transition
metals which are sufficiently different from those
involving third-row metals such as to produce two
different metal arrangements which are conformers of
each other.

$[Pt_3(CO)_3(\mu_2-CO)_3]_2^{2-}$ is the first member of the
series $[Pt_3(CO)_3(\mu_2-CO)_3]_n^{2-}$ ($\underline{n}$ = 2, 3, 4, 5) (209).
The structures of these anions are illustrated in
Figure 113. In the dimer the parallel monomer units
are displaced from an eclipsed $\underline{D}_{3h}$ conformation by $\underline{ca}$
0.51 Å along one of the triangular Pt-Pt edges. In
the trimer the three triangular Pt fragments are
considerably distorted from a regular eclipsed $\underline{D}_{3h}$
conformation by a helical twisting of each fragment
relative to the adjacent ones by $\underline{ca}$ 13° about the
pseudo three-fold axis. The pentamer is formed from
the fusion on a triangular platinum face of two doubly
prismatic $[Pt_3(CO)_3(\mu_2-CO)_3]_3$ units. The bottom of
the five $Pt_3(CO)_3(\mu_2-CO)_3$ layers is virtually eclipsed
with respect to the next layer, the top layer is
twisted by $\underline{ca}$ 8.1^{O} from its neighbour; the central
layer is almost equivalently twisted by $\underline{ca}$ 27.2 and
28.6^{O} relative to its two adjacent inner ones.

It is thought that the observed distortions from
a regular prismatic stacking of Pt atoms represents a
compromise between steric effects (imposed by non-
bonding repulsions mainly between the carbonyl ligands
on adjacent layers) and electronic effects which seem

to favour a regular trigonal-eclipsed metal geometry.
The steric effect is evidenced by the carbonyls
bending away from adjacent layers on the top and
bottom triangles. The individual intratriangular Pt-Pt
distances in the complexes are all within 0.01 $\overset{o}{A}$ of
the mean value of 2.66 $\overset{o}{A}$. The intertriangular Pt-Pt
distances in the three dianions range from 3.03 to
3.10 $\overset{o}{A}$.

The difference in conformation between the nickel
and platinum anions, by which, in the trigonal
prismatic cluster system, each metal in one planar
$M_3(CO)_3(\mu_2-CO)_3$ is only bonded to one corresponding
metal atom in the other identical fragment in
contradistinction to each metal atom in the trigonal
antiprismatic cluster system being coordinated to two
metal atoms in the opposite fragment, presumably
reflects from a viewpoint of total energy minimisation
the inherently greater strength of a given kind of Pt-
Pt bond.

The X-ray structure of $(\eta^5-C_5H_5)_3Ni_3(CO)_2$ (Figure
114) shows that the nickel atoms define an equilateral
triangle with an edge length of 2.39 $\overset{o}{A}$, and that the
two carbonyls lie above and below the Ni_3 plane acting
as μ_3-ligands (210). In agreement with the molecular
formula this complex exhibits the expected paramagnetism
for one unpaired electron.

The Ni-Ni distances in $(\eta^5-C_5H_5)_3Ni_3S_2$ (Figure
115) are relatively long at 2.80 $\overset{o}{A}$. This indicates
relatively weak Ni-Ni bonding, probably the result of
electron donation from the two μ_3-sulphur atoms into
antibonding orbitals of the nickel triangle (211). The
Ni-Ni distances in $Ni_3(PPh_3)_3(S_2C_{10}Cl_6)_3$ are somewhat
shorter at 2.641(4) $\overset{o}{A}$ (212). In this complex the six
sulphur atoms are situated in a trigonal prismatic array
with the Ni atoms 0.60 $\overset{o}{A}$ above the prism faces.

$(\eta^5\text{-}C_5H_5)_3Ni_3NCBu^t$ in isoelectronic with $Co_3(CO)_9S$ and has a similar type of structure (Figure 116). The Ni_3 triangle is irregular with Ni–Ni distances ranging from 2.334(7) to 2.386(5) Å, but the μ_3-N donor ligand caps the triangle symmetrically with an average Ni–N distance of 1.96 Å (213).

$[Ni_3(CO)_3(CF_3C_2CF_3)(C_8H_8)]$ contains an isosceles triangle of Ni atoms (Ni–Ni 2.458(2) Å (twice), 2.703(2) Å). Each metal is bonded to one terminal carbonyl, and the $Ni_3(CO)_3$ species is sandwiched between the two organic ligands.

A bent Ni–Ni–Ni system has been observed in the boron-cage compounds $(C_5H_5Ni)_3B_5H_5CH$ (214). The Ni_3B_5C unit may be described as a monocapped square antiprism (Figure 117) with the nickel atoms occupying three of the positions in one square plane. The average Ni–Ni distance is 2.404(1) Å. Each metal is π-bound to a cyclopentadienyl ring.

In $Pd_3(CNBu^t)_5(SO_2)_2$ (Figure 118) the two SO_2 groups symmetrically bridge edges of the Pd_3 triangle. The metal atom associated with both SO_2 ligands has one terminal isocyanide ligand, and the other two metals have two each. However, there are two short Pd...C contacts (2.81 Å) between the two equivalent metals and two of their coordinated isocyanides, which suggests that the ligands may be bridging very asymmetrically. The two bridged Pd–Pd distances average 2.734(3) Å, and the unbridged distance is 2.760(3) Å (215).

The Pd_3 triangle in $[Pd_3Cl(PPh_2)_2(PEt_3)_3]^+$ is isosceles (Pd–Pd 2.93 (twice), 2.89 Å) (216). The shorter edge is symmetrically bridged by a chlorine ligand, and the two longer edges by diphenylphosphine groups. Each metal is also coordinated to a terminal triethylphosphine ligand (Figure 119). Although the

Pd-Pd distances are longer than in the SO_2 complex there is still considerable metal-metal bonding in the cluster.

The cluster $Pd_3\{C_3Ph(\underline{p}-MeOC_6H_4)_2\}_2(acac)_2$ has crystallographic $\underline{C}_2$ symmetry. The two-fold axis passes through the central metal of the bent Pd-Pd-Pd system (Figure 120), in which the metal-metal bonding distance is 2.662(2) Å and the Pd-Pd-Pd angle is 127°. The structure may be viewed either as having the acyclic $C_3R^1R_2^2$ ligands bridging the two Pd-Pd bonds or as having two palladiacyclobutenyl ligands $(>PdC_3R^1R_2^2)$ π^4-bonded to the central metal atom (217).

Not surprisingly the ligands which are found in palladium triangular clusters are also found in the platinum complexes, and their modes of coordination are similar.

In $Pt_3(SO_2)_3(PPh_3)_3.C_7H_8.SO_2$ (218) all three edges of the platinum triangle are bridged by μ_2-SO_2 groups, and each metal is coordinated to a terminal triphenyl-phosphine ligand (Figure 121). The whole $Pt_3S_3P_3$ system is effectively planar. The Pt-Pt distances have a mean of 2.701 Å which is consistent with that expected for a single bond.

In the molecule $Pt_3(PPh_2)_3(PPh_3)_2Ph$ the diphenyl-phosphine ligands bridge the edges of the triangle, but one edge has been extended to 3.360(1) Å and is non-bridging (219). The Pt-Pt bonding distance is 2.785(1) Å. The unique Pt atom is also coordinated to two phosphido groups and the carbon atom of a σ-bonded phenyl ring (Figure 122).

The three platinum atoms in $Pt_3(CO)_3\left[P(C_6H_{11})_3\right]_4$ define a slightly distorted isosceles triangle with sides of 2.675, 2.736 and 2.714(1) Å. This effect is due to the <u>trans</u> influence of the four terminal

phosphines in equatorial coordination sites; two
ligands are bonded to one metal centre, and one each
to the other two metal atoms. All three edges of the
triangle are bridged by μ_2-CO groups (220).

The Pt_3 triangle in $Pt_3(Bu^tNC)_6$ is genuinely
equilateral with a Pt-Pt distance of 2.531 Å. The
Pt_3C_6 core is similar to that observed in a layer of
the $\left[Pt_3(CO)_6\right]_n^{2-}$ structure, and the bulk of the
isocyanide ligands may prevent polymerisation (221).
Three of the isocyanides bridge the Pt-Pt edges and
the other three are terminal (Figure 123).

Finally, the structure of $Pt_3(PEt_3)_4(PhC_2Ph)_2$
(222) includes a 'V' shaped arrangement of Pt atoms,
with a Pt-Pt distance of 2.905(1) Å and a Pt-Pt-Pt
angle of 144°. The molecule has crystallographic $\underline{C}_2$
symmetry as shown in Figure 124. The acetylenic units
form transverse bridges across each pair of Pt atoms
with the C-C bonds turned towards the underside of the
'V-arrangement' (222).

There are several examples of clusters containing
a tetrahedron of nickel atoms. $Ni_4(CO)_6\left[P(CH_2CH_2CN)_3\right]_4$
has a Ni_4 tetrahedron with 2.51 Å edges. The six μ_2-CO
groups bridge the six edges of the tetrahedron, and
each metal is also bound to a terminal phosphine ligand
(223). The $\left[Ni_4(CO)_4(CF_3C_2CF_3)_3\right]$ cluster has local
three-fold symmetry. The Ni-Ni distance in the basal
plane is 2.669(7) Å and the Ni (apical)-Ni (basal)
distance is shorter at 2.378(8) Å. The three $CF_3C_2CF_3$
ligands coordinate to three triangular faces (224). A
similar mode of bonding for the acetylene ligand is
found in $Ni_4\left[CNBu^t_3\right]_4\left[\mu_3(\eta^2)-C_6H_5C\equiv CC_6H_5\right]_3 \cdot C_6H_6$. Here
the two unique Ni-Ni distances in the tetrahedron are
2.374(2) and 2.686(2) Å. Each metal is also bonded to
one terminal isocyanide ligand. The three diphenylacetylene ligands each coordinate to the apical Ni atom

through one carbon in a σ-bond and to two adjacent basal metal atoms by three-centre μ-type bonds (225). $Ni_4(CNCMe_3)_7 \cdot C_6H_6$ contains a trigonally compressed Ni_4 tetrahedron. Each metal is bonded to one terminal $CNCMe_3$ group and the three remaining ligands bridge either edges or faces. In solution the n.m.r. spectrum shows that the ligands are equivalent (226).

A recent neutron study of $(\eta^5-C_5H_5)_4Ni_4H_3$ confirms that the three μ_3-hydrides cap the three triangular faces with the long Ni-Ni edges of 2.464 Å. Each metal has one π-bonded cyclopentadienyl ligand associated with it. Magnetic measurements on this complex suggest that there are three unpaired electrons.

The only example of a Pd_4 cluster is $[Pd_4(CO)_4(OAc)_4] \cdot (AcOH)_2$. It contains a square plane of Pd atoms with two long and two short Pd-Pd bonds (2.663(1) and 2.909(1) Å respectively), but the Pd-Pd-Pd angles differ markedly from 90° (Figure 125). Each short Pd-Pd bond is μ_2-bridged by two carbonyl groups and each long bond by two chelating acetate groups. Both types of group bridge symmetrically, and the acetates themselves are planar (227).

A square planar metal geometry is favoured in the tetragonal form of the platinum complex $[Pt(CH_3COO)_2]_4$ (228). The very short Pt-Pt distances of 2.492 to 2.498 Å indicate that there is strong metal-metal bonding. The eight bridging acetates are arranged around the square such that four groups are in the plane of the metals and four are alternatively above and below it. This gives the Pt atoms an octahedral coordination geometry (Figure 126). The eight Pt-O bonds in the plane of the metals are rather long (mean 2.162 Å), while those approximately normal to the plane have a mean length of 2.002 Å. The monoclinic form of the complex has the same overall structure (229).

One Ni_5 cluster has been structurally characterised. In $[Ni_5(CO)_{12}]^{2-}$ a $Ni_3(CO)_3(\mu_2\text{-}CO)_3$ species is symmetrically capped by two $Ni(CO)_3$ groups such that the metals define a trigonal bipyramid (Figure 127). The anion has approximate $\underline{C}_{3v}$ symmetry, and the average Ni-Ni (equatorial) distance is 2.36 Å, while the Ni-Ni (equatorial-axial) distances lie in the range 2.743(3) - 2.865(3) Å (230).

The platinum halide 'PtCl$_2$' was found to consist of discrete Pt_6Cl_{12} units (231). The six metal atoms define an octahedron and the twelve chlorines bridge the twelve octahedral edges to give a structure which is analogous to the $[M_6X_{12}]^{2+}$ structure common in the halide chemistry of niobium and tantalum. The Pt-Pt distances are of two types with means of 3.32 and 3.40 Å. These distances are rather long and suggest that bonding occurs largely through the chlorine bridges.

The nickel atoms in the Ni_8 cluster $Ni_8(CO)_8(\mu_4\text{-}PC_6H_5)_6$ define a distorted cube with edge lengths in the range 2.636(3) - 2.681(3) Å (232). Each metal is bound to one terminal carbonyl, and the six phosphorus ligands, which act as four-electron donors, cap the six faces of the cube (Figure 128). Thus the phosphorus atoms define an octahedron with a non-bonding P...P distance of 4.9 Å. The mean Ni-P bond length is 2.183 Å.

Neutron studies on the $[HNi_{12}(CO)_{21}]^{3-}$ and $[H_2Ni_{12}(CO)_{21}]^{2-}$ anions have shown that the hydrides occupy the octahedral holes within these Ni_{12} clusters. It was found that there is more than enough room within a Ni_6 octahedron to accommodate a hydrogen atom with sufficient clearance for it to 'rattle around' in its metal cage.

2.8. Copper, Silver and Gold

The cluster chemistry of copper, silver and gold is markedly different to that of the iron, cobalt and nickel subgroups. There are no carbonyl complexes, and the clusters that do exist are often stabilised by donor ligands coordinated through nitrogen, sulphur or phosphorus. Copper(I) forms small clusters with between three and eight metal atoms. Silver(I) has little tendency to form discrete clusters but rather favours polymeric arrangements containing chains of metal atoms joined by Ag-Ag bonds. Gold also forms polymeric compounds but there are a few larger discrete clusters with between six and eleven metal atoms. The gold atoms in these compounds have mixed oxidation states, either $Au(0)$ or $Au(I)$. Both $Cu(I)$ and $Ag(I)$ halide complexes have 'cubane-like' structures without direct metal-metal bonds.

The structural chemistries of copper and gold are sufficiently different to be discussed separately.

The cation $[Cu_3Cl_2(Ph_2PCH_2PPh_2)_3]^+$ contains a triangle of copper atoms which is capped, above and below, by μ_3-chlorine ligands. These bridges are very asymmetric: one chlorine in involved in two Cu-Cl distances of 2.438(4) and one of 2.678(6) Å, while the other has one Cu-Cl of 2.407(7) and two of 2.598(3) Å. The diphosphine ligands bridge the edges of the Cu_3 triangle (Figure 129).

Tetranuclear copper complexes display both tetrahedral, 'butterfly' and square-planar arrangements. $Cu_4\{(\underline{i}\text{-}C_3H_7O)_2PS_2\}_4$ has an irregular tetrahedral metal

skeleton (Figure 130). In each ligand one sulphur
atom is symmetrically bonded to two Cu atoms at
distances of 2.256(9) and 2.272(16) $\overset{o}{A}$, while the
second S atom is bonded to a third Cu atom at a
distance of 2.272(9) $\overset{o}{A}$. The P-S (bridging) distance
is 2.036(11) $\overset{o}{A}$. The four μ_2-S bridged Cu-Cu bonds are
equivalent with a mean value of 2.74(3) $\overset{o}{A}$. This is
rather shorter than the remaining two Cu-Cu bond
lengths of 2.950(6) $\overset{o}{A}$ (233).

The butterfly configuration of metal atoms is
displayed in 5-methyl-2-[(dimethylamino)methyl]phenyl-
copper (234). The Cu-Cu distances, which average
2.38 $\overset{o}{A}$, are very short. The Cu-Cu distance in copper
metal is 2.56 $\overset{o}{A}$. The phenyl rings act as bridging
groups between pairs of metal atoms (Figure 131). The
Cu-C distances range from 1.97 $\overset{o}{A}$ to 2.16 $\overset{o}{A}$.

In the two related molecules $Cu_4(MeN_3Me)_4$ (235)
and $Cu_4(O_2CCF_3)_4 \cdot 2C_6H_6$ (236) the four copper atoms
define a parallelogram with angles of 67° and 113°.
In the triazene complex (Figure 132) the Cu-Cu distances
around the parallelogram range from 2.64(1) to
2.68(1) $\overset{o}{A}$. In the second complex these Cu-Cu distances
are longer, with a mean of 2.80 $\overset{o}{A}$, and the distance
across the short diagonal is 3.1 $\overset{o}{A}$. The difference in
metal-metal distances between the two compounds must be
due to the more electronegative nature of the oxygen
donor ligands, and because in the trifluoroacetate
complex the copper atoms are three-coordinate since
they also form π-bonds with the benzene solvent
molecules. 2-copper-1-(dimethylaminomethyl)-ferrocene
contains a central square planar of Cu atoms (Cu-Cu
2.443(4) $\overset{o}{A}$) (237). Pairs of these metals are bridged
by a single carbon atom of a cyclopentadienyl ring.
Pairs of adjacent rings then 'sandwich' iron atoms
(Figure 133).

In $Cu_4(CH_2SiMe_3)_4$ (238) and $Cu_4\{N(SiMe_3)_2\}_4$
(Figure 134), the square-planar metal arrangement is
stabilised by the presence of the bulky ligands. In
the former complex the average Cu-Cu distance is
2.417 Å, but in the latter it is significantly longer
at 2.686 Å. This difference in bond lengths may be
attributed to the presence of a lone pair of electrons
in a p orbital on each nitrogen atom in the silylanine
complex. For $Cu_4(CH_2SiMe_3)_4$ the Cu-C-Cu units are
best considered as three-centre two-electron bonds,
but there is also direct Cu-Cu bonding interaction.
In $Cu_4\{N(SiMe_3)_2\}_4$ the Cu-N-Cu systems are bonding for
metal-ligand interactions but antibonding for metal-
metal interactions. This is because the pair of
electrons in the nitrogen p orbitals occupies antibonding molecular orbitals. The distribution of the
bridging ligands in both compounds is similar.

All the clusters containing six copper atoms which
have been investigated crystallographically have a
distorted octahedral metal skeleton. In $[H_6Cu_6(PPh_3)_6]$
[HCO.NMe] (239) two mutually _trans_ faces of the octahedron are enlarged resulting in Cu-Cu distances in the
range 2.632(6) - 2.674(5) Å. This compares to the
range 2.494(6) - 2.595(5) Å for the remaining Cu-Cu
bond lengths. It is thought that the six longer edges
may be bridged by the hydride ligands. Each metal is
terminally bound to the P atom of a triphenylphosphine
ligand. The average Cu-P distance is 2.240(17) Å. In
the structure of $[Cu_6(2-Me_2NC_6H_4)_4Br_2]$ (Figure 135) two
trans edges of the octahedron are bridged by bromine
atoms, while four faces are capped by the organic groups
which bond through the nitrogen atom and through a μ_2-
phenyl carbon. The bromine-bridged Cu-Cu distances
average 2.64 Å, but the other pair of equatorial bonds
are longer, at 2.70 Å, and the metal-metal interaction

is considered to be non- bonding. The other eight Cu-
Cu distances average 2.48 Å in length (240). A similar
structure has been observed for the related complex
$[Cu_6(2-Me_2NC_6H_4)_4(C{\equiv}CC_6H_4Me-4)_2]$ (241). Here also the
dimethylaminophenyl groups span triangular faces of
the octahedron by bridging a Cu (apical)-Cu (equatorial)
edge <u>via</u> a phenyl carbon and bonding to a Cu (equatorial)
atom by nitrogen coordination. Two opposite edges of
the equatorial Cu_4 plane are symmetrically bridged by
4-tolylethynyl ligands which act as one-electron donors.

A cube of copper atoms has been found in several
cluster anions. In $[Cu_8(i-S_2CC(CN)_2)_6][PhMe_3As]_2$ there
is a rhombohedral distortion of the cube and the
average Cu-Cu distance is 2.83 Å (242). The Cu atoms
are arranged inside a icosahedral array of sulphur
atoms and this group of 20 atoms is enclosed by the
carbon atoms of six CS_2 groups located at the vertices
of an octahedron (Figure 136). The same basic structure
is observed when the $[i-S_2CC(CN)_2]^{2-}$ ligand is replaced
by $[S_2C_2(CO)_2]^{2-}$ (243) or $[S_2CC(CO_2Et)_2]^{2-}$ (244). The
Cu-Cu distances in these two complexes are in the range
2.77 - 2.91 Å.

Silver(I) dipropylmonothiocarbamate (245) contains
discrete Ag_6 hexamers. The metals define an octahedron
and the distances between adjacent vertices vary
between 2.943 and 3.281 Å. Each Ag atom is bonded to
two sulphur atoms (mean 2.46 Å) and to one oxygen atom
(2.36 Å). Each carbamate unit lies above a triangular
octahedral face with the sulphur atom bonded symmetrically to two adjacent metal atoms and the oxygen
coordinated to a third.

There are a few discrete gold clusters. Chloro-
(piperidine)gold(I) has a tetrameric solid state
structure with the Au atoms defining a square plane
(Figure 137). The average Au-Au distance is 3.301(5) Å.

This is considerably longer than in gold metal, (2.88 Å)
but there is thought to be significant Au-Au bonding
interaction (246).

In $[Au_6\{P(C_6H_7)_3\}_6]\,[BPh_4]_2$ the octahedron of gold
atoms is significantly distorted. Two <u>trans</u> faces have
edges whose lengths range from 2.932(2) to 2.990(2) Å,
while the remaining Au-Au lengths range from 3.043(2)
to 3.091(2) Å. Each metal is terminally bound to a
$P(C_6H_7)_3$ ligand (247). Despite the relative bulk of
the ligands the variations in metal-metal distances do
not appear to arise from interligand repulsions.

The only Au_9 cluster so far characterised
crystallographically is $[Au_9\{P(p-C_6H_4.Me)_3\}_8]\,[PF_6]_3$
(248). In this compound a central gold atom is
surrounded by eight others which define an icosahedron
from which one equatorial triangle has been removed
(Figure 138). The Au-Au distances involving the
central Au atom are of two different types with bond
lengths of 2.689(3) and 2.729(3) Å. Each peripheral
metal atom is bonded to the central metal, three other
peripheral atoms, at a distance of 2.805 Å, and to the
P atom of a phosphine ligand of 2.30 Å.

The three Au_{11} clusters, $Au_{11}(PPh_3)_7(SCN)_3$ (249),
$Au_{11}[P(p-C_6H_4Cl)_3]_7I_3$ (250) and $Au_{11}[P(pC_6H_4F)_3]_7I_3$
(251), all have the same basic structure (Figure 139).
In $Au_{11}(PPh_3)_7(SCN)_3$ the central gold atom is surrounded,
at an average distance of 2.67 Å, by ten gold atoms
each of which has one terminal ligand associated with
it. The coordination polyhedron may be described
approximately as a combination, by apex sharing of a
pentagonal bipyramid and square pyramid with gold
atoms at all the vertices. The central gold atom has
a formal oxidation state of $\overline{O}$, while the peripheral
Au atoms are in the +1 oxidation state. The minimum
peripheral Au-Au distance is 2.83 Å.

The metal skeleton in the tri-iodido-Au_{11} clusters has been described as a centred icosahedron in which nine of the twelve apices are occupied by Au atoms. A single gold atom substitutes the three remaining apices. Each peripheral atom is bound to the central metal, to three, four or five other peripheral Au atoms, and to either a terminal iodine atom or phosphine group. The Au (central)-Au (peripheral) and Au (peripheral)-Au (peripheral) bond lengths are similar to those found in $Au_{11}(PPh_3)_7(SCN)_3$. An alternative scheme to that proposed by Mason has been given in the assignment of oxidation states. The central gold atom, because of its smaller radius, has been assigned an oxidation state of +3, and the ten peripheral Au atoms an oxidation state of 0. These metals act as one electron donors to give the central Au the 'noble-gas configuration'.

2.9. Zinc, Cadmium and Mercury

Molecular cluster compounds containing zinc, cadmium or mercury are rare. In cases where they do exist they are nearly always bonded to another metallic element to form a mixed metal cluster. These elements show a greater tendency to form structures where there is a polymeric chain of metal atoms.

An example of linear Hg_3 species is in $Hg_3(AlCl_4)_2$ (252) where the two terminal mercury atoms are coordinated to chlorine atoms of the $AlCl_4$ groups. The two Hg-Hg distances are 2.551 and 2.562 Å.

2.10. Mixed Metal Clusters

The structures of mixed metal cluster compounds
are similar to those containing only one type of
metallic element. There are no new modes of ligand
coordination. We will discuss these complexes in an
order based on the number of metal atoms present.

Molybdenum is associated with rhenium in
$\{Re_2Mo(\eta^5-C_5H_5)(CO)_8\}(S)\{SMo(\eta^5-C_5H_5)(CO)_3\}$ (253).
The structure (Figure 140) consists of a triangle of
two Re and one Mo atom with an Re-Re distance of
2.985(1) $\mathring{A}$ and a Re-Mo bond length of 2.909(2) $\mathring{A}$, the
Re-Re-Mo angle is 83.6(1)$^\circ$. A four electron-donating
sulphur atom caps the three metals with Re-S distances
of 2.448(5) and 2.410(5) $\mathring{A}$ and a Mo-S distance of
2.487(5) $\mathring{A}$. A second sulphur atom, which acts as a
six electron donor also bridges the three metal atoms
and is also attached to a terminal $(\eta^5-C_5H_5)Mo(CO)_3$
group.

There is an isosceles WPt_2 triangle in
$Pt_2W\{\mu-C(OMe)Ph\}(CO)_6(PBu^t_2Me)_2$ (254). The Pt-Pt
distance is 2.628(1) $\mathring{A}$ and the two W-Pt distances have
an average value of 2.832 $\mathring{A}$. A carbene species
bridges the Pt-Pt edge. The tungsten atom has four
terminal carbonyls while each Pt atom has a carbonyl
group perturbed towards a semi-bridging interaction
with the W atom (W...C 2.50(3) $\mathring{A}$).

In $HRe_2Mn(CO)_{14}$ the metal atoms adopt a bent
configuration with a Re...Re-Mn angle of 98.1° (255).
The Mn-Re bond length is 2.960 $\mathring{A}$ while the Re...Re
distance of 3.39 $\mathring{A}$ is approximately 0.37 $\mathring{A}$ longer than

a normal Re-Re single bond. The carbonyls bonded to each Re atom are eclipsed with respect to each other rather than the staggered configuration found for carbonyl groups on the central Re atom and the Mn atom. It is presumed that the hydride ligand bridges the long Re...Re edge (Figure 141).

The crystal structure of $Mn_2Fe(CO)_{14}$ shows a linear arrangement of metals without any bridging carbonyls (Figure 142). Each metal atom has an essentially octahedral coordination geometry and the carbonyl groups on the Mn atoms are orientated at 45° to those on iron. The two Mn-Fe distances are 2.80 and 2.83 $\overset{o}{A}$ (256).

In $MnFe(CO)_8(AsMe_2)$ (257) the $AsMe_2$ group bridges the Mn-Fe bond to form a metal triangle. Both the iron and manganese atoms have four terminal carbonyls and it has not been possible to differentiate between the two metals on the basis of the X-ray data.

Manganese forms trinuclear clusters with cadmium and mercury. With cadmium a bent Mn-Cd-Mn metal skeleton is always observed and the Mn atoms are coordinated to five terminal carbonyl ligands. Mn-Cd distances lie in the range 2.68 - 2.79 (258). $(OC)_5Mn-Hg-Mn(CO)_5$ is linear with the carbonyl ligands on the Mn atoms mutually eclipsed (259). The unique Mn-Hg bond length is 2.806 $\overset{o}{A}$.

In $(CO)_4Mn(\mu-CO)(\mu-GeMe_2)Mn(CO)_4$ (260) the $FeMe_2$ group bridges the Mn-Mn bond to form a metal triangle. The Mn-Mn bond of length 2.854(2) $\overset{o}{A}$ is asymmetrically bridged by a carbonyl group (Mn-C 2.159(7) and 2.037(7) $\overset{o}{A}$). The two Mn-Ge distances are 2.432(2) and 2.477(2) $\overset{o}{A}$. In $(CO)_4Mn(\mu-AsMe_2)Mn(CO)_2(\eta^5-C_5H_5)$ there is no bridging carbonyl and two terminal carbonyls on one Mn atom have been replaced by a π-bound C_5H_5 ring (261). The Mn-Mn distance is 2.912(4) $\overset{o}{A}$ and the two

Mn-As distances are 2.350(4) and 2.362(4) $\mathring{A}$. When a
SnX_2 (X = Cl or Br) group is introduced into a dimeric
Mn complex the metal-metal bond is broken and a linear
Mn-Sn-Mn system results. For the $SnCl_2$ complex the
Mn-Sn distance is 2.635(1) $\mathring{A}$, and for $SnBr_2$ it is
2.642(3) $\mathring{A}$ (262).

The complex $(NO)_2Fe(\mu-PMe_2)_2Fe(CO)_2(\mu-PMe_2)Co(CO)_3$
has a bent metal framework. The Fe-Fe is 2.67 $\mathring{A}$ long
and is bridged by two phosphine ligands. The Fe-Co
bond is shorter, at 2.66 $\mathring{A}$, and is bridged by one
phosphine. The molecule is isostructural with
$(NO)_2Fe(\mu-PMe_2)_2Fe(CO)_2(\mu-PMe_2)Fe(CO)_2(NO)$ where the
corresponding Fe-Fe distances are 2.66 $\mathring{A}$ and 2.71 $\mathring{A}$,
respectively. In both molecules all the carbonyl and
nitrosyl ligands are terminal (263). A structurally
related species is the $[Rh\{Fe(PPh_2)(CO)_2(\pi-C_5H_7Me)\}_2]^+$
cation (Figure 143). Here the rhodium atom is the
central metal of the bent metal framework and the
Fe-Rh-Fe angle is 145°. Both Rh-Fe bonds are
asymmetrically bridged by a phosphine and a carbonyl
ligand, and each of the two Fe atoms is also coordinated
to a terminal carbonyl and a π-bound cyclopentadienyl
ring. The average Fe-Rh bond length is 2.665(2) $\mathring{A}$.
In this structure, the Rh atom has a sixteen electron
configuration (264).

There is a series of compounds with the general
formula $MCo_2(CO)_9X$, where M = Fe, Co and X = S, Se, Te.
These molecules have approximate $\underline{C}_{3v}$ symmetry with the
X ligand capping the metal triangle. Each metal is
coordinated to three terminal carbonyl ligands. Data
for these compounds are given in Table 10. Structures
$\underline{a}$ and $\underline{c}$ have unpaired electrons in strongly antibonding
orbitals which are composed of Co 3d atomic orbitals
in the plane of the metals. Replacement of one Co atom
by an Fe atom gives a system with one electron less.

TABLE 10

$\left[MCO_2(CO)_9X\right]$ Clusters

M	X	Structure	M-M (Å)	M-X (Å)	M-X-M (deg)	Ref
Co	S	a	2.637(3)	2.139(4)	76.1(1)	(265)
Fe	S	b	2.554(3)	2.159(4)	72.6(1)	(266)
Co	Se	c	2.616(1)	2.282(1)	69.9(1)	(267)
Fe	Se	d	2.577(1)	2.285(1)	68.7(1)	(267)
Fe	Te	e	2.598(2)	2.466(1)	63.1(1)	(267)

This change results in the expected reduction in bond length. The metal-metal distance is also dependent on the nature of the chalcogen. It increases with the increasing covalent radius of the capping atom.

The complex $FePt_2(CO)_5[P(OPh)_3]_3$ contains a $FePt_2$ triangle. An $Fe(CO)_4$ species bridges the Pt-Pt bond (2.633 Å). One Pt atom is coordinated to two phosphate ligands and the other to one phosphite and a carbonyl (Figure 144). The Fe-Pt distance <u>trans</u> to the equatorial phosphite is longer, at 2.583 Å, than the Fe-Pt distance <u>trans</u> to the equatorial carbonyl which has a length of 2.550 Å. In this complex the Fe atom obeys the 18 electron rule but the Pt atoms have only 16 electrons each (268). In the closely related compound $Fe_2Pt(CO)_9PPh_3$ two $Fe(CO)_4$ species are bonded to a $Pt(CO)PPh_3$ unit by metal-metal bonds. Again a strong <u>trans</u> influence is observed, with the Pt-Fe bond <u>trans</u> to the phosphine longer (2.597(5) Å) than the bond <u>trans</u> to the carbonyl (2.530(5) Å). The Fe-Fe bond is 2.758(8) Å long. The Pt atom has a square planar coordination geometry while the two Fe atoms are pseudo octahedral (269).

In the cation $[(C_5H_5)(C_5H_4)FeAu_2(PPh_3)_2]^+$ the metal atoms define a bent Fe-Au-Au chain and a carbon atom, one of the cyclopentadienyl ligands bridges the Au-Au bond (270).

$Hg\{Fe(CO)_2(NO)(PEt_3)\}_2$ has a linear Fe-Hg-Fe arrangement with the phosphine groups axially bonded to the iron atoms. The ethyl groups are staggered with respect to the equatorial CO and NO ligands, and the two $Fe(CO)_2(NO)$ groups are staggered with respect to each other to give <u>D</u>$_3$d idealised symmetry. The Hg-Fe and Fe-P bonds are 2.534(2) and 2.223(3) Å, respectively, and the Hg-Fe-P angle is 174.8(1)° (271).

In $Fe_2(C_5H_5)_2(CO)_3GeMe_2$ the Fe-Fe bond is bridged

by a carbonyl and a $GeMe_2$ group. One terminal carbonyl
and a cyclopentadienyl are bonded to each metal. The
Fe-Fe bond length is 2.628(1) Å and the Fe-Ge distances
average 2.346(1) Å (272).

The cationic species $[(\pi-C_5H_5)(OC)_2FeSbCl_2Fe(CO)_2$
$(\pi-C_5H_5)]^+$ contains a bent Fe-Sb-Fe system with an
angle at Sb of 135°. The Fe-Sb distance is 2.440 Å
(273).

The two Ru-Pt distances in the metal triangle in
$RuPt_2(CO)_5(PMePh_2)_3$ differ by 0.02 Å [2.707(2) and
2.729(2) Å]. This difference is ascribed to the
asymmetric replacement of an equatorial carbonyl group
on the ruthenium atom and is thought to be largely a
steric effect. The Pt-Pt distance is 2.647(2) Å.
Three of the carbonyls bridge the triangular edges.
Each Pt atom has one terminal phosphine, while the Ru
atom is coordinated to two terminal carbonyls and a
phosphine ligand (274).

In the anion $[Br_2Co_2(CO)_8In]^-$ two $BrCo(CO)_4$
species are linked by an In atom to form a bent Co-In-
Co system with an angle of 123.5^O at iridium (275).
The In-Co distances have an average value of 2.652 Å.

In $Co_2(CO)_7Sn(acac)_2$ the $Sn(acac)_2$ species bridges
the long Co-Co bond (2.626(4) Å) to give a triangle of
metals. This bridge is asymmetric with Sn-Co distances
of 2.564(3) and 2.591(3) Å. One carbonyl ligand also
bridges the Co-Co bond while the other six are terminal,
three being coordinated to each metal (276).

There are a number of tetranuclear mixed metal
clusters. The most common polyhedron displayed by the
metal atoms is the tetrahedron. In the two complexes
$(\mu_2-H)WOs_3(CO)_{12}(\eta^5-C_5H_5)$ and $(\mu_2-H)_3WOs_3(CO)_{11}(\eta^5-C_5H_5)$
the metals define distorted tetrahedra. In the first
compound the tungsten atom is coordinated to three Os
atoms, two terminal carbonyl groups, a π-bound

cyclopentadienyl ring, and an asymmetric μ_2-bridging
carbonyl (Os-C (carbonyl) 2.12(5) and W-C 2.29(5) Å).
The carbonyl bridged W-Os distance (2.914(3) Å) is
intermediate in length between the other two W-Os
bonds (2.908(3) and 2.935(3) Å). The hydride appears
to bridge the long Os-Os bond (2.933(3) Å) which is
0.14 Å longer than the other two. In the second
compound it appears that the bridging carbonyl is
replaced by a bridging hydride and another Os-W edge
is also bridged by a hydride since there are two long
W-Os (3.073(2) and 3.082(3) Å) and one short W-Os
(2.880(3) Å) edge. The long Os-Os edge (2.941(2) Å)
retains its hydride ligand (277).

In $HOs_3Re(CO)_{15}$ (278) the metal atoms define a
planar triangulated rhombus (Figure 145). The metal
skeleton is subject to a fourfold disorder which
principally affects the 'bridgehead' metals and their
associated equatorial ligands. The observed pattern
is consistent with the 'bridgehead' sites being $Re(CO)_4$
and $Os(CO)_3H$ which are mutually linked (Os-Re 2.944(1) Å)
and which are bridged by $Os(CO)_4$ units (Os-Os 2.957(1) Å).
The hydride ligand occupies an equatorial coordination
site on the bridgehead Os atom.

The metal tetrahedron in $H_2FeRu_3(CO)_{13}$ is slightly
distorted (Figure 146). The average Fe-Ru distance is
2.640 Å which is slightly shorter than the value of
2.700 Å for the unbridged Fe-Ru bond. Two Fe-Ru edges
are asymmetrically bridged by μ_2 carbonyl ligands
(mean Fe-C 1.78(5), mean Ru-C 2.28(4) Å). In the Ru_3
triangle two edges are hydride bridged and these Fe-Fe
distances range from 2.885(8) - 2.914(8) Å. The
unbridged Ru-Ru edge is 2.777(7) Å. Each Ru atom is
also coordinated to three terminal carbonyls, and the
Fe atom is bonded to two terminal carbonyl groups
(279).

In $HFeCo_3(CO)_9[P(OMe)_3]_3$ the four metal atoms form a tetrahedron with the apical Fe atom additionally bonded to three terminal carbonyl ligands. Each Co atom is further bonded to one terminal and two bridging CO ligands as well as one phosphite and a hydride ligand (Figure 147). The average Fe-Co distance is 2.560(2) Å and the Co-Co distance is 2.488(12) Å. A neutron study confirms that the hydride lies 0.978(3) Å below the Co_3 plane and bonds to all three Co atoms with Co-H distances of 1.742(3), 1.731(3) and 1.728(3) Å (280).

The cluster $(\eta^5-C_5H_5)RhFe_3(CO)_{11}$ contains a $Fe_3(CO)_9$ unit with Fe-Fe bonds in the range 2.553(3) - 2.594(2) Å. An Rh atom caps this triangle and is also bound to a cyclopentadienyl ring and two carbonyls which bridge two of the Rh-Fe edges. The Rh-Fe bond lengths lie in the range 2.568(3) - 2.615(3) Å (281).

An irregular metal tetrahedron is found in $(\eta^5-C_5H_5)_2Rh_2Fe_2(CO)_8$. One π-bound cyclopentadienyl ligand is associated with each Rh atom. A carbonyl bridges the two Rh atoms [Rh-C 1.96(4) Å]. Two carbonyls are terminally bound to one Fe atom and three to the other. The two remaining carbonyls asymmetrically bridge the Rh-Fe edges. The metal-metal distances are: Fe-Fe 2.539(7), Fe-Rh 2.570(5) and 2.589(5), and Rh-Rh 2.648(4) Å (282).

In the anion $[\{Fe(\pi-C_5H_5)(CO)_2\}_3SbCl]^+$ the metals adopt a 'T' shaped geometry with the Sb atom as the linking species (Figure 148). Each iron atom is also coordinated to two terminal carbonyls and a cyclopentadienyl group (283).

The two complexes $Fe_2(SnMe_2)_2(CO)_8$ (284) and $[(C_5H_5)_2SnFe(CO)_4]_2$ (285) contain planar Fe_2Sn_2 rings. In the first compound the Fe-Sn distances average 2.639 Å, while in the second the average is 2.660 Å.

The cyclopentadienyl ligands are σ-bound to the metals.

The Os_3Co core in $H_3Os_3Co(CO)_{12}$ defines a distorted tetrahedron with approximate $\underline{C}_{3v}$ symmetry. Each metal is bonded to three terminal carbonyl ligands, and the $M(CO)_3$ units adopt a staggered conformation. The three hydrides are thought to bridge the edges of the Os_3 triangle. This is confirmed by the observation of long Os-Os bonds (mean 2.901 Å) but quite short Os-Co bonds (mean 2.694(3) Å) (286).

In $Os_3Pt(\mu\text{-H})_2(CO)_{10}\left[P(C_6H_{11})_3\right]$ the Pt atom is bonded to one terminal carbonyl and the phosphine ligand, while each Os atom is bonded to three terminal carbonyls. Within the tetrahedral metal core the Pt-Os distances range from 2.791(2) to 2.863(2) Å and the Os-Os distances range from 2.741(2) to 2.789(2) Å. From the arrangement of the carbonyls the hydrides are thought to bridge the longest Os-Os and the longest Os-Pt edges. In the related compound $Os_2Pt_2(\mu\text{-H})_2(CO)_8$ $(PPh_3)_2$ the metals define a 'butterfly' configuration. The two Os atoms, which are the 'hinge' metal atoms, are asymmetrically bonded to the 'wingtip' Pt atoms (Os-Pt 2.863(1) and Os-Pt[1] 2.709(1) Å). The Os-Os distance is 2.780(9) Å and the Pt...Pt distance of 3.206(1) Å is non-bonding (287). The hydrides are thought to bridge the two longer Os-Pt edges, and the phosphine ligands are coordinated to the Pt atoms.

The structure of $Ir_2Co_2(CO)_{12}$ is similar to that of $Co_4(CO)_{12}$ with a $Co_3(CO)_9$ species having three μ_2-bridging carbonyls. The metals in this derivative are disordered but the Ir atoms show a preference for the apical site which is not involved in bonding with the bridging carbonyls (Figure 149). The metal-metal distances lie in the range 2.594(6) - 2.693(4) Å (288). Replacement of the apical $Ir(CO)_3$ unit by either

Co(π-xylene) or Co(π-benzene) yields complexes where the metal-metal distances involving the unique Co are slightly longer than in the basal triangle (289).

A 'butterfly' geometry is adopted by the Co_2Pt_2 unit in $Co_2Pt_2(CO)_8(PPh_3)_2$. The Co-Co 'hinge' bond is bridged by a carbonyl ligand as are two Co-Pt bonds. Each Co atom is also associated with two terminal CO groups. Each Pt atom has one terminal phosphite bonded to it and the Pt atom not involved with the CO bridges has one terminal carbonyl ligand (Figure 150). The Co-Co distance is 2.498(2) $\overset{\circ}{A}$, the non-bonding Pt...Pt distance is 2.987(4) $\overset{\circ}{A}$, and the Co-Pt distances range from 2.528(3) to 2.579(3) $\overset{\circ}{A}$ (290).

In $[Co(CO)_4]_3SnBr$ the metal atoms adopt a tetrahedral arrangement (291) but in $[Co(CO)(C_5H_5)SnMe_2]_2$ a planar Co_2Sn_2 ring is observed (292). The average Sn-Co bond in the first complex (2.602(6) $\overset{\circ}{A}$) is longer than in the second (2.542(2) $\overset{\circ}{A}$).

Mixed clusters containing five or more metal atoms are less common but exhibit some interesting structures. In the complex $Cr_2(CO)_6(AsMe_2)_9$ the As atoms form a puckered nine-membered ring six of which are bonded to Cr atoms (mean Cr-As 2.44(1) $\overset{\circ}{A}$) (293). In $(CO)_4Cr(AsMe_2)_4Cr(CO)_4$ the four As and the two Cr atoms form a six-membered ring with the chair conformation. Here the average Cr-As distance is 2.480 $\overset{\circ}{A}$ (294).

Molybdenum and tungsten form ring systems with other metals. The tetramer $[(\eta^5-C_5H_5)_2M(H)Li]_4$ (M = Mo, W) contains an eight-membered M - Li ring (295). Each transition metal which is part of a bent $(\eta^5-C_5H_5)_2M$ system is also coordinated to a hydride ligand. Similar structures have been observed in the cases of $[(\eta^5-C_5H_5)_2MoHMg(C_6H_{11})(Br)_2Mg(Et_2O)]_2$ (296), $[MoH(C_5H_5)(C_5H_4)]_2AlMe_5$ and $[Mo(C_5H_4)_2Al_2Me_3]_2$ (297). In the complex $[Ni_3(CO)_6(M(CO)_5)_2][N(PPh_3)_2]_2$

(M = Mo, W) the nickel atoms form an equilateral
triangle with three μ_2-bridging and three terminal
carbonyl ligands. This triangle is capped on both
sides by $M(CO)_5$ groups, one of which is slightly closer
to the Ni_3 plane than the other (Figure 151). For the
Mo derivative the bond parameters are: Ni-Ni 2.341 Å,
Ni - Mo 3.045(5) - 3.075(5) and 3.132(4) - 3.172(5) Å;
for the W derivative: Ni-Ni 2.339(6) Å, W-Ni 3.032(5) -
3.078(6) and 3.136(5) - 3.182(5) Å (298).

In the complexes $M_2Mn_4(CO)_{18}$ (M = Ga, In) two
$Mn(CO)_4$ groups form a planar four-membered M_2Mn_2 ring
with the two main group metal atoms. Two $Mn(CO)_5$
units sit in a <u>trans</u> configuration to the plane and
coordinate to the Ga or In atoms (Figure 152). Within
the ring the Mn-Mn distances are 3.052(1) and
3.227(1) Å, and the MnMMn angles 76.9° and 76.4°, for
Ga and In, respectively. The Mn-Ga and Mn-In distances
are 2.45(1) and 2.605(1) Å (299). An analogous
structure is observed in $Re_2(CO)_8[\mu_2\text{-InRe}(CO)_5]_2$ where
the Re-Re distance is 3.232(1) Å, the ReInRe angle is
71.1°, and the In-Re distances lie in the range 2.738 -
2.807 Å (300). The structure of $Re_4(CO)_{12}[\mu_3\text{-InRe}(CO)_5]_4$
consists of a tetrahedral array of Re atoms (Re-Re
3.028(5) Å) associated with twelve carbonyl groups, two
bridging each edge. Each triangular Re_3 face is capped
by a μ_3-InRe$(CO)_5$ group with an average Re-(μ_3In)
distance of 2.818(7) Å (301).

In $(\mu_2\text{-H})_2Os_3Re_2(CO)_{20}$ the triangle of Os atoms
has two Re atoms bonded in $\underline{C}_3$-related equatorial sites
on two of the Os atoms (Figure 153). Each Re atom is
bound to five terminal carbonyls. The Os atoms linked
to Re atoms have three terminal carbonyls and the
third Os has four. The Os-Re bond lengths range from
2.942(3) to 2.983(3) Å and the Os-Os distances break
down into two distinct groups; those believed to be

bridged by single μ_2-hydride ligands in which the distance range from 3.059(3) to 3.084(2) Å, and non-bridged bonds where the values are in the range 2.876(3) - 2.880(3) Å (302).

The molecular structure of $Fe_3(CO)_9As_2$ consists of a triangle of $Fe(CO)_3$ units joined by Fe-Fe bonds with an average length of 2.623(4) Å. This triangle is capped above and below by As atoms. The mean Fe-As distance is 2.348(2) Å (303). In $[(C_5H_5)Fe(CO)_2]_2Sn_2$ $Fe_3(CO)_9$ the equatorial $Fe_3(CO)_9$ unit is retained but is now capped on both sides by $SnFe(CO)_2(C_5H_5)$ fragments. The Fe-Fe bond lengths average 2.792(6) Å and the Sn-Fe (equatorial) 2.537(4) Å (304).

The cluster $[(\eta^5-C_5H_5)Co(CO)]_2(GeCl_2)_2Fe(CO)_4$ (Figure 154) contains a planar $FeGe_2Co_2$ ring. The Co-Co bond of length 2.439(5) Å, which is bridged by two carbonyl ligands, is short (305).

In $Ru_3(CO)_9(GeMe_2)_3$ the Ru atoms form an equilateral triangle, coplanar with the three Ge atoms thus forming a six membered heterocyclic ring (Figure 155). Of the three carbonyls bonded to each Ru atom, two are _trans_ to each other and almost perpendicular to the Ru_3Ge_3 plane, whereas the third is in the metal plane. The average Ru-Ru distance is 2.93(1) Å. The $GeMe_2$ groups bridge symmetrically with an Ru-Ge distance of 2.49(1) Å and an Ru-Fe-Ru angle of 71.9(3)° (306).

In the anion $[Co_4Ni_2(CO)_{14}]^{2-}$ the metal atoms are statistically distributed over the vertices of an octahedron that is stretched along a three-fold axis. The metal-metal distances lie in the range 2.487(1) - 2.519(1) Å. Each metal is coordinated to one terminal carbonyl and the other eight ligands cap the triangular faces of the octahedron (307).

The anion $[Rh_5Pt(CO)_{15}]^-$ is structurally related to $Rh_6(CO)_{16}$. One $Rh(CO)_2$ group has been replaced by a Pt(CO) unit but the octahedral cluster skeleton is retained. The average Rh-Pt distance is 2.790 Å and the average Rh-Rh distance is 2.760 Å (308).

The metal skeleton in $Cu_4Ir_2(PPh_3)_2(C\equiv CPh)_8$ is distorted octahedron with the two iridium atoms occupying mutually trans vertices (Figure 156). The average Ir-Cu and Cu-Cu distances are 2.871 Å and 2.740 Å, respectively. Each Ir atom forms an apical bond with a phosphine ligand and four σ-bonds with the four phenylacetylide ligands. Each acetylide is involved in an asymmetric π-linkage with a Cu atom (309).

3. REFERENCES

1. E. O. Fischer and F. Röhrscheid, J. Organometallic
    Chem., 6, 53 (1966).
2. J. C. Huffman, J. G. Stone, W. C. Krusell and
    K. G. Caulton, J. Am. Chem. Soc., 99, 5830
    (1977).
3. K. Brodersen, G. Thiele and H. G. Schnering,
    Z. Anorg. Chem., 337, 120 (1965).
4. F. W. Koknat and R. E. McCarley, Inorg. Chem., 11,
    812 (1972).
5. R. A. Field, D. L. Kepert, B. W. Robinson and
    A. H. White, J. Chem. Soc. Dalton, 1858 (1973).
6. B. Spreckelmeyer and H. G. v. Schnering, Z. Anorg.
    Chem., 386, 27 (1971).
7. V. A. Simon, H. G. Schnering, H. Wohrle and
    H. Schafer, Z. Anorg. Allg. Chem., 339, 155
    (1965).
8. A. Simon and H. G. V. Schnering, Z. Anorg. Chem.,
    361, 235 (1968).
9. D. Bauer and H. G. V. Schnering, Z. Anorg. Chem.,
    361, 259 (1968).
10. C. B. Thaxton and R. A. Jacobson, Inorg. Chem.,
    10, 1460 (1971).
11. R. D. Burbank, Inorg. Chem., 5, 1491 (1966).

12. L. R. Bateman, J. F. Blount and L. F. Dahl,
    J. Am. Chem. Soc., 88, 1082 (1966).

13. A. Simon, Z. Anorg. Allg. Chem., 355, 311 (1967).

14. M. R. Churchill and S. W.-Y. Chang, J. Chem. Soc.
    Chem. Comm., 248 (1974).

15. S. Z. Goldberg, B. Spwack, G. Stanley, R. Eisenberg,
    D. M. Braitsch, J. S. Miller and M. Abkowitz,
    J. Am. Chem. Soc., 99, 110 (1977).

16. H. Schäfer and H. G. V. Schnering, Angew. Chem.,
    76, 833 (1964).

17. P. C. Healy, D. L. Kepert, D. Taylor and
    A. H. White, J. Chem. Soc. Dalton, 646,
    (1973).

18. H. Schäfer, H. G. V. Schnering, J. Tillack,
    T. Kuhnen , H. Wöhrle and H. Baumann, Z. Anorg.
    Chem., 353, 281 (1967).

19. L. J. Guggenberger and A. W. Sleight, Inorg. Chem.,
    8, 2041 (1969).

20. H. G. V. Schnering, Z. Anorg. Chem., 385, 75
    (1971).

21. R. Siepmann and H. G. V. Schnering, Z. Anorg.
    Chem., 357, 289 (1968).

22. H. Schäfer and R. Siepmann, Z. Anorg. Chem., 357,
    273 (1968).

23. R. Siepmann, H. G. V. Schnering and H. Schäfer,
    Angew. Chem. Int. Edn., 6, 637 (1967).

24. K. Jodden, H. G. V. Schnering and H. Schäfer,
    Angew. Chem. Internat. Edn., 14, 570 (1975).

25. W. H. McCarroll, L. Katz and R. Ward, J. Am. Chem.
    Soc., 79, 5410 (1957).

26. P. J. Vergamini, H. Vahrenkamp and L. F. Dahl,
    J. Am. Chem. Soc., 93, 6327 (1971).

27. K. Mennemann and R. Mattes, Angew. Chem. Int. Edn.,
    15, 118 (1976).

98.

28. V. Katovic, J. L. Templeton and R. E. McCalrey,
     J. Am. Chem. Soc., $\underline{98}$, 5706 (1976).

29. J. A. Bertrand, F. A. Cotton and W. A. Dollase,
     Inorg. Chem., $\underline{2}$, 1166 (1963); W. T. Robinson,
     J. E. Fergusson and B. R. Penfold, Proc. Chem.
     Soc., 116 (1963).

30. F. A. Cotton and J. T. Mague, Inorg. Chem., $\underline{3}$,
     1094 (1964).

31. F. A. Cotton and J. T. Mague, Inorg. Chem., $\underline{3}$,
     1403 (1964).

32. M. A. Bush, P. M. Druce and M. F. Lappert,
     J. Chem. Soc. Dalton, 500 (1972).

33. M. J. Bennett, F. A. Cotton and B. M. Foxman,
     Inorg. Chem., $\underline{7}$, 1563 (1968).

34. F. A. Cotton and S. J. Lippard, J. Am. Chem. Soc.,
     $\underline{86}$, 4497 (1964).

35. M. Elder and B. R. Penfold, Inorg. Chem., $\underline{5}$, 1763,
     (1966).

36. B. R. Penfold and W. T. Robinson, Inorg. Chem.,
     $\underline{5}$, 1758 (1966).

37. M. Elder, G. J. Gainsford, M. D. Papps and
     B. R. Penfold, J. Chem. Soc. Chem. Comm.,
     731 (1969).

38. M. B. Hursthouse and K. M. A. Malik, J. Chem. Soc.
     Dalton, 1334 (1978).

39. S. W. Kirtley, J. P. Olsen and R. Bau, J. Am.
     Chem. Soc., $\underline{95}$, 4532 (1973).

40. R. Bau, S. W. Kirtley, T. N. Sorrell and
     S. Winarko, J. Am. Chem. Soc., $\underline{96}$, 998 (1974).

41. W. A. Herrmann, M. L. Ziegler and K. Weidenhammer,
     Angew. Chem. Int. Edn., $\underline{15}$, 368 (1977).

42. R. C. Elder, Inorg. Chem., $\underline{13}$, 1037 (1974).

43. M. R. Churchill, P. H. Bird, H. D. Kaesz, R. Bau
     and B. Fontal, J. Am. Chem. Soc., $\underline{90}$, 7135
     (1968).

44. G. Ciani, G. D'Alfonso, M. Freni, P. Romiti and
      A. Sironi, J. Organometallic Chem., 157,
      199 (1978).

45. G. Ciani, A. Sironi and V. G. Albano, J. Chem.
      Soc. Dalton, 1667 (1977).

46. G. Ciani, G. D'Alfonso, M. Freni, P. Romiti,
      A. Sironi and A. Albinati, J. Organometallic
      Chem., 136, C49 (1977).

47. G. Bertolucci, M. Freni, P. Romiti, G. Ciani,
      A. Sironi and V. G. Albano,
      J. Organometallic Chem., 113, C61 (1976).

48. R. P. White, T. E. Block and L. F. Dahl,
      unpublished work.

49. G. Ciani, A. Sironi and V. G. Albano,
      J. Organometallic Chem., 136, 339 (1977).

50. R. D. Wilson and R. Bau, J. Am. Chem. Soc., 98,
      4687 (1976).

51. G. Bertolucci, G. Ciani, M. Freni, P. Romiti,
      V. G. Albano and A. Abinati,
      J. Organometallic. Chem., 117, C37 (1976).

52. V. G. Albano, G. Ciani, M. Freni and P. Romiti,
      J. Organometallic Chem., 96, 259 (1975).

53. G. Ciani, V. G. Albano and A. Immirzi,
      J. Organometallic Chem., 121, 237 (1976).

54. M. R. Churchill and R. Bau, Inorg. Chem., 7, 2606,
      (1968).

55. R. Hoffmann, B. E. R. Schilling, R. Bau,
      H. D. Kaesz and D. M. P. Mingos, J. Am. Chem.
      Soc., 100, 6088 (1978).

56. M. Laing, P. M. Kieman and W. P. Griffith,
      J. Chem. Soc. Chem. Comm., 221 (1977).

57. F. A. Cotton and J. M. Troup, J. Am. Chem. Soc.,
      96, 4155 (1974).

58. M. R. Churchill, F. J. Hollander and
      J. P. Hutchinson, Inorg. Chem., 16, 2655 (1977).

59. M. R. Churchill and B. G. DeBoer, Inorg. Chem.,
    16, 878 (1977).

60. L. F. Dahl and J. F. Blount, Inorg. Chem., 4,
    1965 (1965).

61. B. F. G. Johnson, J. Lewis, P. R. Raithby and
    G. Süss, J. Chem. Soc. Dalton, in press.

62. G. Raper and W. S. McDonald, J. Chem. Soc. (A),
    3430 (1971).

63. D. J. Dahm and R. A. Jacobson, J. Am. Chem. Soc.,
    90, 5106 (1968).

64. R. E. Benfield, B. F. G. Johnson, P. R. Raithby
    and G. M. Sheldrick, Acta Cryst., B34, 666
    (1978).

65. A. V. Rivera and G. M. Sheldrick, Acta Cryst.,
    B34, 3372 (1978).

66. M. R. Churchill and B. G. DeBoer, Inorg. Chem.,
    16, 2397 (1977).

67. A. G. Orpen, A. V. Rivera, E. G. Bryan, D. Pippard,
    G. M. Sheldrick and K. D. Rouse, J. Chem.
    Soc. Chem. Comm., 723 (1978).

68. V. F. Allen, R. Mason and P. B. Hitchcock,
    J. Organometallic Chem., 140, 297 (1977).

69. D. F. Shriver, D. Lehamn and D. Strope, J. Am.
    Chem. Soc., 97, 1594 (1975).

70. B. F. G. Johnson, J. Lewis, A. G. Orpen,
    P. R. Raithby and G. Süss, J. Organometallic
    Chem., submitted for publication.

71. M. R. Churchill, B. G. DeBoer and F. J. Rotella,
    Inorg. Chem., 15, 1843 (1976).

72. S. Jeannin, Y. Jeannin and G. Larigne, Inorg.
    Chem., 17, 2103 (1978).

73. B. F. G. Johnson, P. R. Raithby and C. Zuccaro,
    J. Chem. Soc. Dalton, submitted for
    publication.

74. J. R. Shapley, M. Tachikawa, M. R. Churchill and
     R. A. Lashewycz, J. Organometallic Chem.,
     162, C39 (1978).

75. M. R. Churchill and B. G. DeBoer, Inorg. Chem.,
     16, 1141 (1977).

76. B. F. G. Johnson, J. Lewis, D. Pippard and
     P. R. Raithby, J. Chem. Soc. Chem. Comm.,
     551 (1978).

77. C. G. Pierpont, Inorg. Chem., 17, 1978 (1976).

78. J. Norton, J. Collman, G. Dolcetti and
     W. T. Robinson, Inorg. Chem., 11, 382 (1972).

79. P. Mastropasqua, A. Riemer, H. Kisch and
     C. Krüger, J. Organometallic Chem., 148,
     C40 (1978).

80. F. A. Cotton, B. E. Hanson and J. D. Jamerson,
     J. Am. Chem. Soc., 99, 6588 (1977).

81. C. G. Pierpont, Inorg. Chem., 16, 636 (1977).

82. G. Gervasio, J. Chem. Soc. Chem. Comm., 25 (1976).

83. A. G. Orpen, D. Pippard, G. M. Sheldrick and
     K. D. Rouse, Acta Cryst., B34, 2466 (1978).

84. J. J. Guy, B. E. Reichert and G. M. Sheldrick,
     Acta Cryst., B32, 3319 (1976).

85. M. Laing, P. Sommerville, Z. Dawoodi, M. J. Mays
     and P. Wheatley, J. Chem. Soc. Chem. Comm.,
     1035 (1978).

86. R. D. Adams and N. M. Golembeski, Inorg. Chem.,
     17, 1969 (1978).

87. B. L. Barnett   C. Krüger, Angew Chem., 83, 969
     (1971).

88. J. F. Blount, L. F. Dahl, C. Hoogzand and
     W. Hubel, J. Am. Chem. Soc., 88, 292 (1966).

89. C. H. Wei and L. F. Dahl, Inorg. Chem., 4, 493
     (1965).

90. R. Mey, J. van der Helm, D. J. Stufkens and
     K. Vrieze, J. Chem. Soc. Chem. Comm., 506 (1978).

102.

91. R. Bau, B. Don, R. Greatrex, R. J. Haines,
R. A. Love and R. D. Wilson, Inorg. Chem.,
14, 3021 (1975).

92. B. F. G. Johnson, J. Lewis, D. Pippard and
P. R. Raithby, Acta Cryst., B34, 3767 (1978).

93. B. F. G. Johnson, J. Lewis, D. Pippard,
P. R. Raithby, G. M. Sheldrick and K. R. Rouse,
J. Chem. Soc. Dalton, in press.

94. G. M. Sheldrick and J. P. Yesinowski, J. Chem.
Soc. Dalton, 873 (1975).

95. M. Catti, G. Gervasio and S. A. Mason, J. Chem.
Soc. Dalton, 2260 (1977).

96. G. Gervasio, D. Osella and M. Valle, Inorg. Chem.,
15, 1221 (1976).

97. M. I. Bruce, M. A. Cairns, A. Cox, M. Green,
M. D. H. Smith and P. Woodward, J. Chem. Soc.,
Chem. Comm., 735 (1970).

98. M. Evans, M. Hursthouse, E. W. Randall,
E. Rosenberg, L. Milone and M. Valle, J. Chem.
Soc. Chem. Comm., 545 (1972).

99. E. G. Bryan, B. F. G. Johnson, J. W. Kelland,
J. Lewis and M. McPartlin, J. Chem. Soc.
Chem. Comm., 254 (1976).

100. K. A. Azam, A. J. Deeming, J. P. Rothwell,
M. B. Hursthouse and L. New, J. Chem. Soc.,
Chem. Comm., 1086 (1978).

101. J. J. de Boer, J. A. van Doom and C. Masters,
J. Chem. Soc. Chem. Comm., 1005 (1978).

102. P. J. Roberts and J. Trotter, J. Chem. Soc. (A),
1479 (1971).

103. K. Yasufuku, K. Aoki and H. Yamazaki, Bull. Chem.
Soc. Japan, 48, 1616 (1975).

104. E. Sappa, A. Tiripicchio and A. M. M. Lanfredi,
J. Chem. Soc. Dalton, 552 (1978).

105.   R. J. Restivo and G. Ferguson, J. Chem. Soc.
        Dalton, 893 (1976).

106.   E. Sappa, L. Milone and A. Tiripicchio, J. Chem.
        Soc. Dalton, 1843 (1976).

107.   S. Aime, L. Milone, E. Sappa and A. Tiripicchio,
        J. Chem. Soc. Dalton, 227 (1977).

108.   J. A. K. Howard, S. A. R. Knox, V. Riera,
        F. G. A. Stone and P. Woodward, J. Chem. Soc.
        Chem. Comm., 452 (1974); J. A. K. Howard,
        S. A. R. Knox, F. G. A. Stone, A. C. Szary
        and P. Woodward, J. Chem. Soc. Chem. Comm.,
        788 (1974).

109.   J. A. K. Howard, S. A. R. Knox, R. J. McKinney,
        R. F. D. Stansfield, F. G. A. Stone and
        P. Woodward, J. Chem. Soc. Chem. Comm., 557
        (1976).

110.   M. R. Churchill, F. R. Scholer and J. Wormald,
        J. Organometallic Chem., $\underline{28}$, C21 (1971).

111.   R. Bau, J. C. Burt, S. A. R. Knox, R. M. Laine,
        R. P. Phillips and F. G. A. Stone, J. Chem.
        Soc. Chem. Comm., 726 (1973).

112.   M. J. Bennett, F. A. Cotton and P. Legzdins,
        J. Am. Chem. Soc., $\underline{90}$, 6335 (1968).

113.   J. D. Edwards, J. A. K. Howard, S. A. R. Knox,
        V. Rieva, F. G. A. Stone and P. Woodward,
        J. Chem. Soc. Dalton, 75 (1976).

114.   G. Gervasio, S. Aime, L. Milone, E. Sappa and
        M. Franchini-Angela, Transition Met. Chem.,
        $\underline{1}$, 96 (1976).

115.   G. Ferraris and G. Gervasio, J. Chem. Soc.
        Dalton, 1813 (1974).

116.   G. Ferraris and G. Gervasio, J. Chem. Soc.
        Dalton, 1057 (1972).

117.   G. Ferraris and G. Gervasio, J. Chem. Soc.
        Dalton, 1933 (1973).

104.

118. C. W. Bradford, R. S. Nyholm, G. J. Gainsford,
      J. M. Guss, P. R. Ireland and R. Mason,
      J. Chem. Soc. Chem. Comm., 87 (1972);
      G. J. Gainsford, J. M. Guss, P. R. Ireland,
      R. Mason, C. W. Bradford and R. S. Nyholm,
      J. Organometallic Chem., 40, C70 (1972).

119. M. R. Churchill and R. A. Laskewycz, Inorg. Chem.,
      17, 1291 (1978).

120. N. Cook, L. Smart and P. Woodward, J. Chem. Soc.
      Dalton, 1744 (1977).

121. R. D. Wilson, S. M. Wu, R. A. Love and R. Bau,
      Inorg. Chem., 17, 1271 (1978).

122. P. F. Jackson, B. F. G. Johnson, J. Lewis,
      M. McPartlin and W. J. H. Nelson, J. Chem.
      Soc. Chem. Comm., 920 (1978).

123. B. F. G. Johnson, J. Lewis, P. R. Raithby and
      C. Zuccaro, Acta Cryst., B34, 3765 (1978).

124. B. F. G. Johnson, J. Lewis, P. R. Raithby,
      G. M. Sheldrick and G. Süss,
      J. Organometallic Chem., 162, 179 (1978).

125. D. B. W. Yawney and R. J. Doedens, Inorg. Chem.,
      11, 838 (1972).

126. M. Manassero, M. Sansoni and G. Longoni, J. Chem.
      Soc. Chem. Comm., 919 (1976).

127. R. J. Doedens and L. F. Dahl, J. Am. Chem. Soc.,
      88, 4847 (1966).

128. M. R. Churchill and R. A. Laskewycz, Inorg. Chem.,
      17, 1950 (1978).

129. B. F. G. Johnson, J. Lewis, B. E. Reichert,
      K. T. Schorpp and G. M. Sheldrick, J. Chem.
      Soc. Dalton, 1417 (1977).

130. R. Mason and K. M. Thomas, J. Organometallic
      Chem., 3, C39 (1972).

131. R. Belford, H. P. Taylor and P. Woodward, J. Chem.
      Soc. Dalton, 2425 (1972).

132. B. F. G. Johnson, J. Lewis, P. R. Raithby,
     G. M. Sheldrick, K. Wong and M. McPartlin,
     J. Chem. Soc. Dalton, 673 (1978).

133. S. Bhaduri, B. F. G. Johnson, J. W. Kelland,
     J. Lewis, P. R. Raithby, S. Rehani,
     G. M. Sheldrick, K. Wong and M. McPartlin,
     J. Chem. Soc. Dalton, in press.

134. E. Sappa, A. Tiripicchio and M. T. Camellini,
     J. Chem. Soc. Dalton, 419 (1978).

135. G. Gerrasio, R. Rossetti and P. L. Stanghellini,
     J. Chem. Soc. Chem. Comm., 387 (1977).

136. B. E. Reichert and G. M. Sheldrick, Acta. Cryst.,
     B33, 173 (1977).

137. J. J. Guy and G. M. Sheldrick, Acta Cryst., B34,
     1725 (1978).

138. J. J. Guy and G. M. Sheldrick, Acta Cryst., B34,
     1722 (1978).

139. A. V. Rivera, G. M. Sheldrick and M. B. Hursthouse,
     Acta Cryst., B34, 3376 (1978).

140. J. M. Fernandez, B. F. G. Johnson, J. Lewis,
     P. R. Raithby and G. M. Sheldrick, Acta
     Cryst., B34, 1994·(1978).

141. A. G. Orpen and G. M. Sheldrick, Acta Cryst.,
     B34, 1992 (1978).

142. J. M. Fernandez, B. F. G. Johnson, J. Lewis and
     P. R. Raithby, J. Chem. Soc. Chem. Comm.,
     1015 (1978).

143. E. H. Braye, L. F. Dahl, W. Hübel and
     D. L. Wampler, J. Am. Chem. Soc., 84, 4633
     (1962).

144. M. R. Churchill, J. Wormald, J. Knight and
     M. J. Mays, J. Am. Chem. Soc., 93, 3073
     (1971).

145. A. Sirigu, M. Bianchi and E. Benedetti, J. Chem.
     Soc. Chem. Comm., 596 (1969).

106.

146. R. Mason and W. R. Robinson, J. Chem. Soc. Chem.
     Comm., 468 (1968).

147. C. R. Eady, B. F. G. Johnson, J. Lewis,
     M. C. Malatesta, P. Machin and M. McPartlin,
     J. Chem. Soc. Chem. Comm., 945 (1976).

148. M. McPartlin, C. R. Eady, B. F. G. Johnson and
     J. Lewis, J. Chem. Soc. Chem. Comm., 883
     (1976).

149. M. R. Churchill and J. Wormald, J. Am. Chem. Soc.,
     93, 5670 (1971).

150. R. Mason, K. M. Thomas and D. M. P. Mingos,
     J. Am. Chem. Soc., 95, 3802 (1973).

151. A. G. Orpen and G. M. Sheldrick, Acta Cryst.,
     B34, 1989 (1978).

152. A. V. Rivera, G. M. Sheldrick and M. B. Hursthouse,
     Acta Cryst., B34, 1985 (1978).

153. C. R. Eady, P. D. Gavens, B. F. G. Johnson,
     J. Lewis, M. C. Malatesta, M. J. Mays,
     A. G. Orpen, A. V. Rivera, G. M. Sheldrick
     and M. B. Hursthouse, J. Organometallic
     Chem., 149, C43 (1978).

154. C. R. Eady, J. M. Fernandez, B. F. G. Johnson,
     J. Lewis, P. R. Raithby and G. M. Sheldrick,
     J. Chem. Soc. Chem. Comm., 421 (1978).

155. J. M. Fernandez, B. F. G. Johnson, J. Lewis and
     P. R. Raithby, Acta Cryst., B34, 3086 (1978).

156. C. R. Eady, B. F. G. Johnson, J. Lewis, R. Mason,
     P. B. Hitchcock and K. M. Thomas, J. Chem.
     Soc. Chem. Comm., 385 (1977).

157. J. J. Guy and G. M. Sheldrick, Acta Cryst., B34,
     1718 (1978).

158. P. W. Sutton and L. F. Dahl, J. Am. Chem. Soc.,
     89, 261 (1967).

159. V. Batzel, U. Muller and R. Allmann,
     J. Organometallic Chem., 102, 109 (1975).

160. V. Batzel, Z. Naturforsch., 31b, 342 (1976).

161. F. Klanberg, W. B. Askew and L. F. Guggenberger,
     Inorg. Chem., 7, 2265 (1968).

162. M. D. Brice, R. J. Dellaca, B. R. Penfold and
     J. L. Spencer, J. Chem. Soc. Chem. Comm.,
     72 (1971).

163. C. H. Wei and L. F. Dahl, Inorg. Chem., 6, 1229
     (1967).

164. G. Bor, G. Gervasio, R. Rossetti and
     P. L. Stanghellini, J. Chem. Soc. Chem.
     Comm., 841 (1978).

165. T. W. Matheson and B. R. Penfold, Acta Cryst.,
     B33, 1980 (1977).

166. F. W. B. Einstein and R. D. G. Jones, Inorg.
     Chem., 11, 395 (1972).

167. Y. S. Ng and B. R. Penfold, Acta Cryst., B34,
     1978 (1978).

168. R. S. McCallum and B. R. Penfold, Acta Cryst.,
     B34, 1688 (1978).

169. P. D. Frisch and L. F. Dahl, J. Am. Chem. Soc.,
     94, 5082 (1972).

170. F. A. Cotton and J. D. Jamerson, J. Am. Chem.
     Soc., 98, 1273 (1976).

171. E. F. Paulus, E. O. Fischer, H. P. Fritz and
     H. Schuster-Woldan, J. Organometallic
     Chem., 10, C3 (1967).

172. O. S. Mills and E. F. Paulus, J. Organometallic
     Chem., 10, 331 (1967).

173. O. S. Mills and E. F. Paulus, J. Organometallic
     Chem., 11, 587 (1968).

174. W. D. Jones, M. A. White and R. G. Bergman,
     J. Am. Chem. Soc., 100, 6772 (1978).

175. T. Toan, R. W. Broach, S. A. Gardner and
     M. D. Rausch, Inorg. Chem., 16, 279 (1977).

108.

176. F. H. Carve, F. A. Cotton and B. A. Frenz,
        Inorg. Chem., 15, 380 (1976).

177. C. H. Wei, Inorg. Chem., 8, 2384 (1969).

178. V. Albano, P. L. Bellon, P. Chini and
        V. Scatturin, J. Organometallic Chem., 16,
        461 (1969).

179. V. Albano, P. Chini and V. Scatturin, J. Chem.
        Soc. Chem. Comm., 163 (1968).

180. V. G. Albano, M. Sansoni, P. Chini and
        S. Martinengo, J. Chem. Soc. Dalton, 651
        (1973).

181. V. G. Albano, P. L. Bellon and M. Sansoni,
        J. Chem. Soc. (A), 678 (1971).

182. E. R. Corey, L. F. Dahl and W. Beck, J. Am. Chem.
        Soc., 85, 1202 (1963).

183. V. G. Albano, P. L. Bellon and G. Ciani,
        J. Chem. Soc. Chem. Comm., 1024 (1969).

184. V. G. Albano, G. Ciani, S. Martinengo, P. Chini
        and G. Giovdano, J. Organometallic Chem.,
        88, 381 (1975).

185. V. G. Albano, P. Chini, G. Ciani, S. Martinengo
        and M. Sansoni, J. Chem. Soc. Dalton, 463
        (1978).

186. V. G. Albano, P. Chini, S. Martinengo, M. Sansoni
        and D. Strumolo, J. Chem. Soc. Chem. Comm.,
        299 (1974).

187. G. Ciani, L. Garlaschelli, M. Manassero and
        U. Sartorelli, J. Organometallic Chem., 129,
        C25 (1977).

188. R. S. Gall, N. G. Connelly and L. F. Dahl, J. Am.
        Chem. Soc., 96, 4017 (1974).

189. G. Huttner and H. Lorenz, Chem. Ber., 108, 973
        (1975).

190. L. F. Dahl and D. L. Smith, J. Am. Chem. Soc.,
        84, 2450 (1962).

191. R. C. Ryan and L. F. Dahl, J. Am. Chem. Soc.,
    97, 6904 (1975).

192. C. H. Wei and L. F. Dahl, Cryst. Struct. Comm.,
    4, 583 (1975).

193. E. Keller and H. Vahrenkamp, Angew. Chem. Intern.
    Edn., 16, 731 (1977).

194. V. G. Albano and P. L. Bellon, J. Organometallic
    Chem., 19, 405 (1969).

195. V. G. Albano, P. Chini, S. Martinengo, M. Sansoni
    and D. Strumolo, J. Chem. Soc. Dalton, 459
    (1978).

196. V. G. Albano, A. Ceriotti, P. Chini and G. Ciani,
    S. Martinengo and W. M. Anker, J. Chem. Soc.
    Chem. Comm., 859 (1975).

197. S. Martinengo, G. Ciani, A. Sironi and P. Chini,
    J. Am. Chem. Soc., 100, 7096 (1978).

198. V. G. Albano, M. Sansoni, P. Chini, S. Martinengo
    and D. Strumolo, J. Chem. Soc. Dalton, 970
    (1976).

199. J. L. Vidal, R. A. Fiato, L. A. Cosby and
    R. L. Pruett, Inorg. Chem., 17, 2574 (1978).

200. D. F. Chodosh, R. H. Crabtree, H. Felkins and
    G. E. Morris, J. Organometallic Chem., 161,
    C67 (1978).

201. M. R. Churchill and J. P. Hutchinson, Inorg. Chem.,
    17, 3528 (1978).

202. P. Chini, G. Ciani, L. Garlaschelli, M. Manassero,
    S. Martinengo, A. Sironi and F. Canziani,
    J. Organometallic Chem., 152, C35 (1978).

203. G. Ciani, M. Manassero, V. G. Albano,
    F. Canziani, G. Giovdano, S. Martinengo
    and P. Chini, J. Organometallic Chem., 150,
    C17 (1978).

204. V. Albano, P. L. Bellon and V. Scatturin, J. Chem.
    Soc. Chem. Comm., 730 (1967).

110.

205. G. F. Stuntz, J. R. Shapley and C. G. Pierpont,
     Inorg. Chem., 17, 2596 (1978).

206. P. F. Heveldt, B. F. G. Johnson, J. Lewis,
     P. R. Raithby and G. M. Sheldrick, J. Chem.
     Soc. Chem. Comm., 340 (1978).

207. C. G. Pierpont, G. F. Stuntz and J. R. Shapley,
     J. Am. Chem. Soc., 100, 616 (1978).

208. J. C. Calabrese, L. F. Dahl, A. Cavalieri,
     P. Chini, G. Longoni and S. Martinengo,
     J. Am. Chem. Soc., 96, 2616 (1974).

209. J. C. Calabrese, L. F. Dahl, P. Chini, G. Longoni
     and S. Martinengo, J. Am. Chem. Soc., 96,
     2614 (1974).

210. A. Hock and O. S. Mills, Advances in Chemistry
     of Coordination Compounds, MacMillan,
     New York, 1961.

211. H. Vahrenkamp, V. A. Uchtman and L. F. Dahl,
     J. Am. Chem. Soc., 90, 3272 (1968).

212. W. P. Basman and H. G. M. van der Linden, J. Chem.
     Soc. Chem. Comm., 714 (1977).

213. N. Kamijyo and T. W. Watanabe, Bull. Chem. Soc.
     Japan, 47, 373 (1974).

214. C. G. Salentine, C. E. Strouse and
     M. F. Hawthorne, Inorg. Chem., 15, 1832 (1976).

215. S. Otsuka, Y. Tatsuno, M. Miki, M. Matsumoto,
     H. Yoshioka and K. Nakatsu, J. Chem. Soc.
     Chem. Comm., 445 (1973).

216. G. W. Bushnell, K. R. Dixon, P. M. Moroney,
     A. D. Rattray and Ch'eng Wan, J. Chem. Soc.
     Chem. Comm., 709 (1977).

217. P. M. Bailey, A. Keasey and P. M. Maitlis,
     J. Chem. Soc. Dalton, 1825 (1978).

218. D. C. Moody and R. R. Ryan, Inorg. Chem., 16,
     1054 (1977).

219. N. J. Taylor, P. C. Chieh and A. J. Carty,
          J. Chem. Soc. Chem. Comm., 448 (1975).

220. A. Abinati, G. Carturan and A. Musco, Inorg.
          Chim. Acta., 16, L3 (1976).

221. M. Green, J. A. K. Howard, M. Murry, J. L. Spencer
          and F. G. A. Stone, J. Chem. Soc. Dalton,
          1509 (1977).

222. N. M. Boag, M. Green, J. A. K. Howard,
          J. L. Spencer, R. F. D. Stansfield,
          F. G. A. Stone, M. D. O. Thomas,
          J. Vicente and P. Woodward, J. Chem. Soc.
          Chem. Comm., 930 (1977).

223. M. J. Bennett, F. A. Cotton and B. H. C. Winquist,
          J. Am. Chem. Soc., 89, 5366 (1967).

224. J. L. Davidson, M. Green, F. G. A. Stone and
          A. J. Welch, J. Am. Chem. Soc., 97, 7490
          (1975).

225. M. G. Thomas, E. L. Muetterties, R. O. Day and
          V. W. Day, J. Am. Chem. Soc., 98, 4645 (1976).

226. V. W. Day, R. O. Day, J. S. Kristoff,
          F. J. Hirsekom and E. L. Muetterties,
          J. Am. Chem. Soc., 97, 2571 (1975).

227. I. I. Moiseev, T. A. Stromnova, M. N. Vargaftig,
          G. J. Mazo, L. G. Kuz'mina and
          Y. T. Struchkov, J. Chem. Soc. Chem. Comm.,
          27 (1978).

228. M. A. A. F. de C. T. Carrondo and A. C. Skapski,
          Acta Cryst., B34, 1857 (1978).

229. M. A. A. F. de C. T. Carrondo and A. C. Skapski,
          Acta Cryst., B34, 3576 (1978).

230. G. Longoni, P. Chini, L. D. Lower and L. F. Dahl,
          J. Am. Chem. Soc., 97, 5034 (1975).

231. K. Brodersen, G. Thiele and H. G. Schnering,
          Z. Anorg. Chem., 337, 120 (1965).

112.

232. L. D. Lower and L. F. Dahl, J. Am. Chem. Soc.,
     $\underline{98}$, 5046 (1976).

233. S. L. Lawton, W. J. Rohrbaugh and G. J. Kokotailo,
     Inorg. Chem., $\underline{11}$, 612 (1972).

234. G. van Koten and J. G. Noltes, J. Organometallic
     Chem., $\underline{84}$, 129 (1975).

235. J. E. O'Connor, G. A. Janusonis and E. R. Corey,
     J. Chem. Soc. Chem. Comm., 445 (1968).

236. P. F. Rodesiler and E. L. Amma, J. Chem. Soc.
     Chem. Comm., 599 (1974).

237. A. N. Nesmeyanov, Yu. T. Struchkov, N. N. Sedova,
     V. G. Andrianov, Yu. V. Valgin and
     V. A. Sazonova, J. Organometallic Chem.,
     $\underline{137}$, 217 (1977).

238. J. A. J. Jarvis, B. T. Kilboum, R. Pearce and
     M. F. Lappert, J. Chem. Soc. Chem. Comm.,
     475 (1973).

239. S. A. Bezmann, M. R. Churchill, J. A. Osborn
     and J. Wormald, Inorg. Chem., $\underline{11}$, 1818
     (1972).

240. J. M. Guss, R. Mason, K. M. Thomas, G. van Koten
     and J. G. Noltes, J. Organometallic Chem.,
     $\underline{40}$, C79 (1972).

241. R. W. M. den Hoedt, J. G. Noltes, G. van Koten
     and A. L. Spek, J. Chem. Soc. Dalton, 1800
     (1978).

242. L. E. McCandlish, E. C. Bissell, D. Coucouvanis,
     J. P. Fackler and K. Knox, J. Am. Chem.
     Soc., $\underline{90}$, 7357 (1968).

243. F. J. Hollander and D. Coucouvanis, J. Am. Chem.
     Soc., $\underline{96}$, 5646 (1974).

244. F. J. Hollander, M. L. Caffery and D. Coucouvanis,
     Abstracts of 167th A.C.S. National Meeting,
     Los Angeles, 1974.

245. P. Jennische and R. Hesse, Acta. Chem. Scand.,
          25, 423 (1971).

246. J. J. Guy, P. G. Jones, M. J. Mays and
          G. M. Sheldrick, J. Chem. Soc. Dalton,
          8 (1977).

247. P. Bellon, M. Manassero and M. Sansoni, J. Chem.
          Soc. Dalton, 2423 (1973).

248. P. Bellon, F. Cariati, M. Manassero, L. Naldini
          and M. Sansoni, J. Chem. Soc. Chem. Comm.,
          1423 (1971).

249. M. McPartlin, R. Mason and L. Malatesta,
          J. Chem. Soc. Chem. Comm., 334 (1969).

250. V. G. Albano, P. L. Bellon, M. Manassero and
          M. Sansoni, J. Chem. Soc. Chem. Comm., 1210
          (1976).

251. P. Bellon, M. Manassero and M. Sansoni,
          J. Chem. Soc. Dalton, 1481 (1972).

252. R. D. Ellison, H. A. Levy and K. W. Fung, Inorg.
          Chem., 11, 833 (1972).

253. P. J. Vergamini, H. Vahrenkamp and L. F. Dahl,
          J. Am. Chem. Soc., 93, 6326 (1971).

254. T. V. Ashworth, M. Berry, J. A. K. Howard,
          M. Laguna and F. G. A. Stone, J. Chem. Soc.
          Chem. Comm., 45 (1979).

255. H. D. Kaesz, R. Bau and M. R. Churchill, J. Am.
          Chem. Soc., 89, 2775 (1967).

256. P. A. Agron, R. D. Ellison and H. A. Levy,
          Acta Cryst., 23, 1079 (1967).

257. H. Vahrenkamp, Chem. Ber., 106, 2570 (1973).

258. W. Clegg and P. J. Wheatley, J. Chem. Soc.
          Dalton, 90 (1973); ibid, 424 (1974); ibid,
          511 (1974).

259. W. Clegg and P. J. Wheatley, J. Chem. Soc. (A),
          3572 (1971).

114.

260. K. Triplett and M. D. Curtis, J. Am. Chem. Soc.,
     97, 5747 (1975).

261. H. Vahrenkamp, Chem. Ber., 107, 3867 (1974).

262. H. Preut, W. Wolfes and H.-J. Haupt, Z. Anorg.
     Chem., 412, 121 (1975).

263. E. Keller and H. Vahrenkamp, Angew Chem. Int.
     Edn., 16, 542 (1977).

264. R. J. Haines, R. Mason, J. A. Zubieta and
     C. R. Nolte, J. Chem. Soc. Chem. Comm.,
     990 (1972).

265. C. H. Wei and L. F. Dahl, Inorg. Chem., 6, 1229
     (1967).

266. D. L. Stevenson, C. H. Wei and L. F. Dahl,
     J. Am. Chem. Soc., 93, 6027 (1971).

267. C. E. Strouse and L. F. Dahl, J. Am. Chem. Soc.,
     93, 6032 (1971).

268. V. G. Albano, G. Ciani, M. I. Bruce, G. Shaw and
     F. G. A. Stone, J. Organometallic Chem., 42,
     C99 (1972).

269. R. Mason and J. A. Zubieta, J. Organometallic Chem.,
     66, 289 (1974).

270. V. G. Andrianov, Yu. T. Struchkov and
     E. R. Rossinskaya, J. Chem. Soc. Chem. Comm.,
     338 (1973).

271. F. S. Stephens, J. Chem. Soc. Dalton, 2257 (1972).

272. R. D. Adams, M. D. Brice and F. A. Cotton, Inorg.
     Chem., 13, 1080 (1974).

273. F. W. B. Einstein and R. D. G. Jones, Inorg. Chem.,
     12, 1690 (1973).

274. A. Modinos and P. Woodward, J. Chem. Soc. Dalton,
     1534 (1975).

275. P. D. Cradwick, J. Organometallic Chem., 27, 251
     (1971).

276. R. D. Ball and D. Hall, J. Organometallic Chem.,
     56, 209 (1973).

277. M. R. Churchill, F. J. Hollander, J. R. Shapley
      and D. S. Foose, J. Chem. Soc. Chem. Comm.,
      534 (1978).

278. M. R. Churchill and F. J. Hollander, Inorg.
      Chem., 16, 2493 (1977).

279. C. J. Gilmore and P. Woodward, J. Chem. Soc. (A),
      3453 (1971).

280. R. G. Toller, R. D. Wilson, R. K. McMillan,
      T. F. Koetzle and R. Bau, J. Am. Chem. Soc.,
      100, 3871 (1978).

281. M. R. Churchill and M. V. Veidis, J. Chem. Soc.
      (A), 2995 (1971).

282. M. R. Churchill and M. V. Veidis, J. Chem. Soc.
      (A), 2170 (1971).

283. T.-Toan and L. F. Dahl, J. Am. Chem. Soc., 93,
      2654 (1971).

284. C. J. Gilmore and P. Woodward, J. Chem. Soc.
      Dalton, 1387 (1972).

285. P. G. Harrison, T. J. King and J. A. Richards,
      J. Chem. Soc. Dalton, 2097 (1975).

286. S. Bhaduri, B. F. G. Johnson, J. Lewis,
      P. R. Raithby and D. J. Watson, J. Chem. Soc.
      Chem. Comm., 343 (1978).

287. L. J. Farrugia, J. A. K. Howard, P. Mitrprachachon,
      J. L. Spencer, F. G. A. Stone and P. Woodward,
      J. Chem. Soc. Chem. Comm., 260 (1978).

288. V. G. Albano, G. Ciani and S. Martinengo,
      J. Organometallic Chem., 78, 265 (1977).

289. P. H. Bird and A. R. Frazer, J. Organometallic
      Chem., 73, 103 (1974).

290. J. Fischer, A. Mitschler, R. Weiss, J. Dehard
      and J. F. Wennig, J. Organometallic Chem.,
      91, C37 (1975).

291. R. D. Ball and D. Hall, J. Organometallic Chem.,
      52, 293 (1973).

116.

292. J. Weaver and P. Woodward, J. Chem. Soc. Dalton,
 1060 (1973).

293. P. S. Elmes, B. M. Gatehouse, D. J. Lloyd and
 B. O. West, J. Chem. Soc. Chem. Comm., 953
 (1974).

294. F. A. Cotton and T. R. Webb, Inorg. Chim. Acta,
 10, 127 (1974).

295. R. A. Forder and K. Prout, Acta Cryst., B30,
 2318 (1974).

296. M. L. H. Green, G. A. Moser, J. Packer, F. Petit,
 R. A. Forder and K. Prout, J. Chem. Soc.
 Chem. Comm., 839 (1974).

297. R. A. Forder and K. Prout, Acta Cryst., B30,
 2312 (1974).

298. J. K. Ruff, R. P. White and L. F. Dahl, J. Am.
 Chem. Soc., 93, 2159 (1971).

299. H. Preut and H.-J. Haupt, Chem. Ber., 107, 2860
 (1974).

300. H. Preut and H.-J. Haupt, Chem. Ber., 108, 1447
 (1975).

301. H.-J. Haupt, F. Newman and H. Preut,
 J. Organometallic Chem., 99, 439 (1975).

302. M. R. Churchill and F. J. Hollander, Inorg. Chem.,
 17, 3546 (1978).

303. L. F. J. Delbaere, L. J. Kruczynski and
 D. W. McBride, J. Chem. Soc. Dalton, 307
 (1973).

304. T. J. McNeese, S. S. Wreford, D. L. Sipton and
 R. Bau, J. Chem. Soc. Chem. Comm., 390
 (1977).

305. M. Elder and W. L. Hutchson, J. Chem. Soc.
 Dalton, 15 (1972).

306. J. A. K. Howard and P. Woodward, J. Chem. Soc.
 (A), 3648 (1971).

307. V. G. Albano, G. Ciani and P. Chini, J. Chem.
     Soc. Dalton, 432 (1974).

308. A. Famagalli, S. Martinengo, P. Chini,
     A. Albinati, S. Bruckner and B. T. Heaton,
     J. Chem. Soc. Chem. Comm., 195 (1978).

309. O. M. Abu Salah, M. I. Bruce, M. R. Churchill
     and S. A. Bezman, J. Chem. Soc. Chem.
     Comm., 858 (1972).

118.

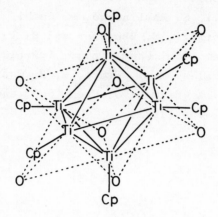

Figure 1

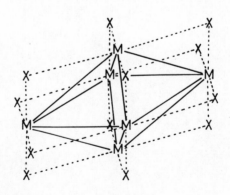

Figure 2

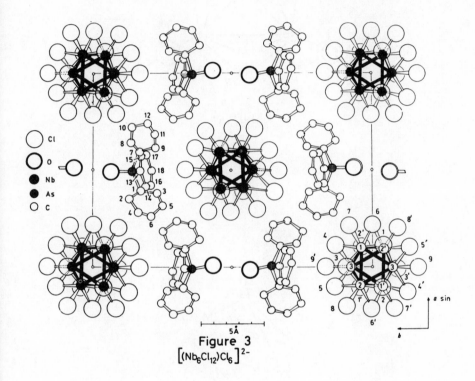

Figure 3

$\left[(Nb_6Cl_{12})Cl_6\right]^{2-}$

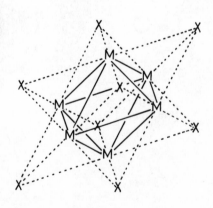

Figure 4

Figure 5

Cl
Cl
Cl
Mo
Cl
Cl
Mo
Mo
Cl
Cl
Cl
Mo
Mo
Cl
Cl
Cl
Cl
Cl

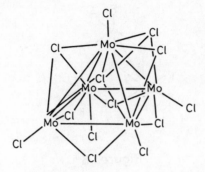

Figure 6

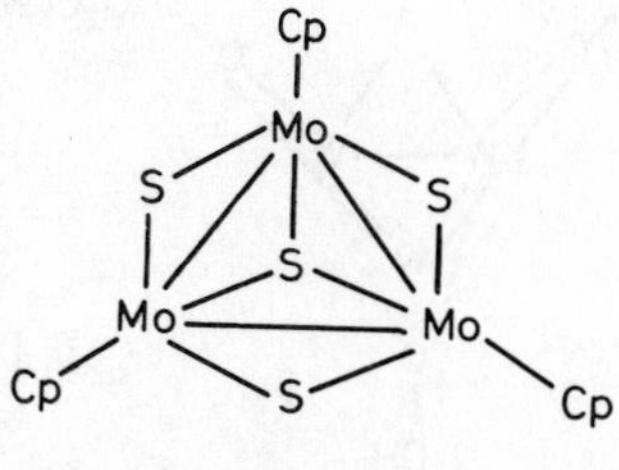

Cp
Mo
S
S
S
Mo
Mo
Cp
S
Cp

Figure 7

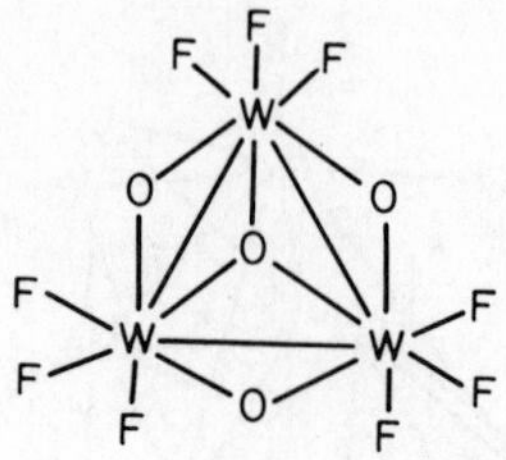

Figure 8

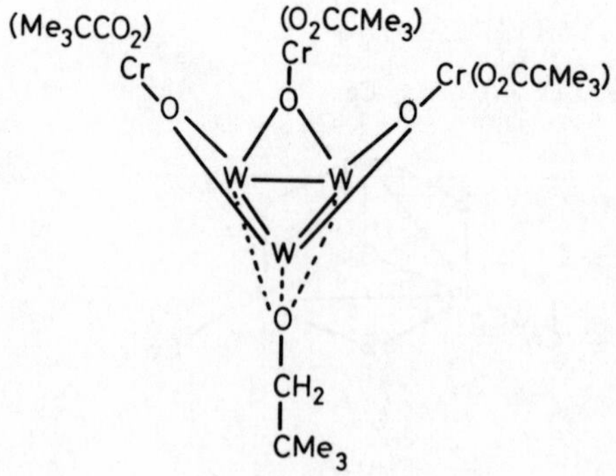

Figure 9

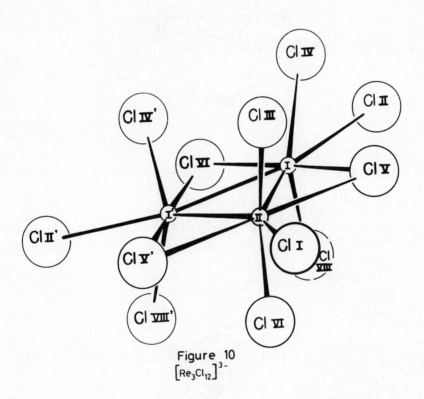

Figure 10
$[Re_3Cl_{12}]^{3-}$

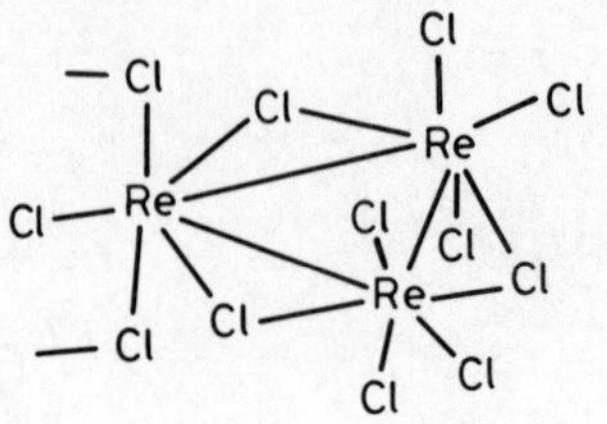

124.

Figure 11

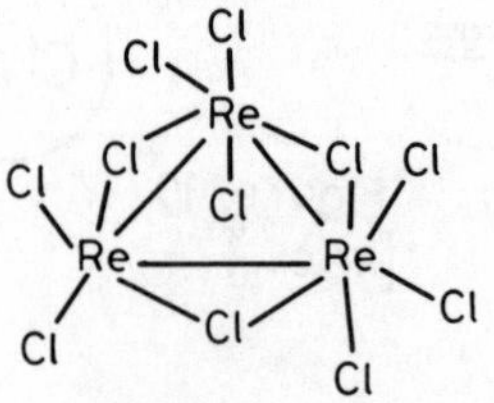

Figure 12

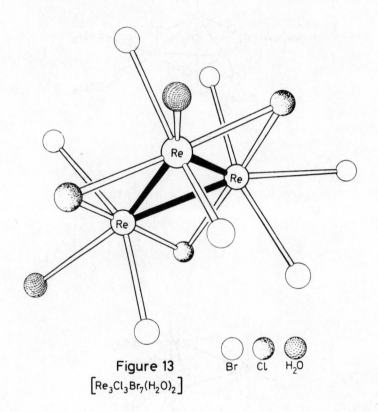

Figure 13

$[Re_3Cl_3Br_7(H_2O)_2]$

Br   Cl H_2O

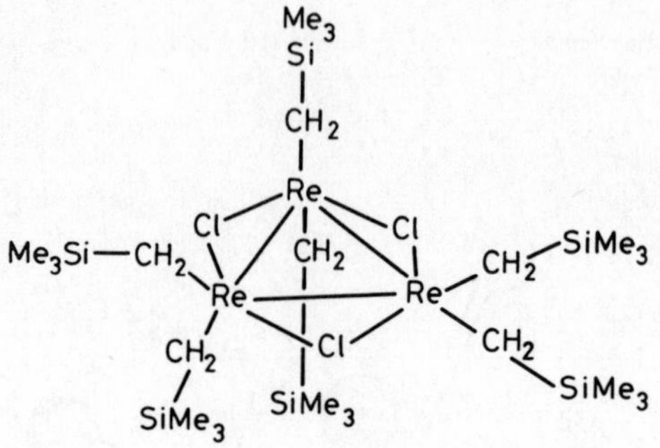

Figure 14

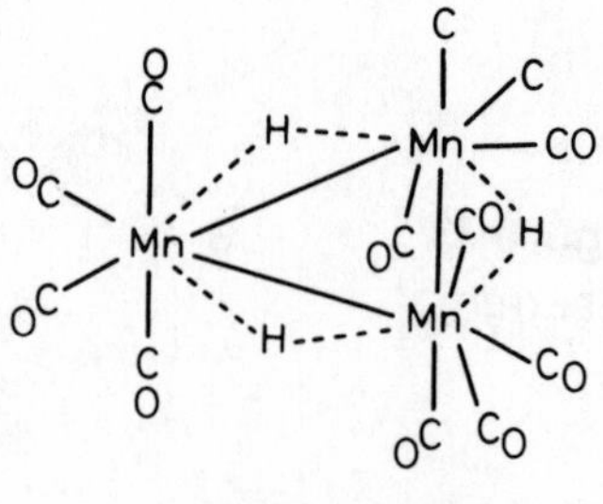

Figure 15

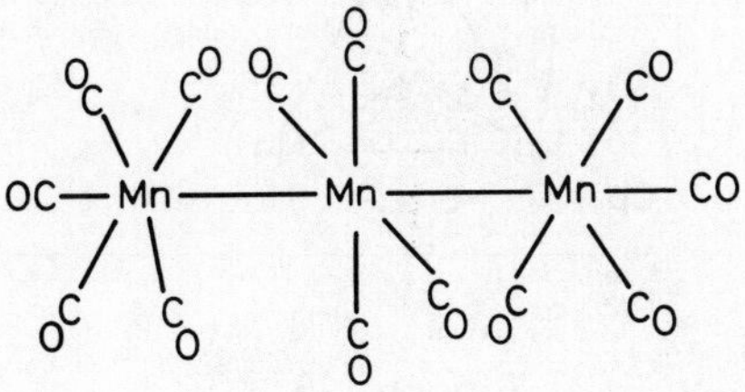

Figure 16

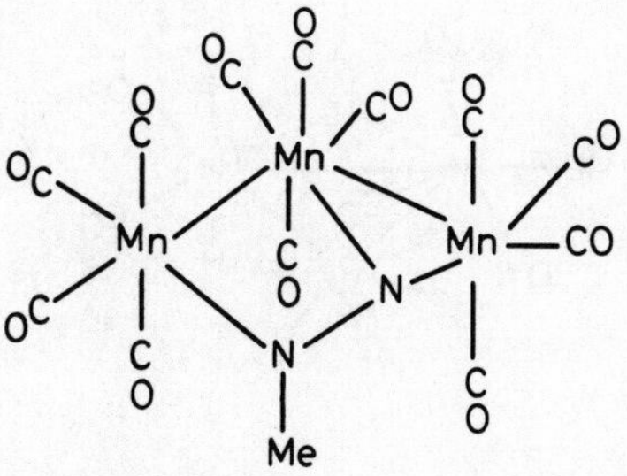

Figure 17

128.

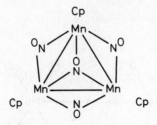

Figure 18

Figure 19

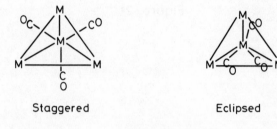

Staggered   Eclipsed

Figure 20

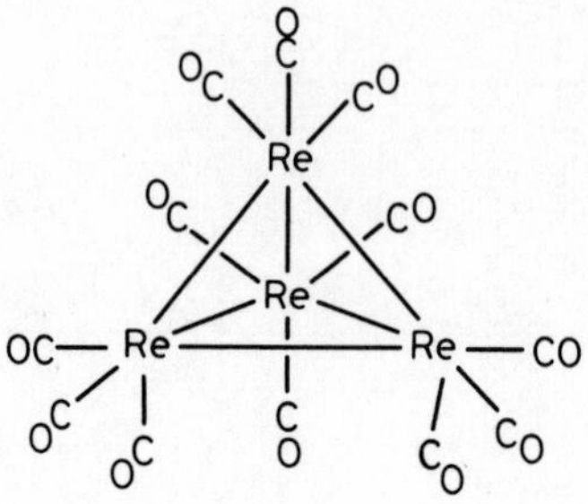

Figure 21

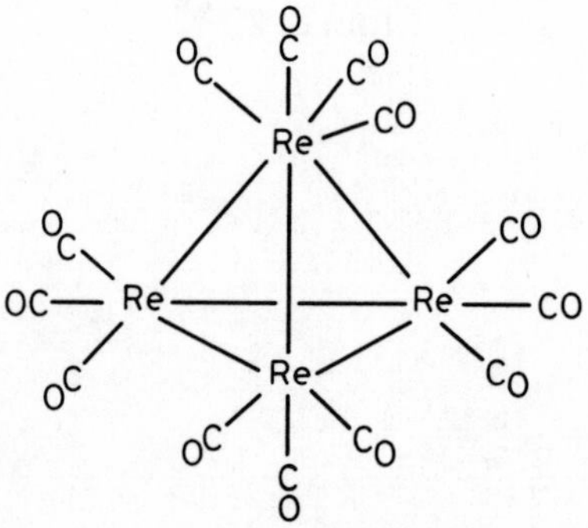

Figure 22

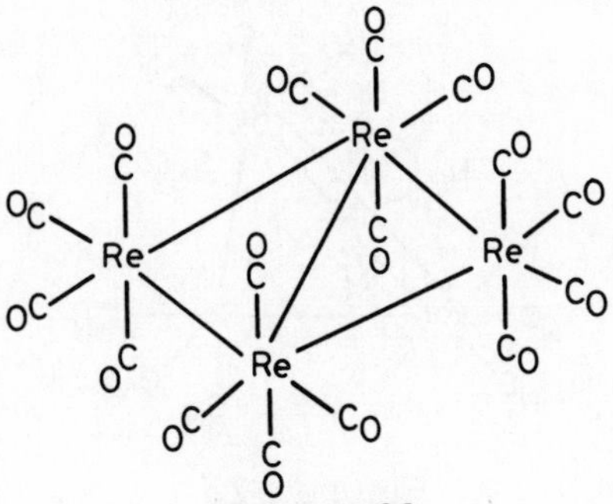

Figure 23

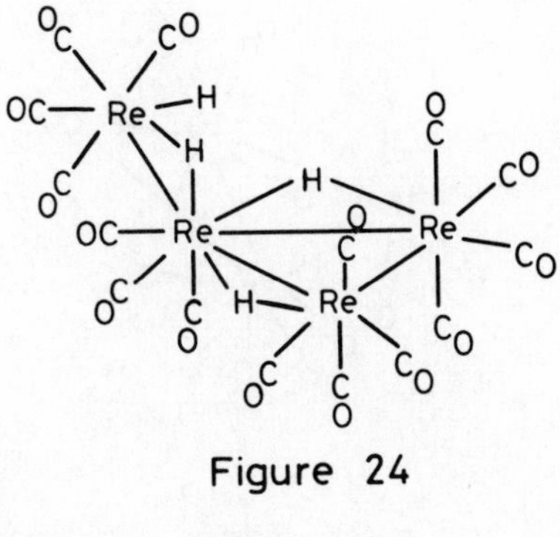

Figure 24

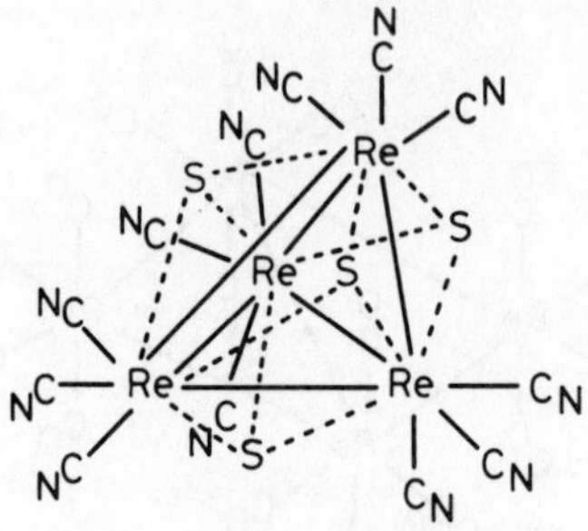

Figure 25

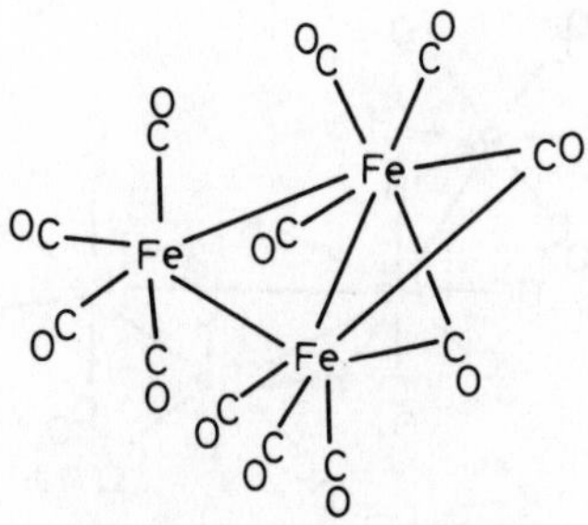

Figure 26

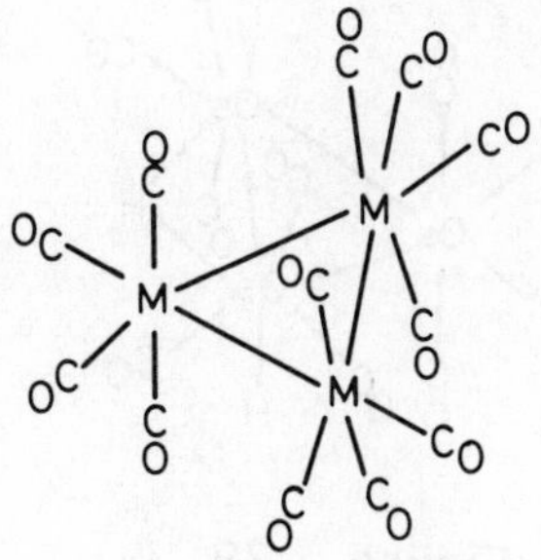

M = Ru, Os

Figure 27

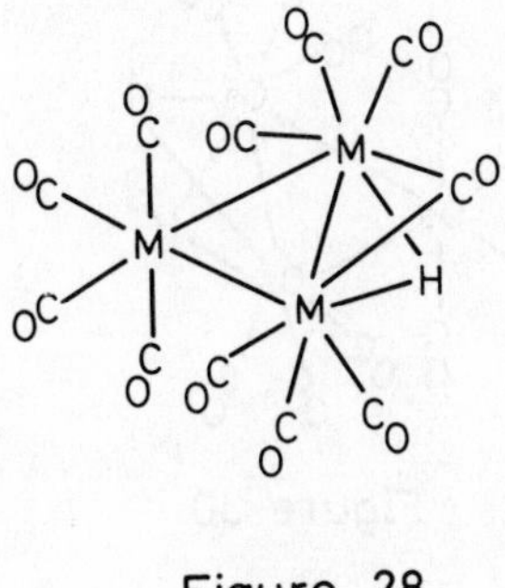

M = Fe, Ru

Figure 28

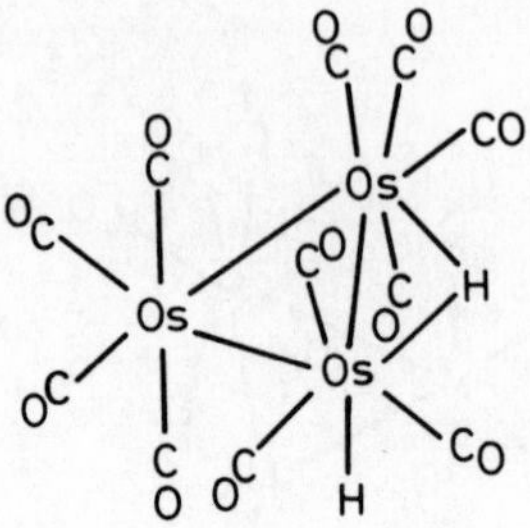

Figure 29

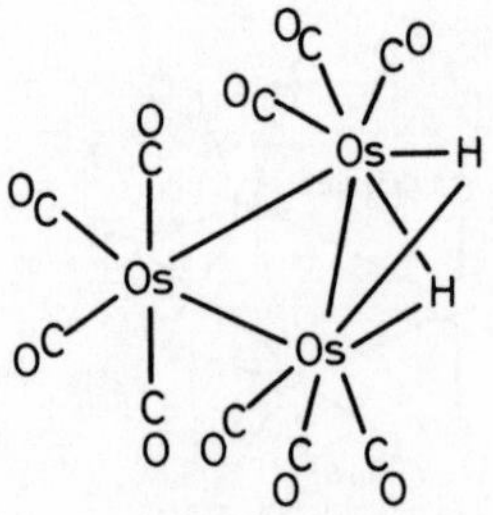

Figure 30

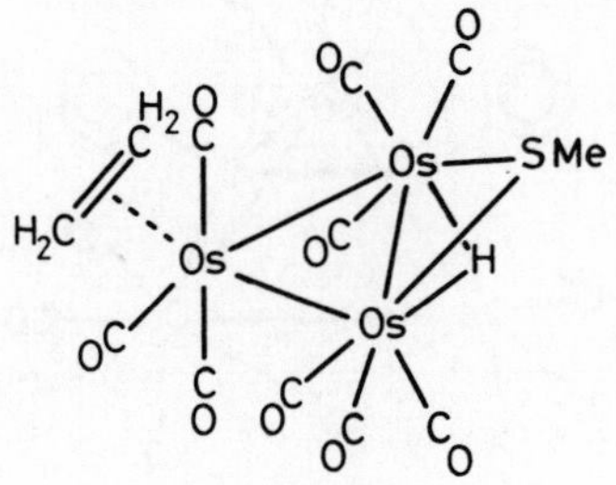

Figure 31

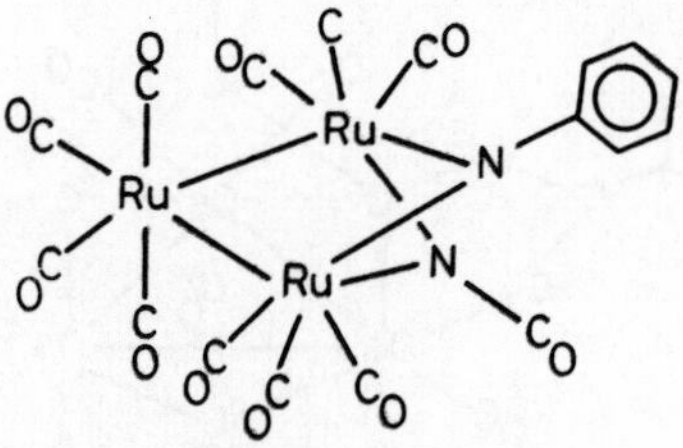

Figure 32

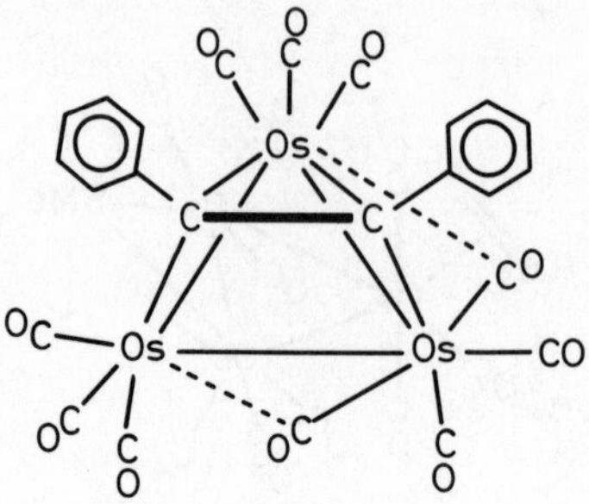

Figure 33

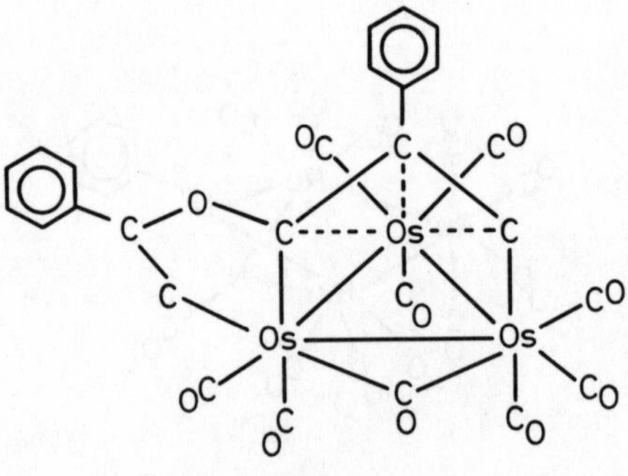

Figure 34

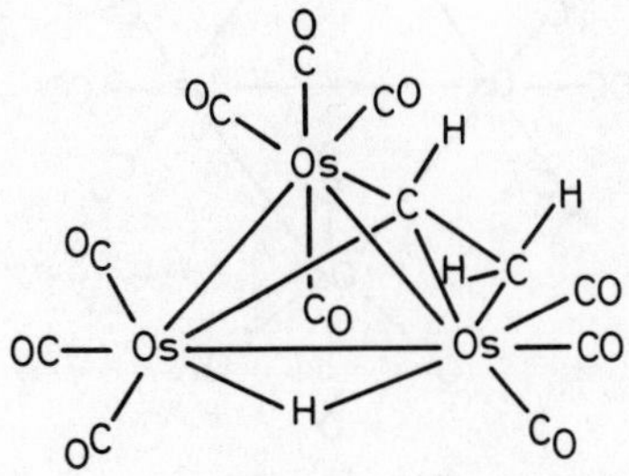

Figure 35

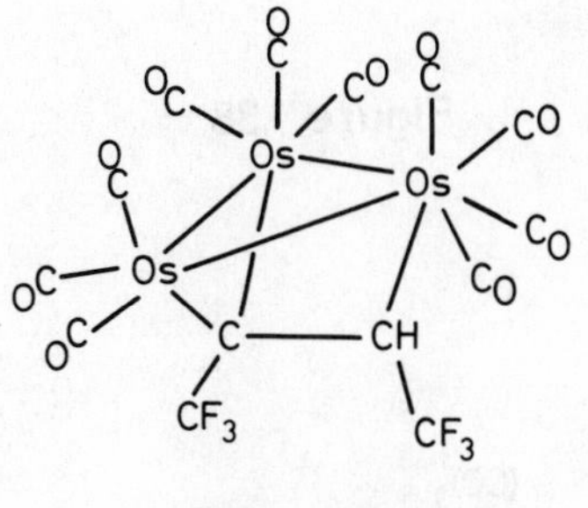

Figure 36

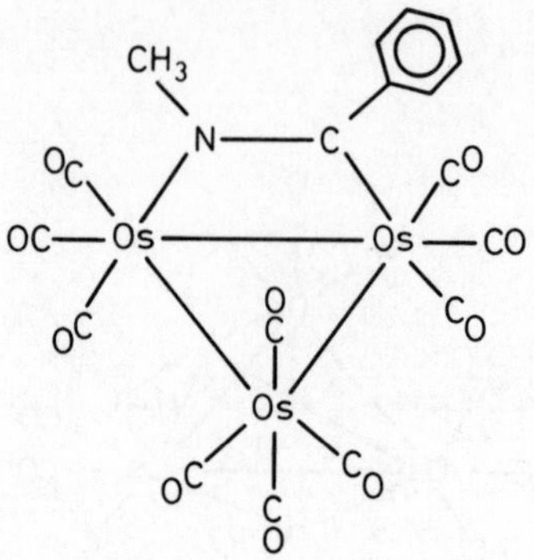

Figure 37

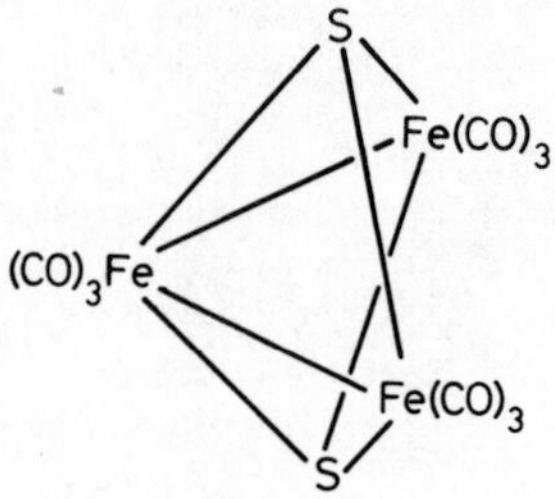

Figure 38

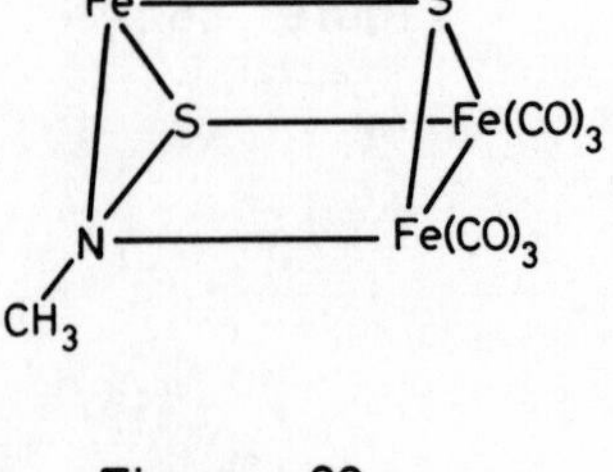

Figure 39

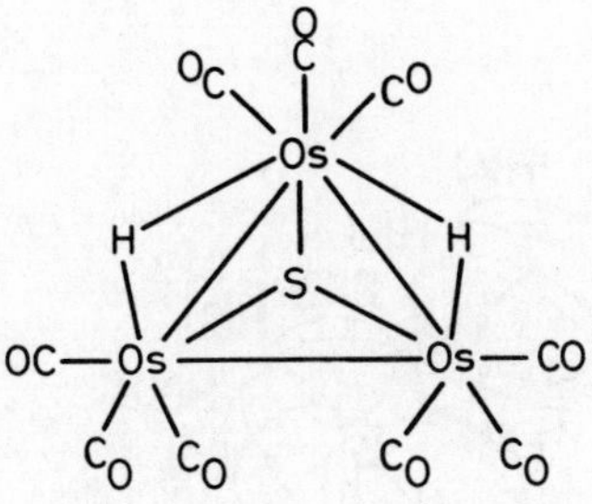

Figure   40

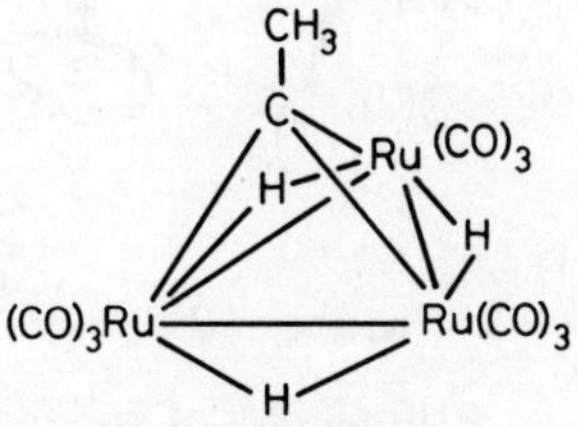

Figure   41

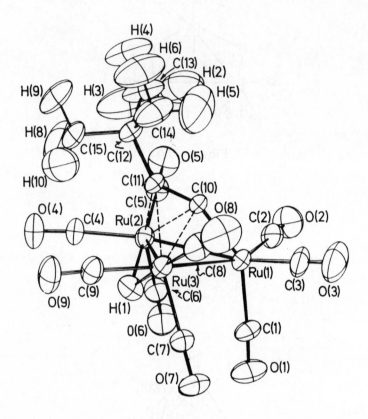

Figure 42

$HRu_3(CO)_9(C \equiv CBu^t)$

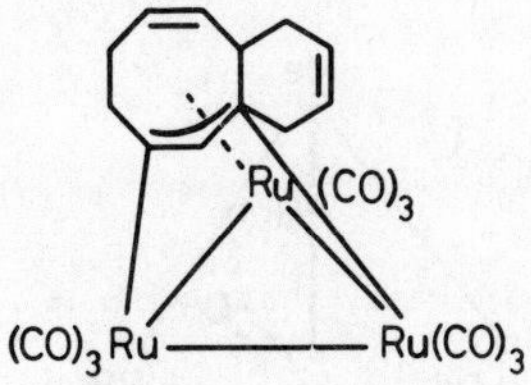

Figure 43

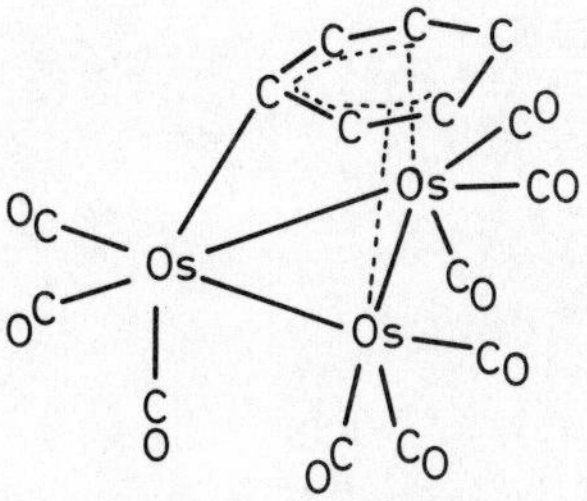

Figure 44

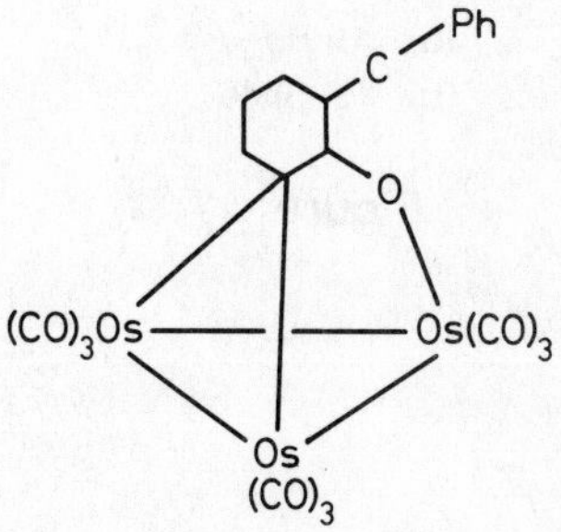

Figure 45

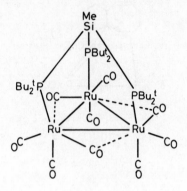

Figure 46

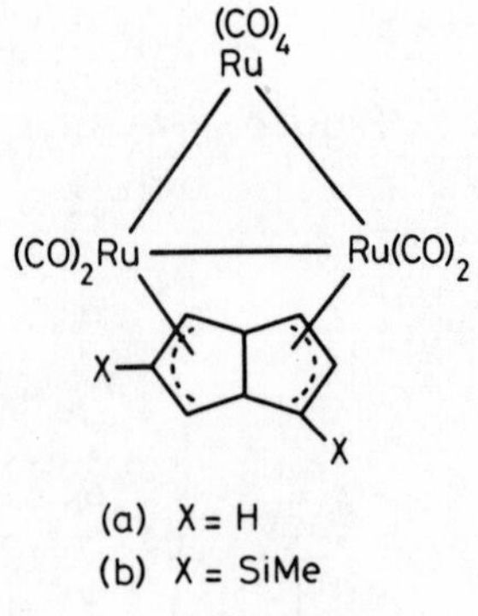

(a) X = H
(b) X = SiMe

Figure 47

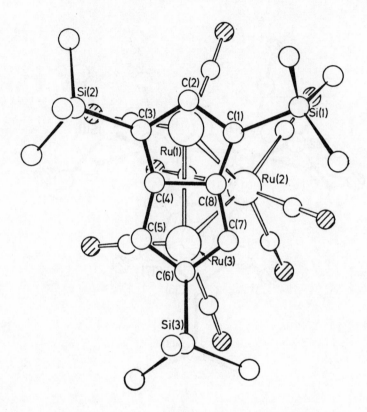

Figure 48

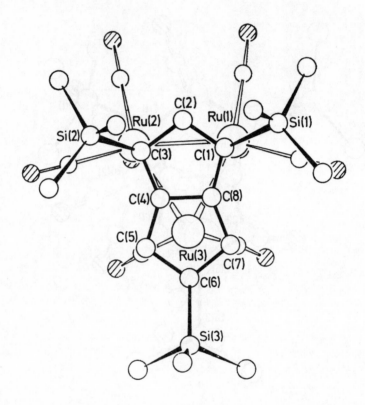

Figure 49

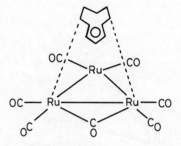

Figure 50

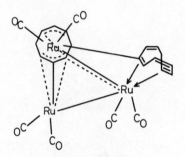

Figure 51

Figure 52

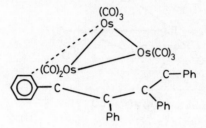

Figure 53

Figure 54

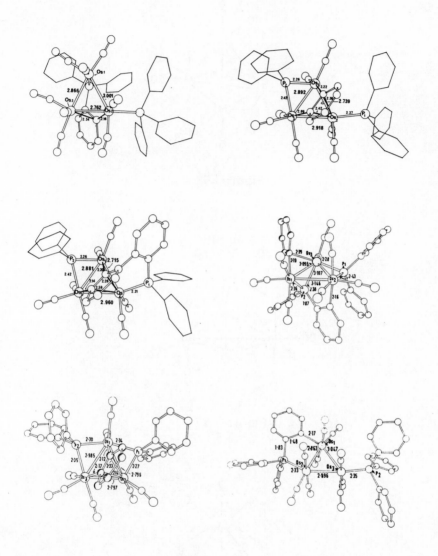

Figure 55

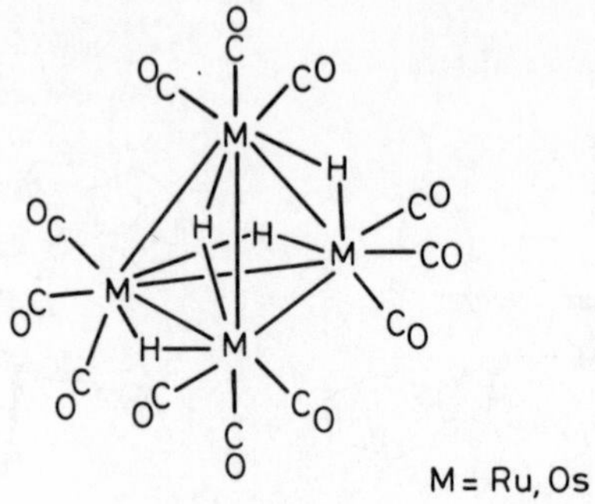

148.

M = Ru, Os

Figure 56

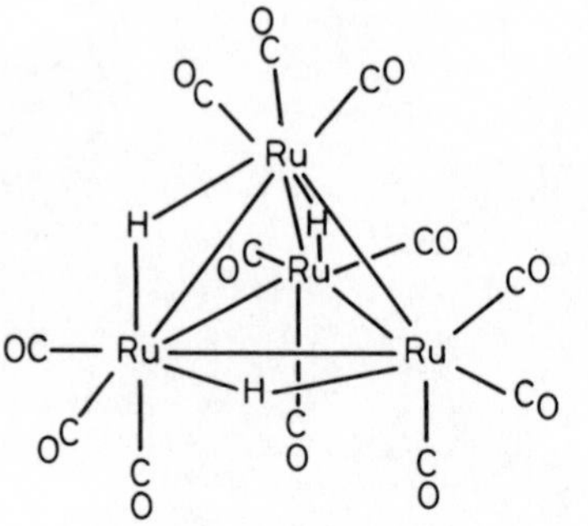

Figure   57

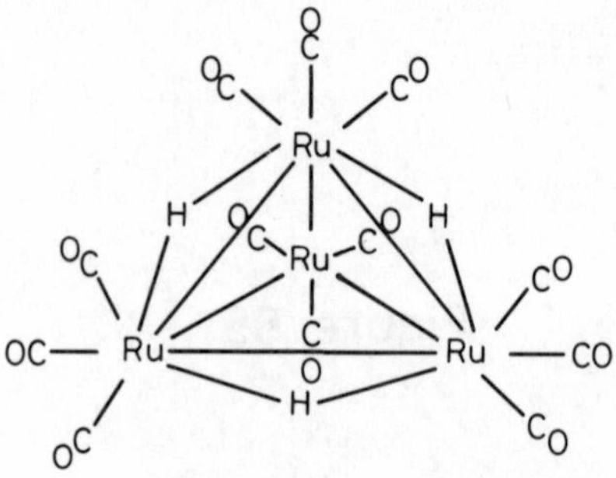

Figure 58

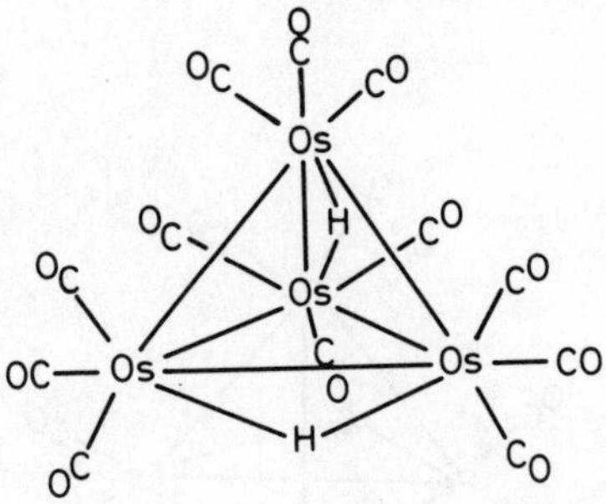

Figure 59

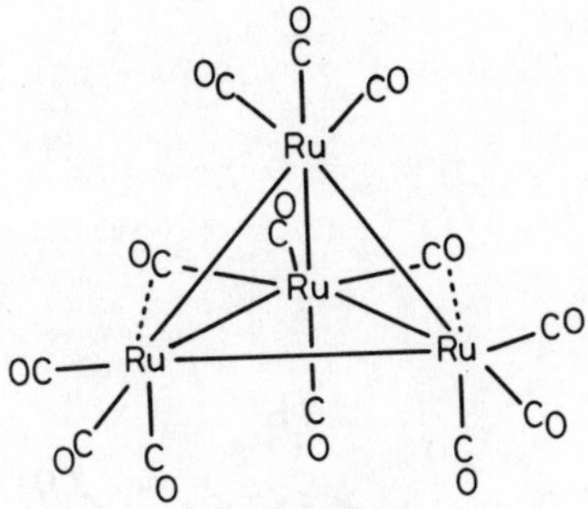

Figure 60

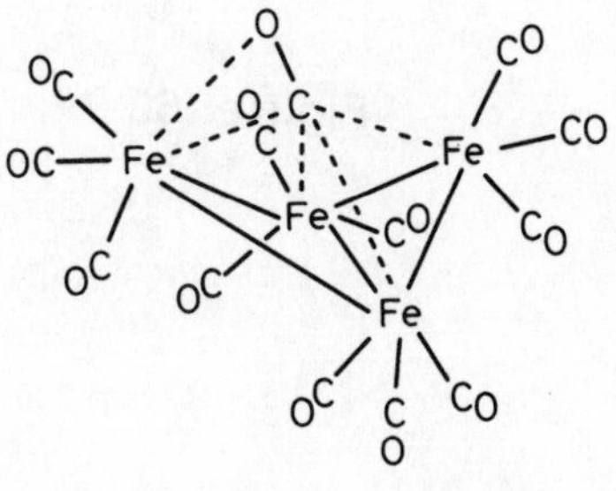

Figure 61

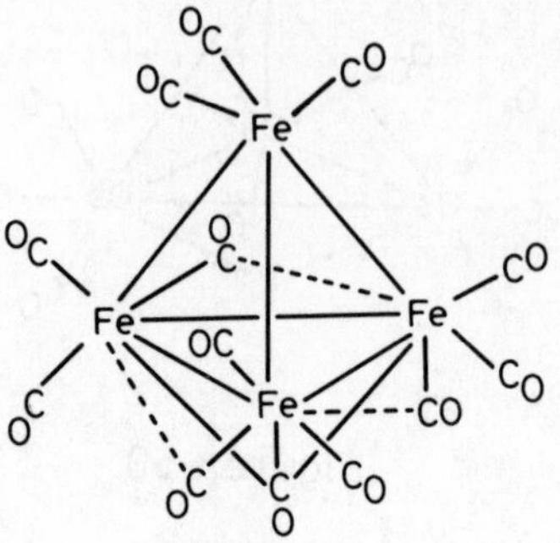

Figure 62

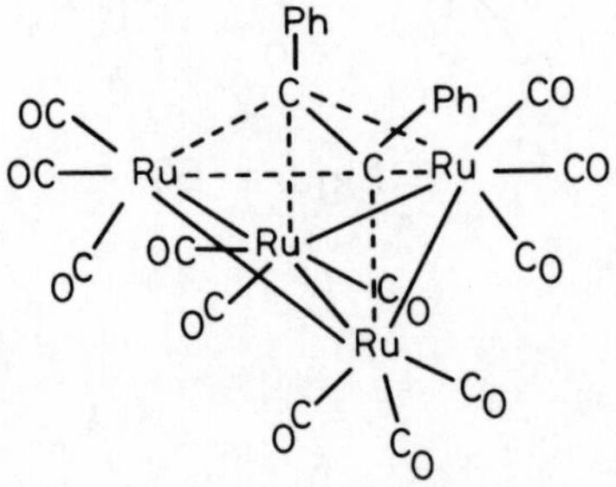

Figure 63

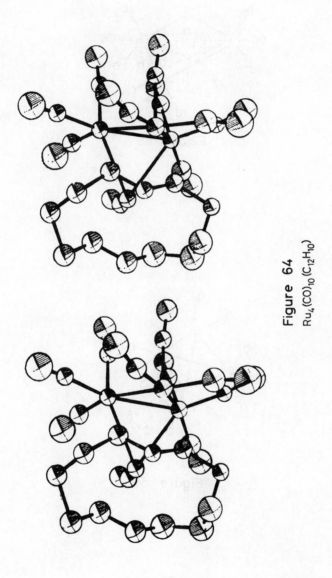

Figure 64

$Ru_4(CO)_{10}(C_{12}H_{10})$

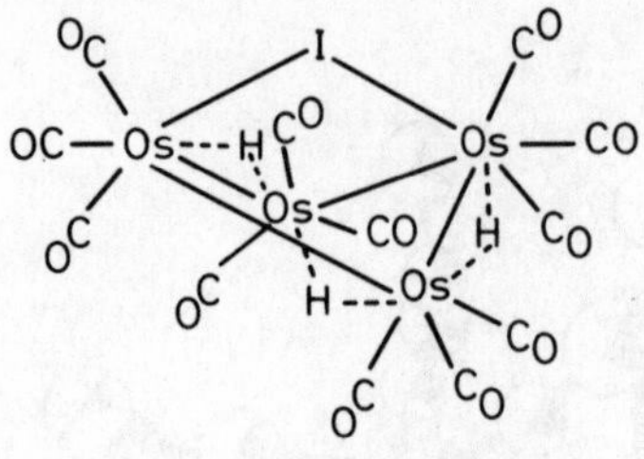

Figure 65

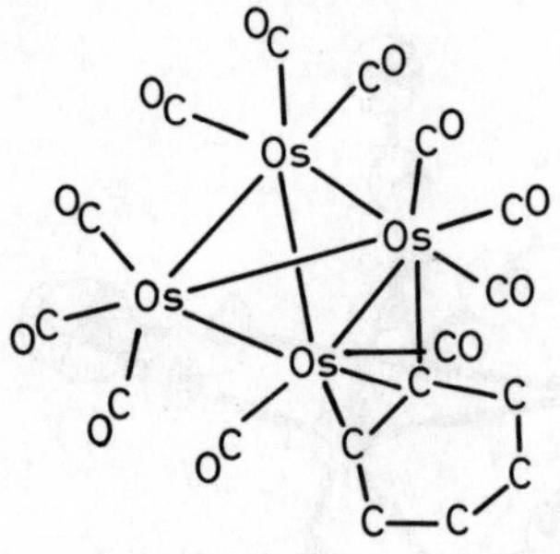

Figure 66

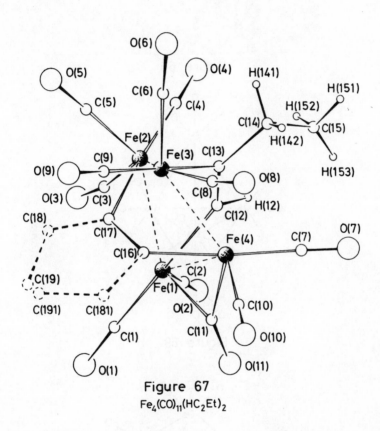

Figure 67

$Fe_4(CO)_{11}(HC_2Et)_2$

154.

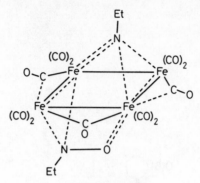

Figure 68

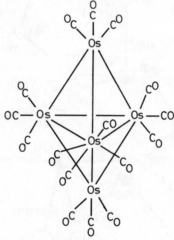

Figure 69

Figure 70

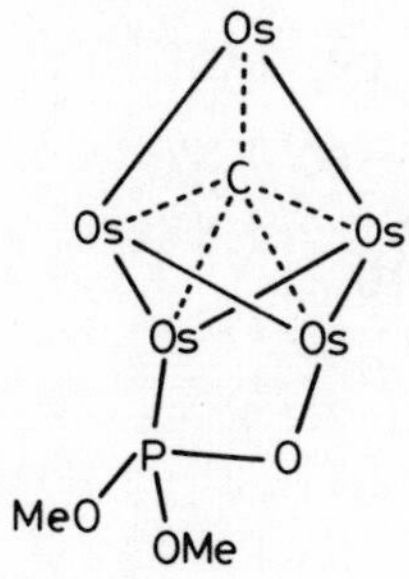

$HOs_5C(CO)_{14}[OP(OMe)_2]$

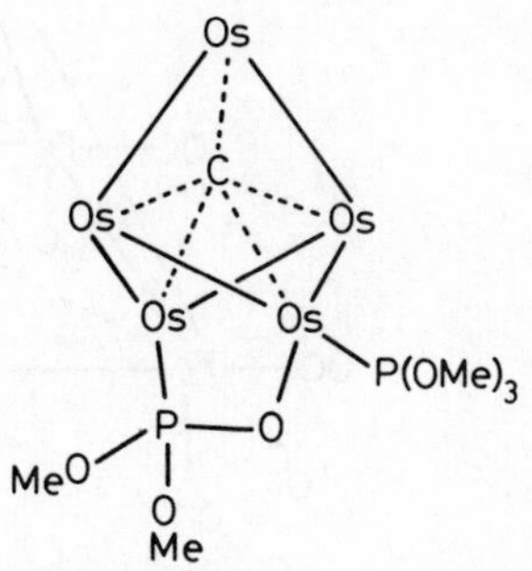

$HOs_5C(CO)_{12}[OP(OMe)_2][P(OMe)_3]$

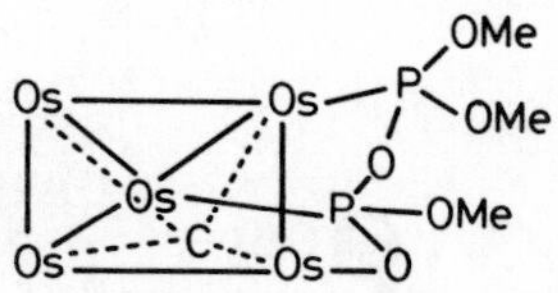

$HOs_5C(CO)_{13}[OP(OMe)OP(OMe)_2]$

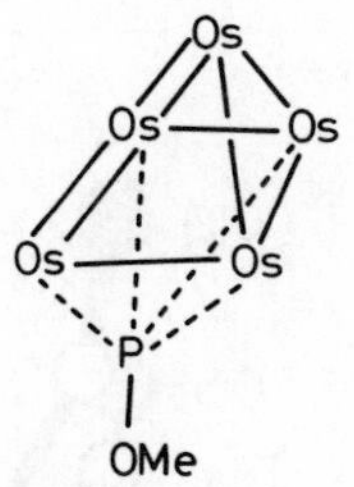

$Os_5(CO)_{15}POMe$

Figure 71

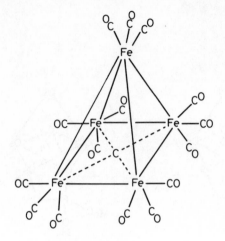

Figure 72

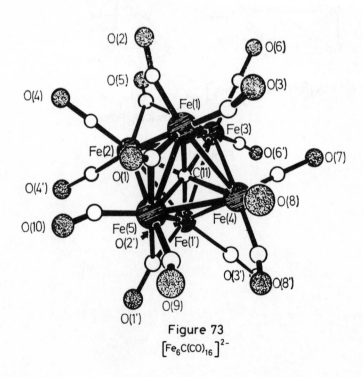

Figure 73

$$[Fe_6C(CO)_{16}]^{2-}$$

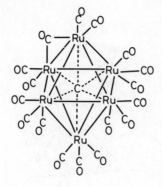

Figure 74

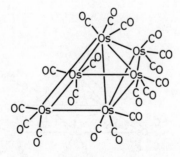

Figure 75

Figure 76

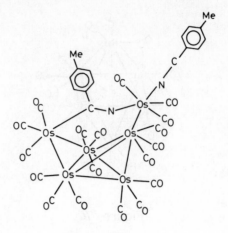

Figure 77

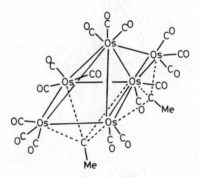

Figure 78

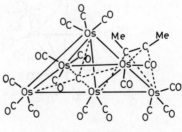

Figure 79

159.

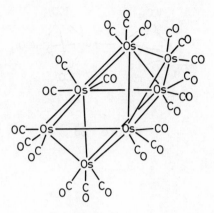

Figure 80

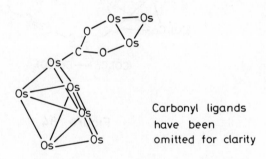

Carbonyl ligands
have been
omitted for clarity

Figure 81

CH₃

Figure 82

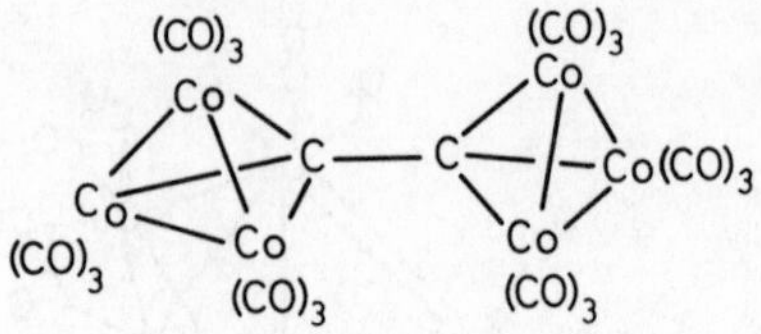

160.

Figure 83

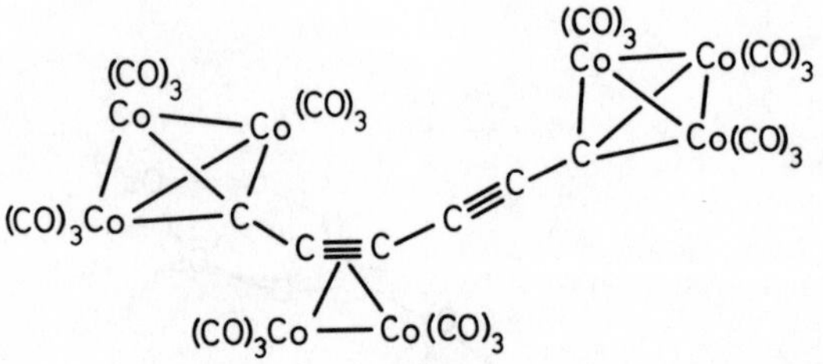

Figure 84

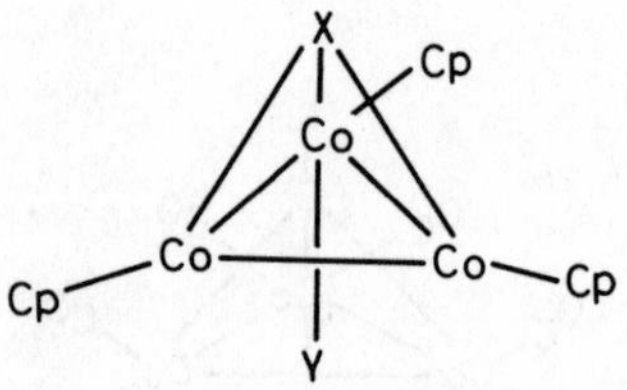

Figure 85

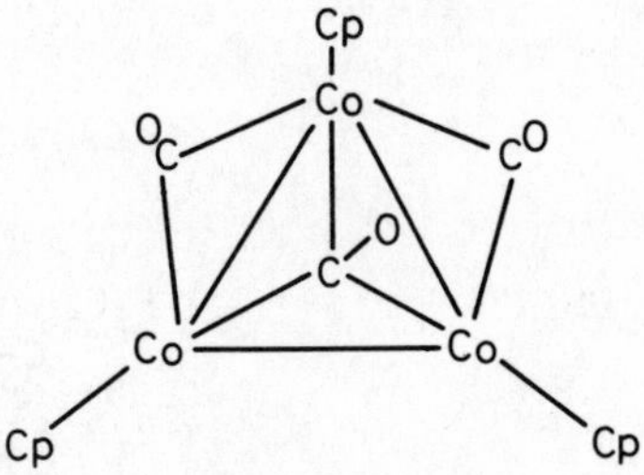

Figure 86

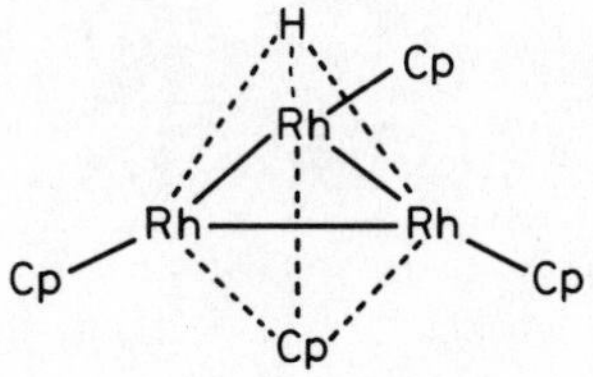

Figure 87

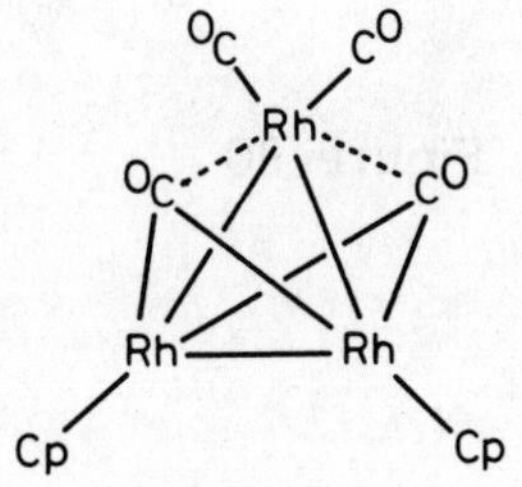

Figure 88

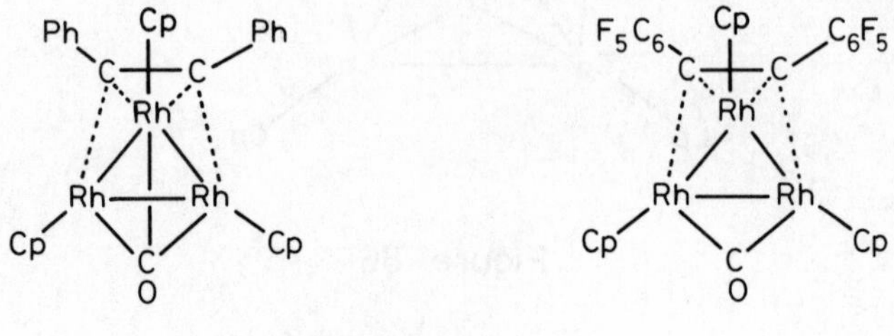

Figure 89

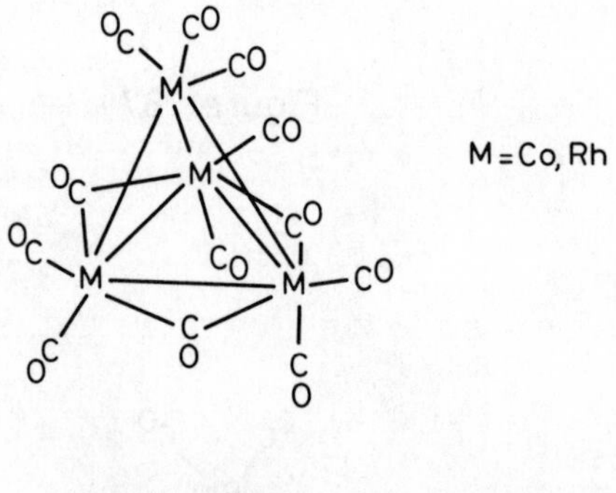

M = Co, Rh

Figure 90

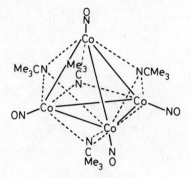

Figure 91

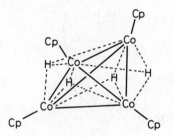

Figure 92

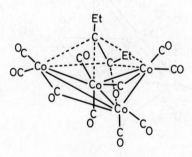

Figure 93

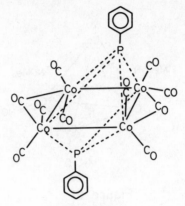

Figure 94

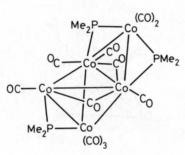

Figure 95

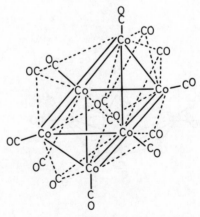

Figure 96

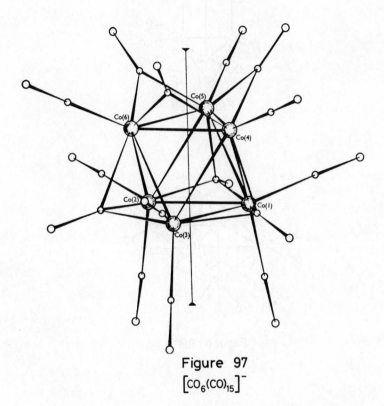

Figure 97

$\left[Co_6(CO)_{15} \right]^{-}$

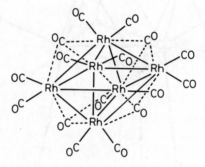

Figure 98

Figure 99

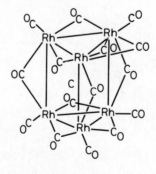

Figure 100

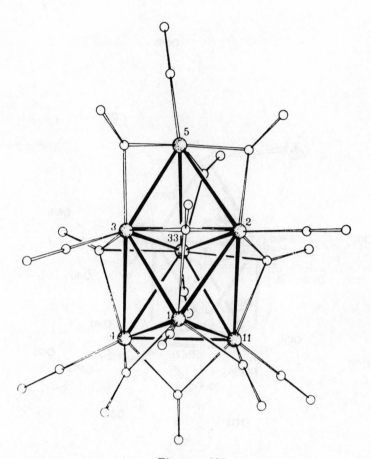

Figure 101

$[Rh_7(CO)_{16}]^{3-}$

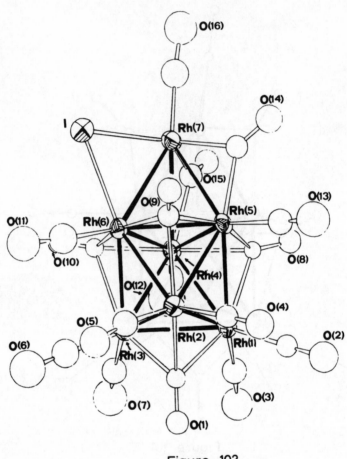

Figure 102

$[Rh_7(CO)_{16}I]$

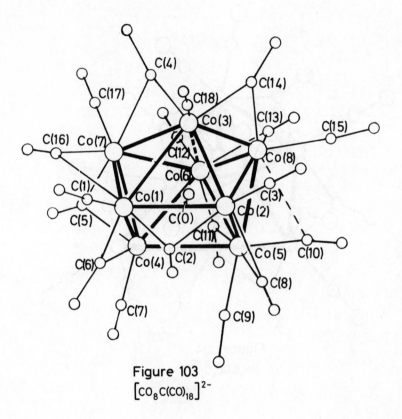

Figure 103

$\left[Co_8C(CO)_{18}\right]^{2-}$

170.

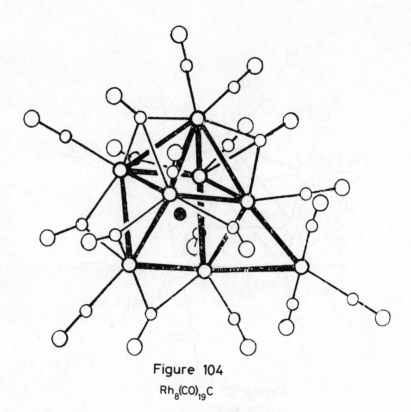

Figure 104

$Rh_8(CO)_{19}C$

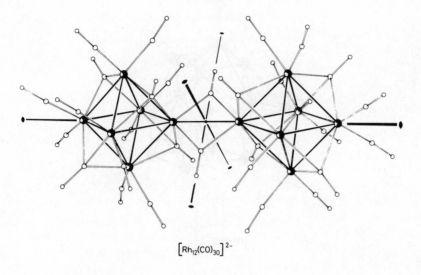

$$\left[Rh_{12}(CO)_{30}\right]^{2-}$$

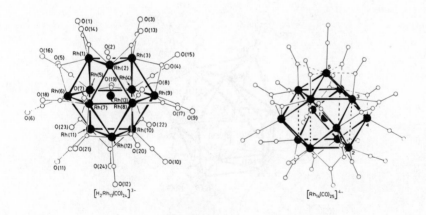

$$\left[H_2Rh_{13}(CO)_{24}\right]^{3-}$$

$$\left[Rh_{14}(CO)_{25}\right]^{4-}$$

Figure 105

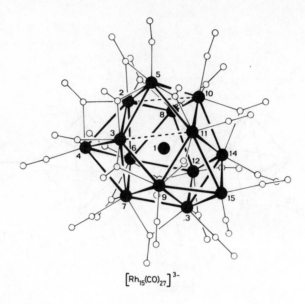

$[Rh_{15}(CO)_{27}]^{3-}$

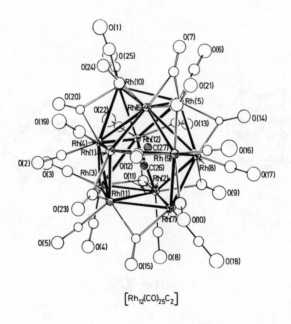

$[Rh_{12}(CO)_{25}C_2]$

Figure 105

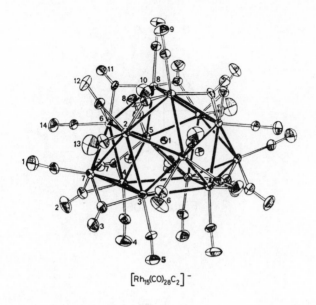

$$[Rh_{15}(CO)_{28}C_2]^-$$

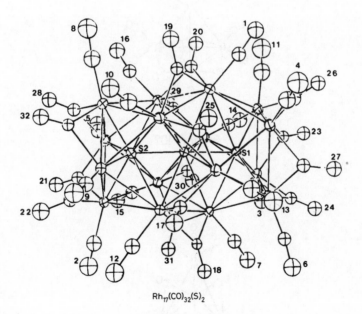

$$Rh_{17}(CO)_{32}(S)_2$$

Figure 105

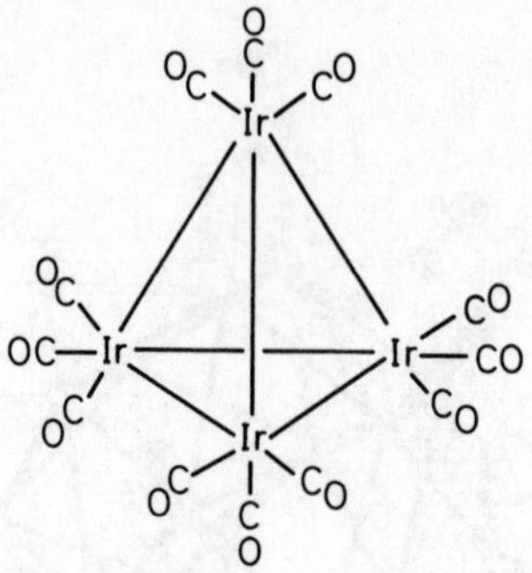

Figure 106

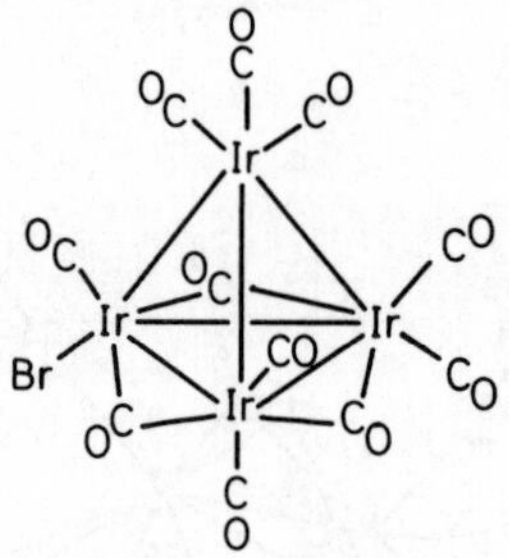

Figure 107

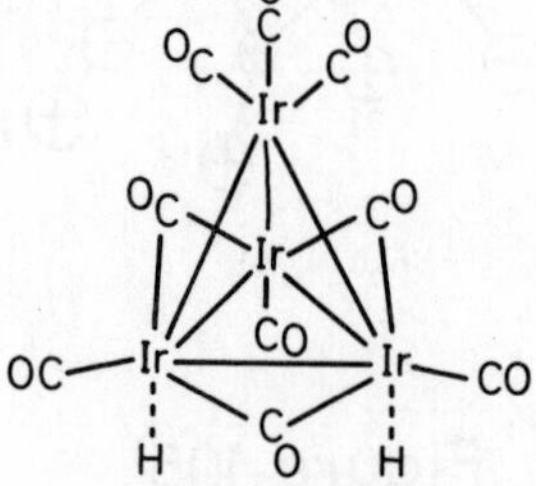

Figure 108

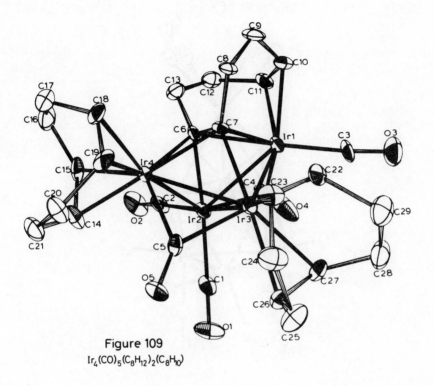

Figure 109

$Ir_4(CO)_5(C_8H_{12})_2(C_8H_{10})$

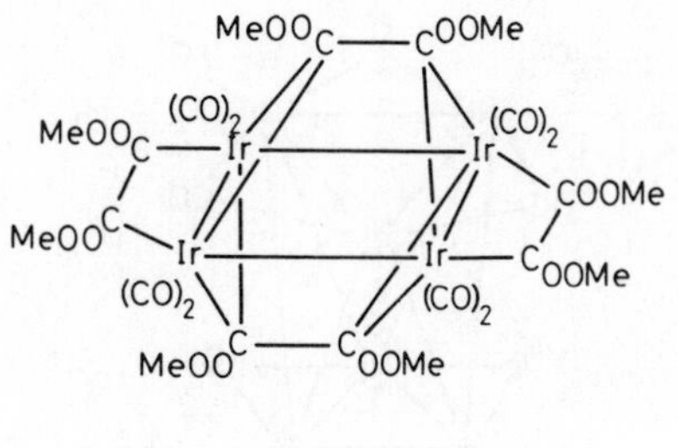

Figure 110

176.

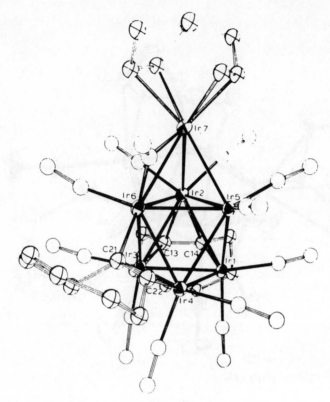

Figure 111

$Ir_7(CO)_{12}(C_8H_{12})(C_8H_{11})(C_8H_{10})$

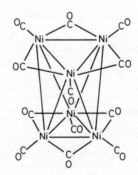

Figure 112

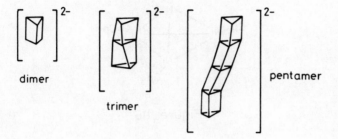

dimer trimer pentamer

Figure 113

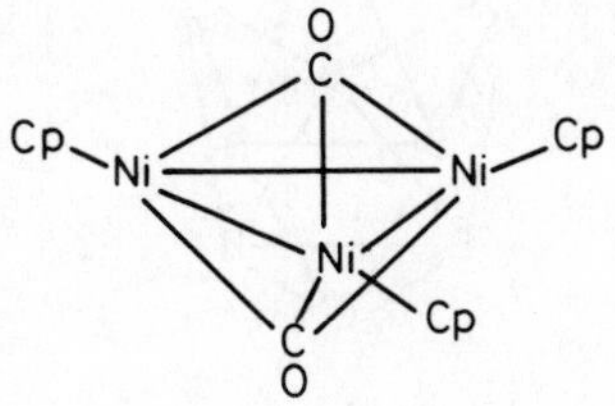

Figure 114

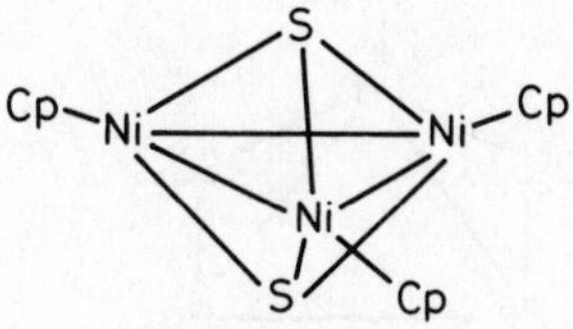

Figure 115

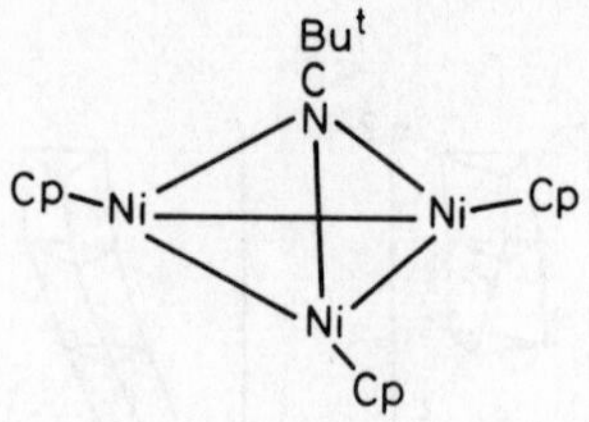

Figure 116

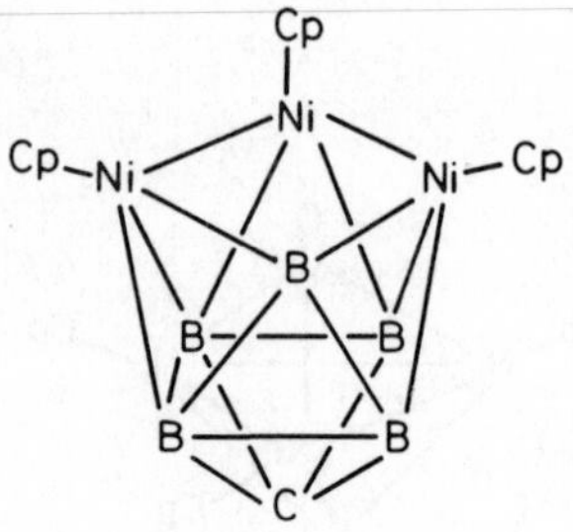

Figure 117

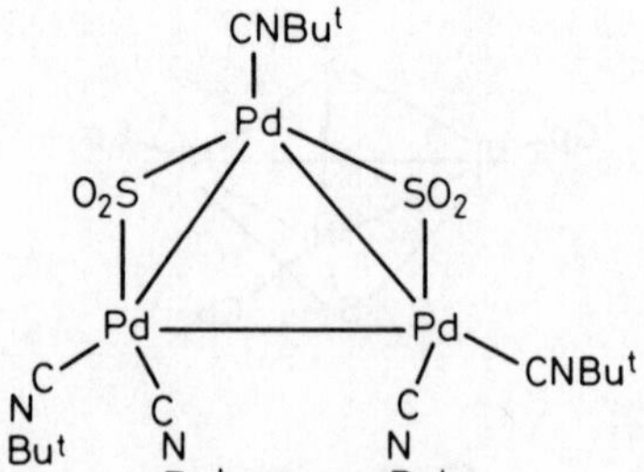

Figure 118

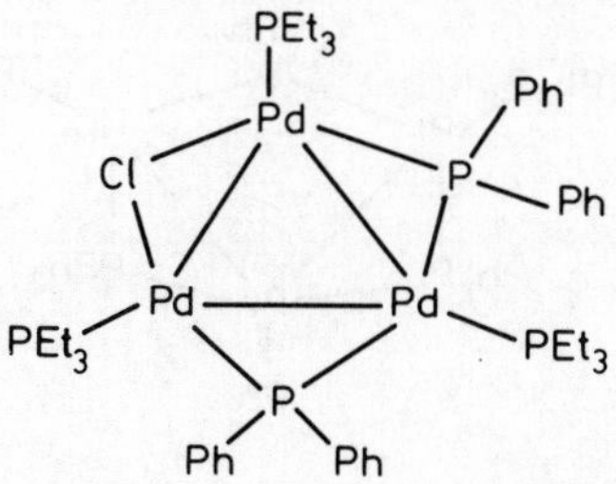

Figure 119

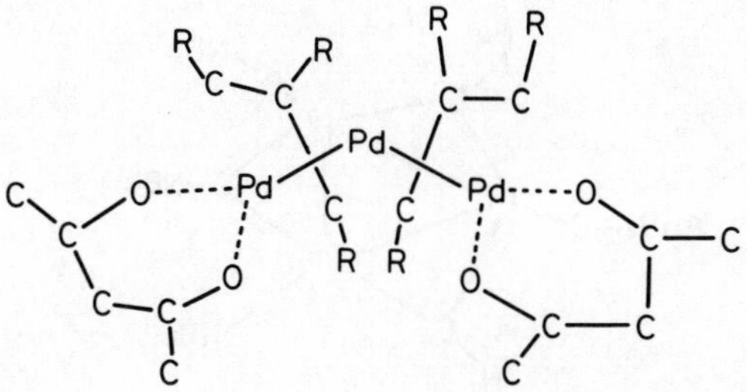

Figure 120

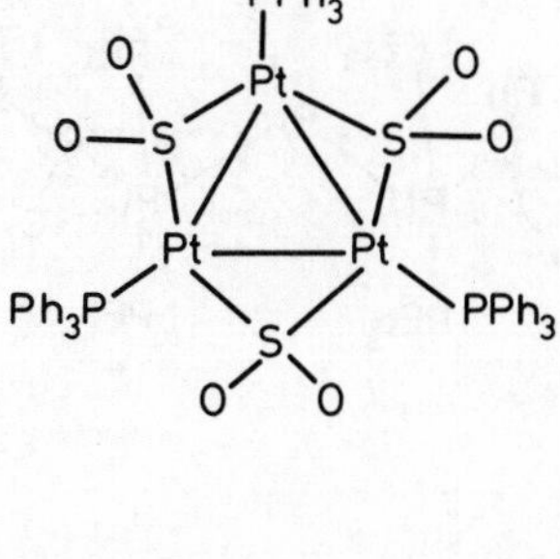

Figure 121

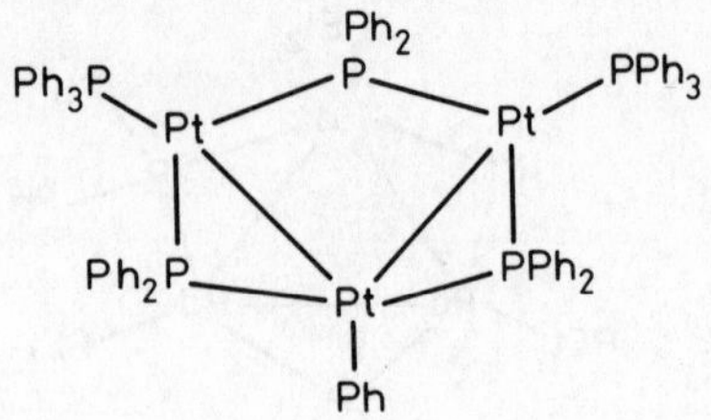

Figure 122

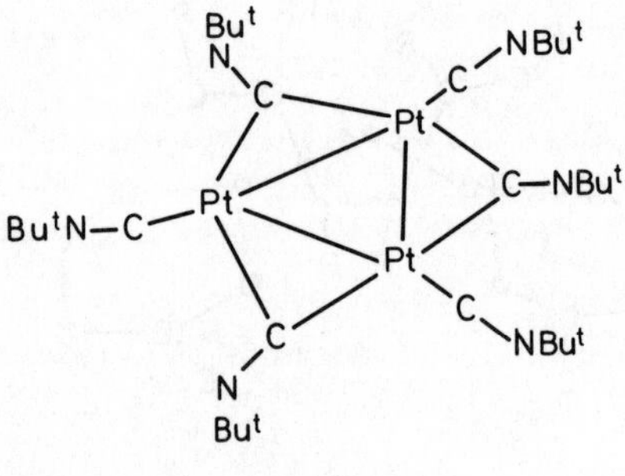

Figure 123

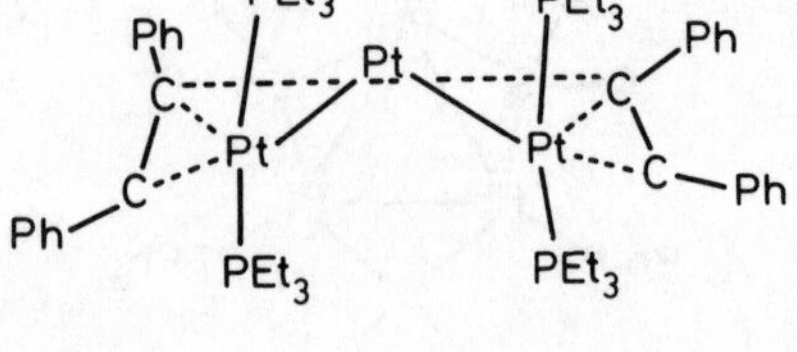

Figure 124

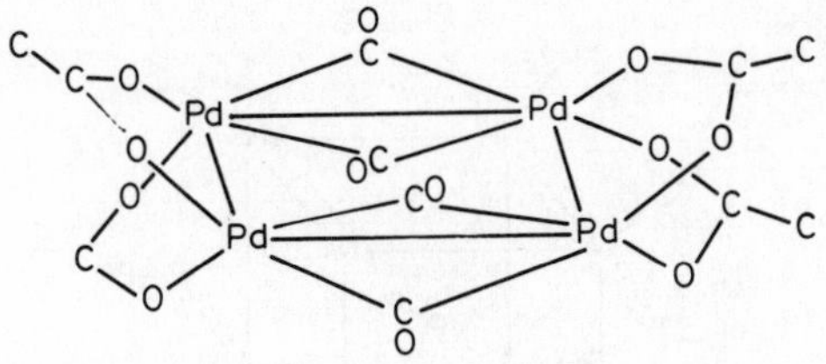

Figure 125

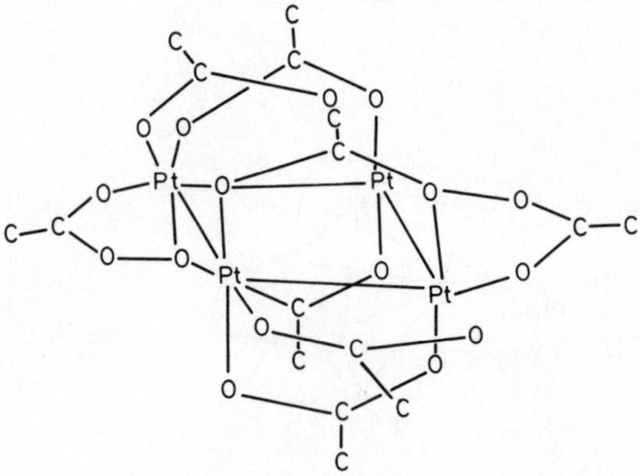

Figure 126

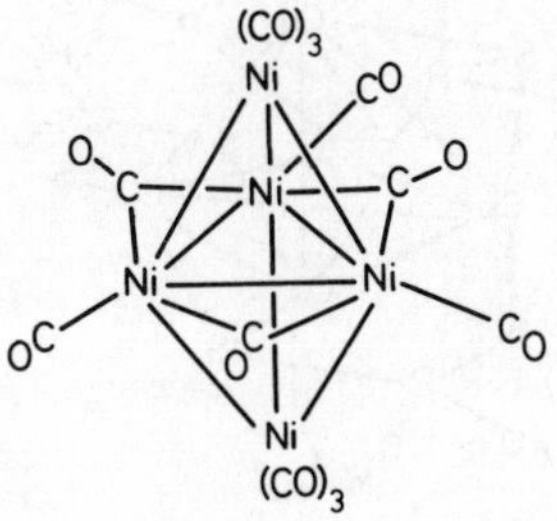

Figure 127

182.

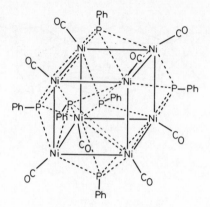

Figure 128

Figure 129

Figure 130

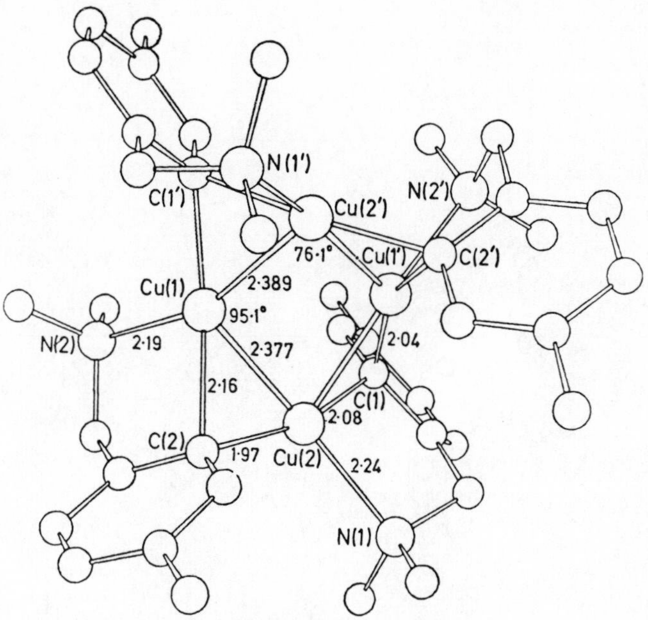

Figure 131

5 methyl-2 [(dimethylamine) methyl] phenyl copper (II)

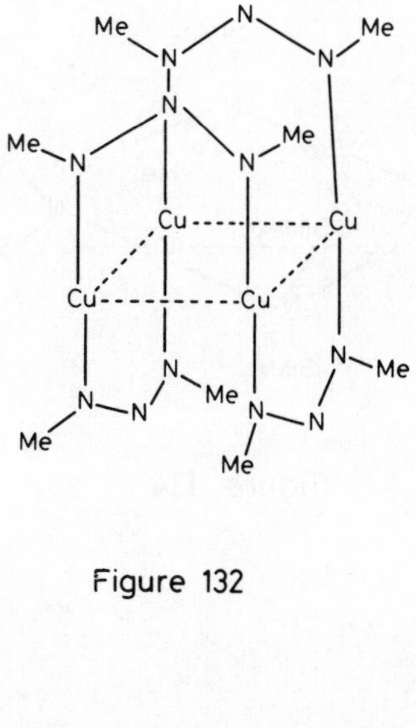

Figure 132

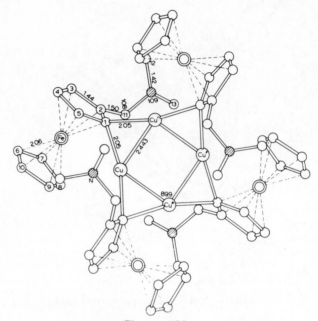

Figure 133

2-copper-1-(dimethylaminomethyl) ferrocene

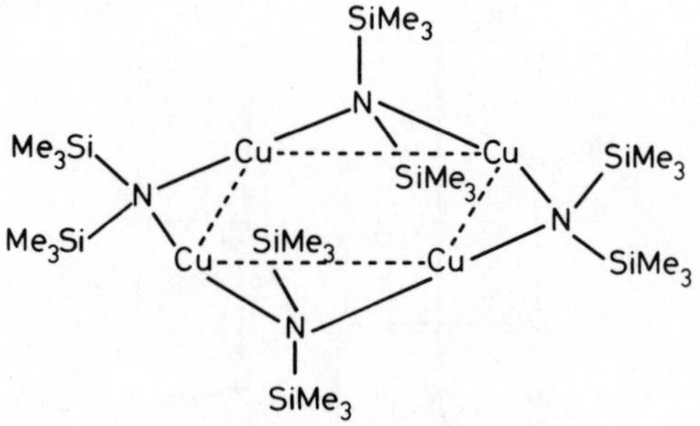

Figure 134

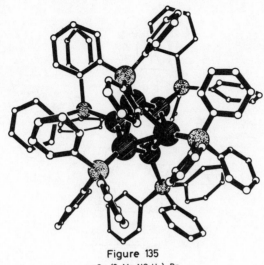

Figure 135
$Cu_6(2-Me_2NC_6H_4)_4 Br_2$

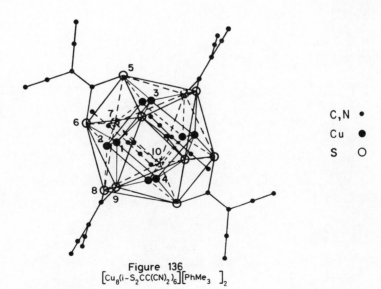

C,N •

Cu ●

S ○

Figure 136
$[Cu_8(i-S_2CC(CN)_2)_6][PhMe_3]_2$

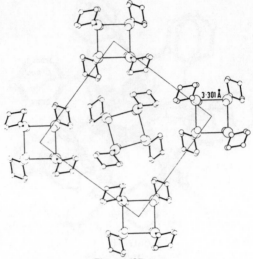

Figure 137
Chloro(piperidine) gold(I)

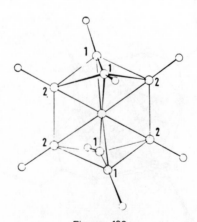

Figure 138

$\left[Cu_9 \{P(p-C_6H_4Me)_3\}_8\right]\left[PF_6\right]_3$

187.

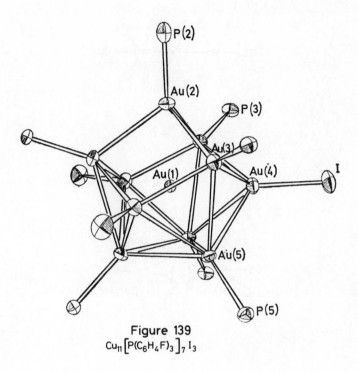

Figure 139

$Cu_{11}\left[P(C_6H_4F)_3\right]_7 I_3$

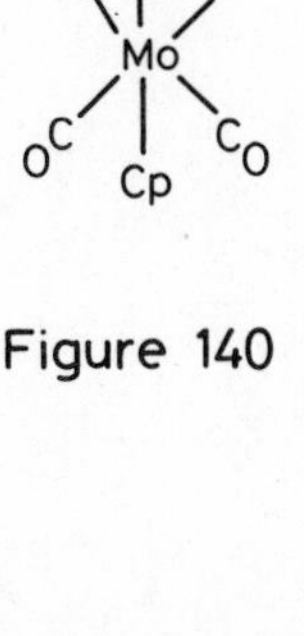

Figure 140

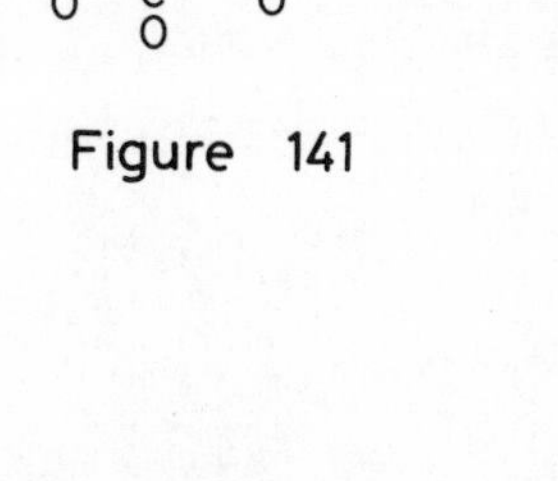

Figure 141

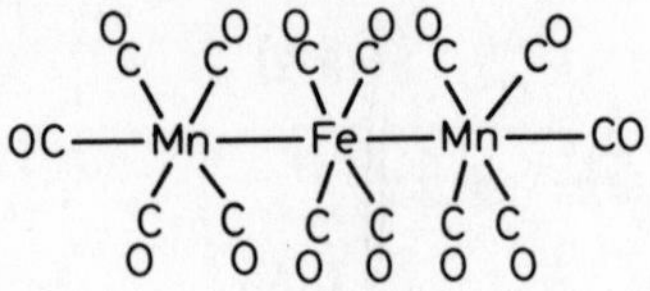

188.

Figure 142

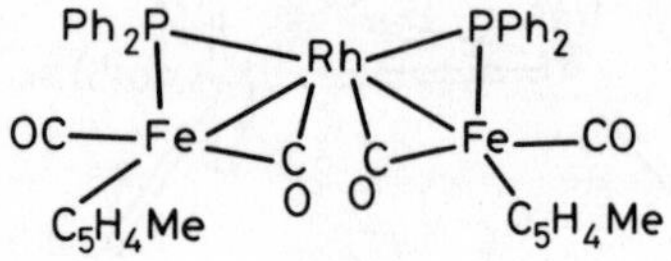

Figure 143

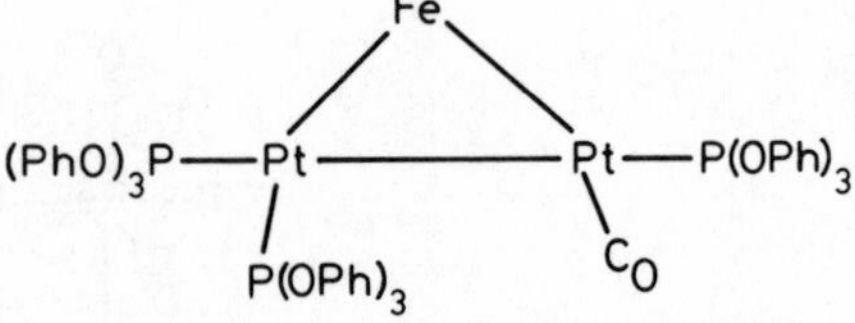

Figure 144

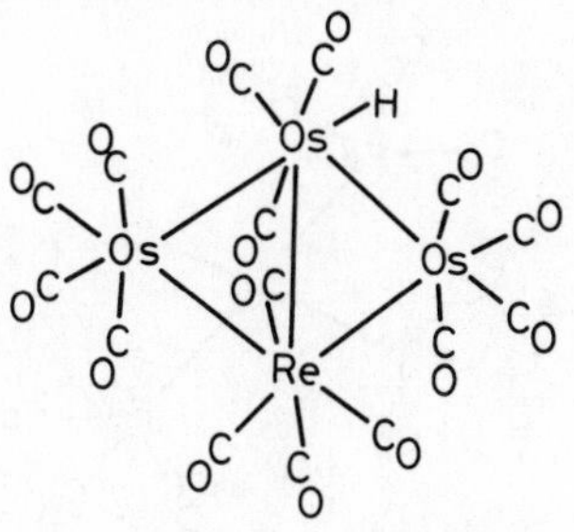

Figure 145

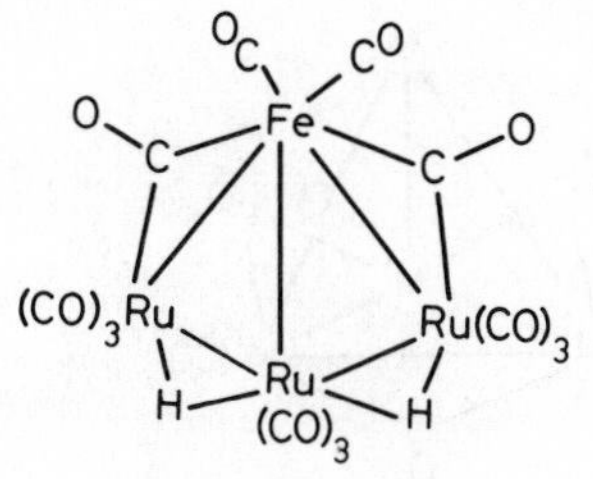

Figure 146

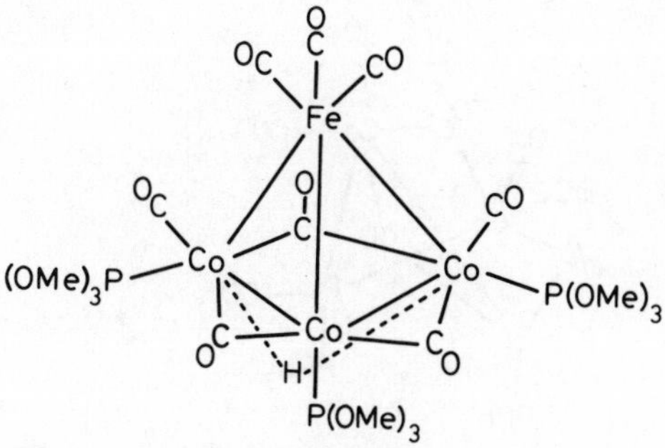

Figure 147

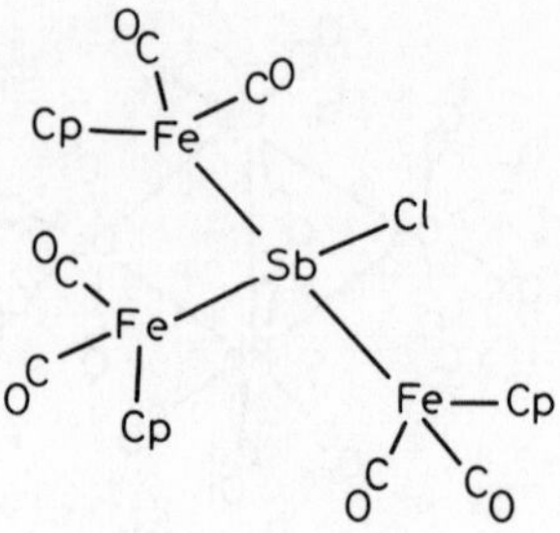

Figure 148

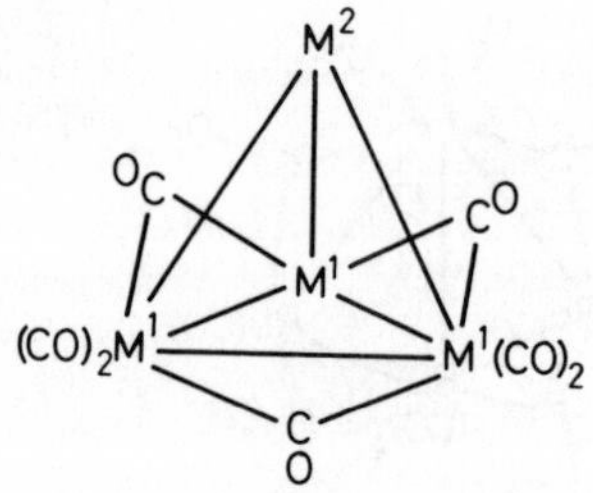

Figure 149

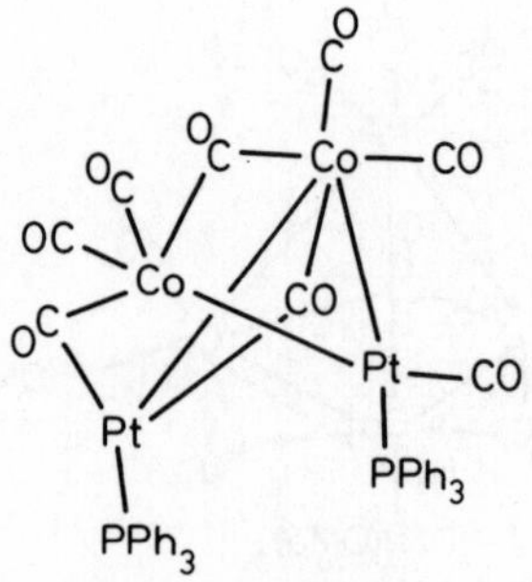

Figure 150

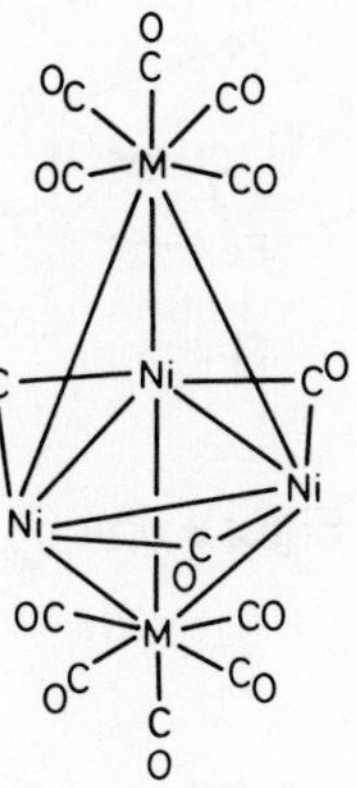

Figure 151

M = Mo, W

Figure 152

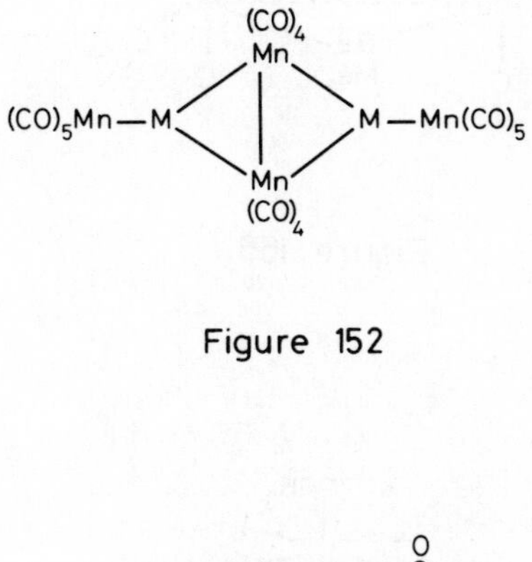

M = Ga, In

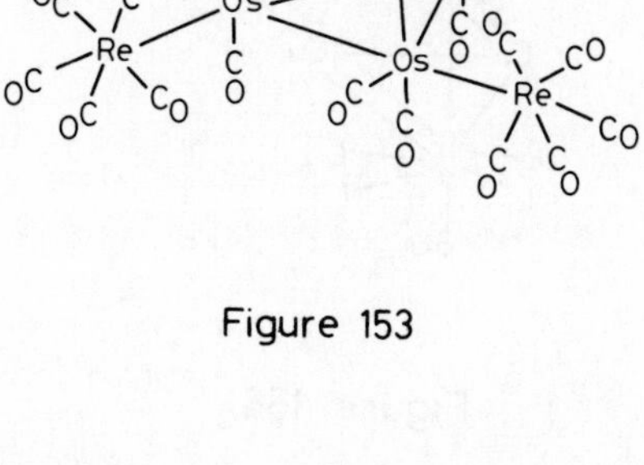

Figure 153

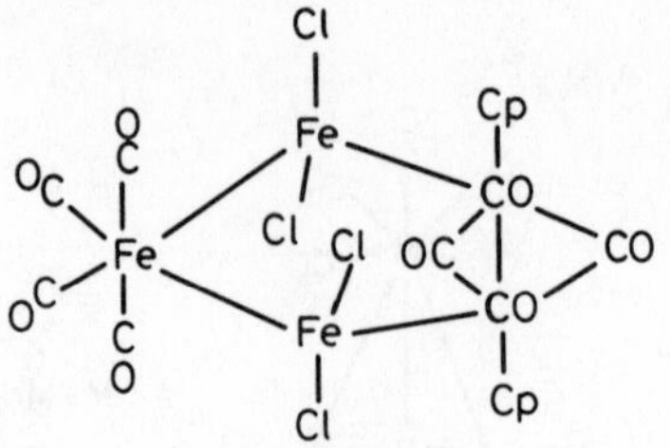

Figure 154

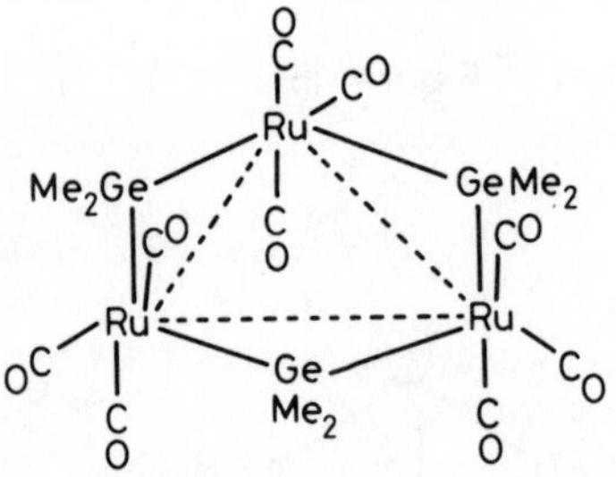

Figure 155

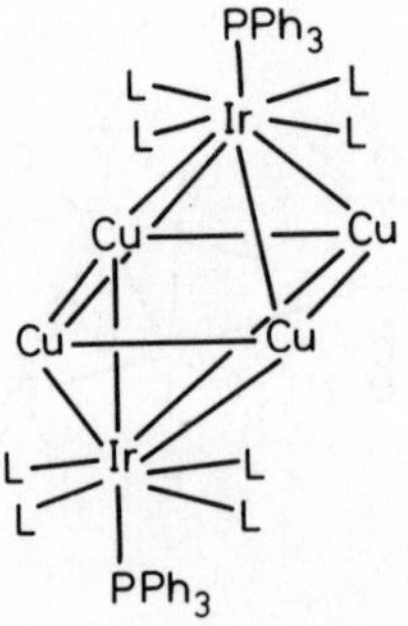

Figure 156

Transition Metal Clusters
Edited by Brian F.G. Johnson
© 1980 John Wiley & Sons Ltd.

CHAPTER 3

Some Bonding Considerations

K. Wade

Department of Chemistry, University of Durham,
Science Laboratories, Durham, U.K.

1.   INTRODUCTION

The purpose of this chapter is to consider certain
features of the bonding in transition metal clusters.
In particular, the intention is to see whether simple
treatments of their metal-ligand and metal-metal
bonding, generally developed for less complicated
systems, allow one to get a feeling for the manner in
which the valence shell electrons are distributed among
the intricate networks of atoms commonly found in
clusters (1-8), and to appreciate how their structures
may vary with the number of electrons they contain.
Most attention will be paid to metal carbonyl clusters
and to the relationship between them and other
categories of cluster, such as boron clusters and metal-
hydrocarbon η-complexes (5), though brief reference will
also be made to metal halide clusters and clusters formed
by the copper group metals.

As indicated elsewhere in this book, particularly
in Chapter 2, the structures of transition metal
clusters vary widely in complexity. At one extreme,
the shapes defined by their skeletal atoms (triangles,
tetrahedra, trigonal prisms, cubes) may be such as to
allow the bonding to be described quite satisfactorily
in terms of localized 2-centre electron pair bonds.

At the other extreme are complicated close-packed arrangements of the metal atoms reminiscent of those present in the metals themselves, for which 2-centre electron pair bonding descriptions are clearly inadequate. Between these extremes, many intermediate-sized clusters adopt shapes (octahedra, square pyramids, capped tetrahedra etc.) for which 2-centre bonding descriptions are generally less helpful than molecular orbital (MO) treatments, and it is on these intermediate-sized clusters that attention will be most concentrated, with the object of illustrating the value of simple MO-based arguments. If one looks at the units, e.g. metal carbonyl fragments $M(CO)_n$, of which metal carbonyl clusters are composed, considers what atomic orbitals (AO's) and electrons they make available for cluster skeletal bonding, and considers the symmetry-dictated interactions between these AO's when the fragments are brought together, one can get some feeling for the link between electron numbers and shape that is a pronounced feature of cluster chemistry, and that allows one to see family relationships between metal clusters and other cluster and ring systems. Electron-counting schemes are outlined below that allow one to rationalise or predict the connectivity, and often the 3-dimensional arrangement, of the skeletal atoms. Chapter illustrates a different approach to cluster bonding which treats clusters $M_x(CO)_y$ effectively by placing a bare metal cluster M_x, itself treated as a fragment of the bulk metal, within a $(CO)_y$ shell.

2. THE METAL-LIGAND BONDING IN METAL CARBONYL CLUSTERS

Before considering the metal-metal skeletal
bonding in metal carbonyl clusters, it is convenient to
note the type of metal-ligand bonding interactions they
contain, since these are relevant not only to such
features as ligand mobility and reactivity, but also to
the question of which metal AO's will be available for
metal-metal bonding.

Carbon monoxide is a remarkably versatile ligand.
It can bond to one, two or three atoms of a cluster with
apparently equal ease, a facility that makes it a very
mobile ligand (see Chapter 7). Though it normally
coordinates to metals through only its carbon atom,
it can simultaneously coordinate through its oxygen
atom also. It can act both as a source of, and as a
sink for, electrons. In the former capacity it can
supply either two or four electrons. It can function
as an inert group that occupies a spare valency, though
lost readily enough to make that spare valency available
again, or it can itself undergo reactions while
remaining coordinated. The manner in which it can
bond to metals is discussed in most general inorganic
texts, but a reminder of the salient features is given
here.

198.

The main ways in which a carbonyl ligand can coordinate to one or more of the metal atoms of a cluster are as follows (see Fig. 1):

(i) Terminal attachment to one metal atom, by a linear or nearly linear M-C-O unit (angle MCO 165-180°).

(ii) Doubly bridging (μ_2) between two metal atoms, coordinating exclusively (though not necessarily symmetrically) through the carbon atom (i.e. monohapto, η^1), the C-O axis being perpendicular to the M-M axis, or nearly so.

(iii) Doubly bridging in such an unsymmetrical way as to imply dihapto coordination to one of the metal atoms, with some M....O bonding.

(iv) Triply bridging (μ_3) between three metal atoms, the C-O axis being perpendicular to the M_3 plane, or nearly so.

(v) Triply bridging (μ_3) between three metal atoms, while coordinating dihapto to a fourth.

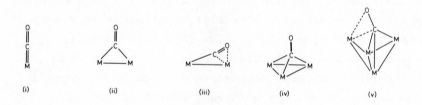

Figure 1 Ways in which carbonyl ligands can coordinate

Modes of coordination intermediate between these types 1(i) - (v) are also possible, as is attachment of a second Lewis acid to the oxygen of a carbonyl ligand. The bonding interactions that allow this versatility are outlined below.

2.1. Terminal Carbonyls

When a CO molecule coordinates to a single metal
atom, it does so by a metal-carbon bond much shorter
than the single M-C bonds of metal alkyls. For
example, the carbonyl ligands of the molybdenum complex
$(\eta^5-C_5H_5)Mo(CO)_3Et$ (9) are attached to the metal atom
by bonds of length 1.97(3) Å, whereas the single Mo-C
bond to the ethyl group is 2.38(3) Å long. Moreover,
the CO bonds in carbonyl complexes are apparently
invariably longer than that in free CO, though this
has been established unambiguously in relatively few
cases because of the difficulty of measuring these
short bonds accurately enough, and the CO stretching
frequency and force constant decrease on coordination.

These features can be rationalised in valence bond
terms by describing the MCO bonding as a resonance
hybrid of the canonical forms $M=C=O$ and $\bar{M}-C\equiv\overset{+}{O}$, indicating
an M-C bond order between 1 and 2, and a C-O bond order
between 2 and 3. The more commonly-used bonding
description draws attention to the metal and ligand
orbitals involved (Figure 2). The 'lone-pair' electrons
on carbon can be donated to a suitable vacant metal
orbital (e.g. the $\underline{d}_{z^2}$ AO, with possibly some contribution
from the metal $\underline{s}$ or $\bar{\underline{p}}_z$ AO), while the ligand→metal
electron drift assumed to occur in this process is
compensated for by back π-donation from metal to
ligand, from filled metal $\underline{d}_{xz}$ and $\underline{d}_{yz}$ AO's, possibly
reinforced by contributions from the $\underline{p}_x$ and $\underline{p}_y$ AO's,
into the ligand π^* orbitals. The synergic nature of
the bonding allows each component to reinforce the

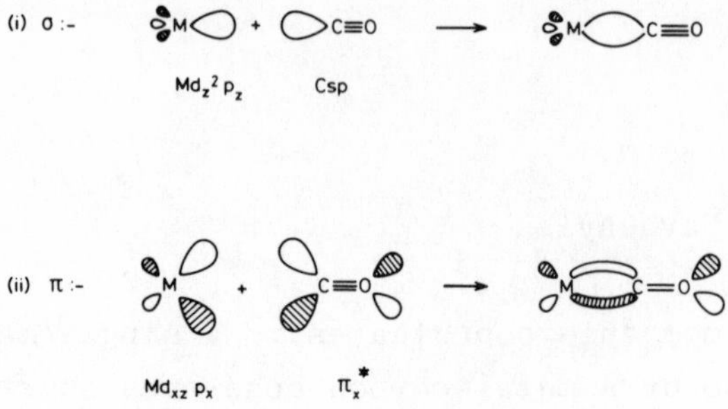

200.

(i) σ :- $Md_z^2 p_z$ Csp

(ii) π :- $Md_{xz} p_x$ π_x^*

Figure 2

The σ and π metal-carbon bonding interactions
of terminally coordinated carbonyl groups.

other, it being commonly assumed that the ligand remains
essentially neutral.

Closer consideration of the orbitals involved,
however, shows that on coordination, carbon monoxide
normally acquires a significant negative charge (10-13).
This is because the σ-donation from ligand to metal
involves less ligand→metal charge transfer than is
commonly supposed, while the back π-donation involves
more metal-ligand charge transfer than is commonly
supposed. The 'lone-pair' electrons on the carbon atom
of CO occupy an orbital that is essentially a carbon
sp hybrid, nonbonding orbital (it is actually very
feebly C-O antibonding). They hardly move away from
the ligand on coordination because the metal $d_z^2-p_z$
hybrid acceptor orbital, overlapping the lone pair
orbital, requires them to occupy essentially the same
region that they originally occupied, which region
becomes the M-C σ-bonding region. However, the back
π-donation results in significant electron transfer to
the ligand, since the acceptor orbitals involved, the
ligand π* MO's, have substantial oxygen p orbital
character and so require some increase in electron
density on the oxygen atom. Moreover, there are two

such π^* orbitals.

This picture of the bonding, leading to negatively charged carbonyl ligands and positively charged metal atoms, is supported by theoretical and spectroscopic studies (11-14) on various carbonyl complexes, although uncertainty remains about the relative contributions of the σ and π components to the metal-ligand bond strength. The canonical form $\overset{-}{M}-C\equiv\overset{+}{O}$ by which metal-carbonyl bonding is still occasionally represented thus has three defects. It suggests a weaker M-C bond, and a stronger C-O bond, than are present, and implies erroneously that the ligand acquires an overall positive charge.

The CO bond orders of terminally coordinated carbonyls, as indicated by their stretching frequencies and force constants, are generally in the range 2.4 - 2.8, though they may fall to about 2.2 - 2.3. For example, the CO force constants for $Cr(CO)_6$ and $Ni(CO)_4$ are 17.2 and 17.9 millidynes/$\overset{o}{A}$ respectively (c.f. 19.8 md/$\overset{o}{A}$ for CO itself) implying CO bond orders of about 2.65 and 2.75 respectively. Decreasing C-O bond orders in the isoelectronic series $Ni(CO)_4$, $Co(CO)_4^-$ and $Fe(CO)_5^{2-}$, as the ligands become progressively more negatively charged, are reflected in the infra-red-active CO stretching frequencies of these tetracarbonyl species (2060, 1890 and 1790 cm^{-1} respectively; free CO has ν(CO) 2143 cm^{-1}). A similar effect is apparent in the series $Mn(CO)_6^+$ (2090 cm^{-1}), $Cr(CO)_6$ (2000 cm^{-1}) and $V(CO)_6^-$ (1860 cm^{-1}) (14).

Although the metal-ligand bonding may thus vary from one system to another in the extent to which electronic charge is transferred to the ligand, these systems show no variation in the number of electrons formally donated from ligand to metal. For electron-book-keeping purposes, the terminal carbonyl ligand is

202.

invariably regarded as a 2-electron ligand, as that is the number it contributes to the metal atom's valence shell.

2.2. Monohapto Doubly Bridging Carbonyl

When a carbonyl ligand adopts a μ_2 mode of coordination (Fig. 1(ii)) in di- or poly-nuclear metal carbonyls, it can still function as a 2-electron ligand because the 'lone-pair' orbital on carbon can overlap with a suitable in-phase (metal-metal bonding) combination of empty metal orbitals, while an out-of-phase (metal-metal antibonding) combination of filled metal orbitals can interact with, and so release electronic charge into, the CO π^* orbital in the M_2CO plane (Figure 3). The net result is to reduce the CO

Figure 3

The bonding of a μ_2-carbonyl ligand.

bond order to about 2 (μ_2 carbonyl ligands generally have $\nu(C-O)$ in the range 1750 - 1850 cm^{-1}, overlapping with that of ketones, but below the range 1900 - 2150 cm^{-1} in which most (though by no means all - c.f. above) terminal carbonyl stretching frequencies lie. Since the metal-ligand bonding effectively involves two pairs of electrons, the bridging carbonyl can be regarded as bonded by a single M-C bond to each of the bridged metal atoms in a cyclopropanone-type arrangement (Figs. 1(ii) and 3(iii); the bridged metal atoms are usually, though not invariably, directly linked by a metal-metal bond).

Although the two M-C single bonds of a μ_2-carbonyl group are together likely to have more energy than the one M=C multiple bond of a terminal carbonyl ligand, this does not mean that a doubly-bridging carbonyl group is necessarily bound more strongly than a terminal carbonyl group, because the bridging ligand will have lost more C-O bond energy, relative to uncoordinated CO, than the terminal ligand has. The ease with which CO ligand scrambling occurs between bridging and terminal sites (15) suggests that there is little difference between the strength of attachment in the two types of site (16), a conclusion supported by an assessment (17) of the bond energies of the dinuclear iron carbonyl $Fe_2(CO)_9$, which has six terminal and three doubly-bridging carbonyl ligands.

2.3. Doubly Bridging Dihapto Carbonyl Ligands

The ligand geometry shown in Fig. 1(iii) is to be found in the dinuclear manganese complex $Mn_2(CO)_5(Ph_2PCH_2PPh_2)_2$ (18). In this, the carbonyl ligand is effectively terminally bound to one manganese atom (M^1) while coordinating η^2 to the other (M^2), presumably in a manner like that of a dihapto alkene complex (Fig. 4). Altogether, this carbonyl ligand thus contributes four electrons, two to the valence

204.

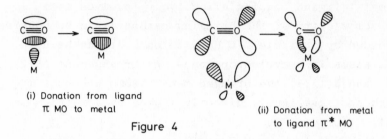

(i) Donation from ligand
π MO to metal

Figure 4

(ii) Donation from metal
to ligand π* MO

Bonding interactions possible for a carbonyl ligand
bonding dihapto to a single metal atom

shell of each metal atom. Though this mode of
coordination is rarely found among isolable
crystalline compounds, it may not be uncommon among
coordinatively unsaturated reaction intermediates.

2.4. Triply-bridging Carbonyls

In the μ_3 mode of coordination (Figure 1(iv)),
the 'lone-pair' orbital on carbon can combine with a
suitable in phase combination of metal orbitals
(Fig. 5(i)), and receive electron density into both
CO π* MO's from suitable E-symmetry combinations of
metal orbitals (Figs. 5(ii) and (iii)). The metal-
ligand bonding can thus be regarded as involving
three pairs of electrons, enough to allow the
alternative description of three metal-carbon single
bonds holding the carbon atom to the metal triangle,
with the CO bond order formally reduced to one
(Fig. 1(iv)). In fact, although the C-O bond lengths
of triply-bridging carbonyl groups are not known
accurately enough to assess their effective bond orders,

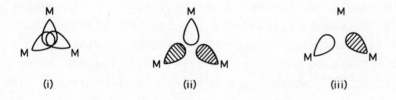

(i)   (ii)   (iii)

Figure 5

The combinations of metal orbitals with which the
'lone pair' $\left[(i)\right]$ and $\pi^*\left[(ii) \text{ and } (iii)\right]$ ligand orbitals
can interact to bond a μ_3 carbonyl ligand

their stretching frequencies, on average lower still than those of doubly bridging carbonyl groups, are consistent with a CO bond order just below two.

On progressing from terminal coordination through μ_2 to μ_3 coordination, a carbonyl ligand thus experiences progressive lengthening of its CO bond and accumulates a progressively greater negative charge. An increase in the total energy of the metal-carbon bonds is believed to occur in the same sequence, offsetting the concomitant loss of CO bond energy, so the strength of attachment of the ligand to the cluster probably changes little with its mode of coordination.

An example of a <u>quadruply</u>-bridging dihapto carbonyl ligand has been reported (19). In the anion $[Fe_4(CO)_{13}H]^-$, which has a butterfly-shaped arrangement of its metal atoms, a unique carbonyl ligand triply bridges one wing of the metal skeleton while π-bonding to the fourth metal atom (Fig. 1(v)).

2.5. The Basicity of Carbonyl Ligands

A further mode of coordination of the carbonyl group that may be of considerable importance in certain reactions of metal carbonyl clusters involves attachment of a Lewis acid to the oxygen atom of the carbonyl ligand, a bonding situation possible for both terminal and bridging carbonyl ligands (Figure 6) (20-22). The capacity of the oxygen atoms of carbonyl ligands to coordinate to Lewis acids by use of their 'lone pair' electrons will clearly increase with the accumulation of negative charge on the ligand. Accordingly, triply-bridging carbonyl ligands are expected to be more basic than doubly-bridging carbonyl ligands, which in turn

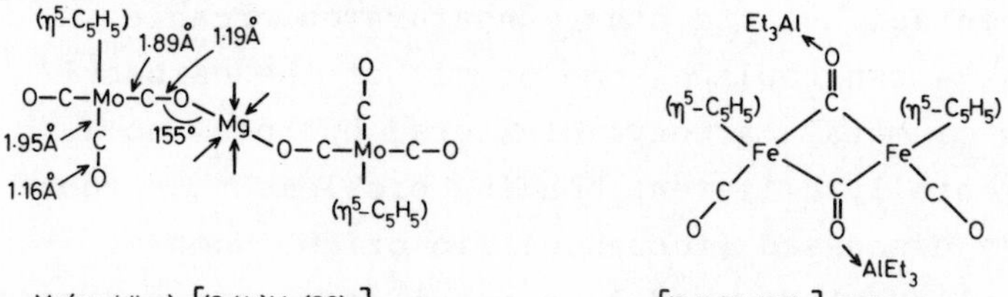

$Mg(pyridine)_4[(C_5H_5)Mo(CO)_3]_2$ (20) $[(C_5H_5)Fe(CO)_2]_2.2AlEt_3$ (21)

Figure 6 Carbonyl ligands acting as Lewis bases by use of their oxygen 'lone pair' electrons.

are expected to be more basic than terminal carbonyls. Among terminal carbonyls, those with low CO stretching frequencies, indicative of relatively negatively charged ligands, are the most likely to function as Lewis bases. Addition of a Lewis acid to a di- or poly-nuclear metal carbonyl can cause carbonyl groups to move from terminal to bridging positions to facilitate coordination at oxygen (23).

The metal-carbon and carbon-oxygen bond distances in the compound $Mg(pyridine)_4[(\eta^5-C_5H_5)Mo(CO)_3]_2$ (20) (Fig. 6) indicate that coordination through the oxygen 'lone-pair' lengthens the C-O bond but shortens the Mo-C bond attaching the carbonyl ligand to the metal, showing that draining electronic charge from oxygen strengthens the metal→ligand back π-bonding.

2.6. Other Ligands

Various other ligands, particularly those with similar bonding characteristics to carbon monoxide - π-acid ligands such as NO, phosphines, arsines,

cyclopentadienyl and other unsaturated organic
ligands - can replace some or all of the carbonyl
groups of metal carbonyl clusters, but pose no
fundamentally different bonding problems than those
already discussed (though nitric oxide, when
terminally coordinated to a metal atom, can act as a
source of either three electrons or one electron,
depending on whether the MNO unit is linear or bent).

     Many metal carbonyl hydride clusters are also
known. Hydrogen can coordinate to the same types of
terminal or bridging site as CO, the important bonding
difference being that hydrogen has no AO's of suitable
symmetry and energy with which to form multiple bonds.
Its 1s AO will interact with either a σ-bonding AO of
a particular metal atom of the type shown in Fig. 2(i),
or an in-phase (i.e. metal-metal bonding) combination
of metal orbitals as in Figs. 3(i) and 4(i). Hydrogen
as a ligand can thus be thought of as protonating what
would otherwise be a 'lone pair' on a single metal
atom, or bond pairs associated with the edges or faces
of cluster polyhedra, forming 2-centre M-H, 3-centre
M_2H or 4-centre M_3H bonds respectively. Hydrogen as a
ligand is also capable of protonating the oxygen 'lone
pair' of a carbonyl ligand (24), and there is increasing
evidence that occasionally hydrogen atoms may move into
cavities within the polyhedral skeletons of metal
clusters.

3. METAL-METAL BONDING IN METAL CARBONYL CLUSTERS

In cluster chemistry just as elsewhere in chemistry,
the number of electrons associated with a group of
atoms determines the way(s) those atoms may arrange
themselves in space. Knowing how many electrons are
associated with a particular group of cluster atoms
can allow one to rationalise the shape of the cluster
they form, or to predict that shape if it has not been
determined.

There are two methods in common use in cluster
chemistry by which the number of valence shell electrons
can be used to deduce or rationalise structures. One
is essentially a valence-bond approach that allocates
the skeletal (metal-metal) bonding electrons to
localised 2-centre metal-metal bonds, and is useful
and adequate for many of the smaller clusters. The
other, instead of assigning the skeletal electrons to
individual bonds, notes the relationship between the
polyhedral shape of a cluster and the total number of
skeletal electrons. It draws heavily on analogies
with other categories of cluster compound, notably
boron hydride clusters, for which 2-centre bond schemes
have proved unsatisfactory, and is useful for
rationalising the shapes of medium-sized metal carbonyl

clusters (from 4 to 7 skeletal atoms) and mixed
clusters. Both approaches are outlined below.

3.1. The 18 Electron Rule and 2-centre Bond Schemes

     The simplest arguments based on electron numbers
assume that the skeletal atoms are held together by a
network of 2-centre electron pair bonds, and that the
cluster atoms use all their valence shell AO's, either
for skeletal metal-metal bonding, for metal-ligand
bonding, or to accommodate nonbonding 'lone-pair'
electrons. Each skeletal atom in a cluster is allocated
one AO for each 2-centre skeletal bond it forms. The
remaining valence shell AO's are used to bond ligands
or to accommodate nonbonding electron pairs. Since a
transition metal atom has nine valence shell AO's
(five $\underline{d}$, one $\underline{s}$ and three $\underline{p}$ AO's), it can accommodate a
total of eighteen electrons in these orbitals, and the
majority of organometallic and carbonyl complexes of
transition metals (at least those formed by the Cr, Mn,
Fe and CO subgroup metals) apparently use all nine
valence shell orbitals and so conform to the '18
electron rule' (25,26). For metals of the nickel and
copper groups, particularly the heavier metals, the
higher energies of the $\underline{p}$ orbitals make them less
readily available for bonding use, and so these metals
tend to form complexes in which one or even two valence
shell AO's are left vacant. For these, 16 or 14
electrons respectively will be associated with the
valence shell.
     The vast majority of known metal carbonyl clusters
contain metals that are expected to use all nine valence
shell AO's, however, and so the familiar 18 electron

rule can be applied. As trivial examples, the mononuclear carbonyls $Cr(CO)_6$, $Fe(CO)_5$ and $Ni(CO)_4$ have
the numbers of carbonyl groups needed to provide the
12, 10 and 8 electrons (respectively) needed to bring
the valence shell count to 18 in each case. In the
dinuclear complexes $Mn_2(CO)_{10}$, $Fe_2(CO)_9$ and $Co_2(CO)_8$
there are 34 valence shell electrons, enough to
allocate 17 to the valence shall of each metal atom
in the absence of metal-metal bonding. Whether the
carbonyl groups are bridging or terminal is immaterial
to the electron count (bridging of the type shown in
Fig. 1(iii) is so rare as to be ignored in predicting
likely structures). A single metal-metal bond in each
of these dinuclear complexes completes the rare gas
18 electron configuration.

Treated similarly, the trinuclear complexes
$M_3(CO)_{12}$ (M = Fe, Ru or Os) contain 3 x 8 + 2 x 12 = 48
electrons, enough to allocate 16 to each metal atom
assuming no metal-metal bonding. Each metal atom
therefore needs to acquire two extra electrons to
achieve a total of 18, which it can do by forming two
metal-metal 2-centre bonds, so the singly-bonded
triangular structures shown in Figure 7 are appropriate.

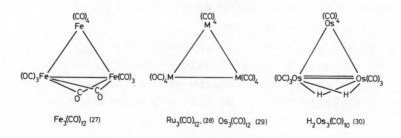

Figure 7

Some triangular metal carbonyl clusters

Again, since the electron count is the same, whether
the carbonyl ligands are bridging or terminal, this
approach allows no prediction to be made about the
positions of the carbonyl ligands, except that they
must be distributed in a manner that does not
coordinatively saturate any metal atom and so prevent
it from bonding to the other metal atoms. For example,
it could not have been predicted that $Fe_3(CO)_{12}$ would
have two bridging carbonyl ligands (27), whereas
$Ru_3(CO)_{12}$ (28) and $Os_3(CO)_{12}$ (29) have none (Figs. 7(i)
and (ii)), though hypothetical structures that
allocated five terminal carbonyl groups to one of the
metal atoms of $Fe_3(CO)_{12}$, $Ru_3(CO)_{12}$ or $Os_3(CO)_{12}$ could
have been excluded from consideration.

The presence of multiple metal-metal bonds in
certain clusters can be inferred from the electron
count. The osmium carbonyl hydride $Os_3(CO)_{10}H_2$, for
example, contains 46 electrons associated with the
metal valence shells, indicating that _four_ metal-metal
bonds linking these three atoms are needed to give each
metal atom a share in 18 electrons, suggesting a cyclo-
propene-type structure with one Os=Os double bond
(Fig. 7(iii)). A neutron diffraction study (30)
compound has shown that its hydrogen atoms are
associated with the multiple bond, which may therefore
be regarded as a protonated double bond like that in
the BH_2B bridge of diborane.

The tetranuclear cobalt group clusters $M_4(CO)_{12}$
(M = Co, Rh or Ir), with 60 valence shell electrons,
_i.e._ 15 per metal atom, require each metal to form
three M-M single bonds, enough for a bond along each
of the six edges of their M_4 tetrahedra (7).

From these examples, one can see that the
connectivity of a cluster building unit such as a
metal carbonyl unit $M(CO)_n$ - the number of 2-centre

M-M bonds it needs to form to obey the 18 electron rule - is equal to the number of electrons by which its own valence shell electron count falls short of 18.

The use of the 18 electron rule to deduce how many 2-centre electron pair metal-metal bonds may be present in a cluster may be put on a general basis as follows. For a cluster of formula $M_x(CO)_y$, there will be a total of $(vx + 2y)$ valence shell electrons, if v is the number of valence shell electrons appropriate for the element M ($v = 6$ for Cr, Mo or W; 7 for Mn, Tc or Re; 8 for Fe, Ru or Os; and 9 for Co, Rh and Ir). If there were no metal-metal bonds in the cluster $M_x(CO)_y$, then 18x valence shell electrons altogether would be needed. The deficit, $[18x-(vx + 2y)]$, is the number of electrons needed in metal-metal bonds to give each metal atom a share in 18 electrons. The number of 2-centre M-M bonds is thus

i.e. $\frac{1}{2}[(18-v)x - 2y]$, $\frac{1}{2}[18x-(vx + 2y)]$,

More generally still, for clusters of x transition metal atoms, not necessarily all the same, accompanied by assorted ligands, the number of metal-metal 2-centre bonds can be deduced simply by calculating the total number of valence shell electrons present (including those provided by the ligands), subtracting that number from 18x, and dividing the remainder by two. For $CoNi_2(\eta^5-C_5H_5)_3(CO)_2$, for example, the total number of valence shell electrons available is $9 + 2 \times 10 + 3 \times 5 + 2 \times 2 = 48$. Subtracting this from 54 (18x) leaves 6, implying three metal-metal bonds, and so the triangular structure actually adopted (a less likely alternative might have been an open-chain structure with one double and one single metal-metal bond). Similarly, $Co_4(\eta^5-C_5H_5)_4H_4$ contains $4 \times 9 + 4 \times 5 + 4 = 60$ valence shell electrons, and so requires $\frac{1}{2}[18 \times 4 - 60] = 6$ metal-metal bonds to hold its four cobalt atoms together, the appropriate number for the tetrahedral structure adopted.

With the exception of $Os_3(CO)_{10}H_2$, which contains an Os=Os double bond, the examples cited so far show a direct correspondence between the number of metal-metal 2-centre electron pair bonds and the number of metal-metal contacts. This is by no means generally true for metal carbonyl and related clusters, however, though most low nuclearity clusters show this correlation. Further exceptions are provided by such tetrahedral clusters as $Re_4(CO)_{12}H_4$ (31) and $Ni_4(\eta^5-C_5H_5)_4H_3$ (32), for which the electron counts would suggest 8 and $4\frac{1}{2}$ metal-metal 2-centre electron pair bonds respectively. Although the shortage of electrons in the former might in principle be relieved by the allocation of two Re=Re double bonds to two of the four tetrahedral edges, and four Re-Re single bonds to the remaining four edges, the distortion to D_{2d} symmetry this would require is not observed. An alternative description, in terms of four 3-centre Re_3 metal-metal bonds in the faces of the tetrahedron, is a much more appropriate way of giving each metal atom in this cluster a share in 18 electrons. As the hydrogen atoms in $Re_4(CO)_{12}H_4$ cap the four tetrahedral faces, the skeletal bonding may strictly be regarded as involving four 4-centre Re_3H bonds. For the nickel cluster $Ni_4(\eta^5-C_5H_5)_4H_3$, the odd number of electrons present could be accommodated on a 2-centre bond scheme by assuming the presence of three single bonds and three half bonds, but both these examples reveal the predictive limitations of the 2-centre bond approach.

Further limitations of the approach are apparent if one attempts to describe larger clusters using it. For example, there are many 86-electron octahedral clusters known, of which the rhodium carbonyl $Rh_6(CO)_{16}$

is an example (Figure 8) (34), whose skeletal bonding can be described in terms of 2-centre electron pair bonds only if one invokes resonance of the eleven such bonds dictated by the 18 electron rule ($\frac{1}{2}\{6 \times 18 - 86\}$) between the twelve octahedral edges. These clusters contain just two electrons too many to allow a satisfactory 2-centre bonding description. However, in their reactions they show no sign that two electrons can readily be dispensed with - they are not readily oxidised to 84-electron clusters - and it is significant that the osmium cluster $Os_6(CO)_{18}$, which does contain 84 valence shell electrons, adopts a different structure, a bicapped tetrahedron (Figure 8) (35) which admittedly

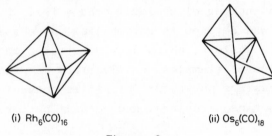

(i) $Rh_6(CO)_{16}$ (ii) $Os_6(CO)_{18}$

Figure 8

The polyhedral skeletons of $Rh_6(CO)_{16}$ (34) and $Os_6(CO)_{18}$ (35)

has the 12 2-centre links that would have been expected, but is less symmetrical than the O_h octahedron.

Describing the bonding in clusters in terms of localised 2-centre electron pair bonds is thus of limited value. An alternative approach is outlined below that allows one to deal more satisfactorily with the shapes and bonding of many systems that in 2-centre bond terms appear exceptional.

3.2. The Polyhedral Skeletal Electron Pair Theory (PSEPT)

This approach gets its name from the clear link that exists between the polyhedral shapes of many cluster compounds and the numbers of electron pairs available to hold their skeletal atoms together. It retains the assumption that each metal atom uses all nine of its valence shell AO's, but instead of assigning the skeletal electron pairs to 2-centre bonds, it allows the probable structure to be deduced from the total number of skeletal bond pairs. Since the approach was devised to rationalise the shapes of boron hydride clusters, which provide simpler and more complete series of examples to illustrate the approach than metal clusters do, (5,38,39), it is convenient to outline the borane structural and bonding pattern here.

3.2.1. Boron Hydride Clusters

The simplest series of borane clusters is the closo (closed cage) series, of formulae $B_nH_n^{2-}$ or isoelectronic variations (carboranes) in which BH is replaced by CH^+, e.g. $CB_{n-1}H_n^-$ or $C_2B_{n-2}H_n$. These are clusters of n BH (or CH) units whose polyhedral structures have exclusively triangular faces (Fig. 9). The hydrogen atoms point radially outwards, away from the centres of the pseudo-spherical polyhedra. Since a BH unit can supply two electrons, and a CH unit 3 electrons, for skeletal bonding (the number remaining

after two have been allocated to the BH or CH bond),
<u>closo</u> boranes or carboranes have (n + 1) skeletal bond
pairs holding their n skeletal atoms together.   The
polyhedra in Figure 9 are the shapes that make best use

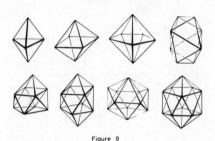

Figure 9

The polyhedra that form the bases for the structures of
<u>closo</u> borane anions $B_nH_n^{2-}$ and carboranes $C_2B_{n-2}H_n$

of these electrons (the structures of the <u>closo</u> borane
anions and carboranes are evidently the thermodynamically
preferred shapes, since these clusters are commonly
prepared by high temperature syntheses and have high
thermal stability).

Molecular orbital treatments have shown that all
the polyhedra in Figure 9 have appropriate symmetries
to generate (n + 1) skeletal bonding MO's from the
three AO's (one inward-pointing, radially orientated
<u>sp</u> hybrid AO, and two tangentially orientated <u>p</u> AO's -
see Figure 10) that each skeletal atom contributes,
with an appreciable energy gap before the next
available orbital, <u>i.</u>e. between HOMO and LUMO.   Since
the combinations of AO's that generate the MO's are
symmetry-dictated, it follows that these same polyhedra
represent suitable structures for whole series of
clusters which, like the <u>closo</u> boranes and carboranes,
consist of groups of n skeletal atoms, each providing
three AO's similar to those shown in Figure 10.
and held together by a total of (n + 1) skeletal bond pairs.

218.

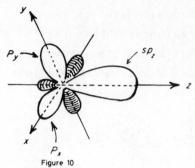

Figure 10

The radial sp hybrid AO and tangential p AO's that
a BH unit can use for skeletal bonding in borane clusters

Before considering examples of <u>closo</u> structures
among metal carbonyl clusters, it is worth noting
further features of the borane-carborane structural
pattern. In addition to the <u>closo</u> borane anions and
carboranes, there are other series of boron clusters
that contain more skeletal bonding electrons, and so
adopt more open structures, than the <u>closo</u> series. One
such is the <u>nido</u> (nest-like) series, of general
formula B_nH_{n+4} (or $C_xB_{n-x}H_{n+4-x}$). Their structures
contain n BH (or CH) units effectively held together by
(n + 2) skeletal bond pairs, since they are all
formally derivable from hypothetical parent anions
$B_nH_n^{4-}$ by protonation (the extra four hydrogen atoms
of the neutral boranes B_nH_{n+4} tend to occupy $B \cdots H \cdots B$
bridging positions in which they are clearly intimately
involved in the skeletal bonding - see Figures 11 and
12) or by isoelectronic replacement of B by C^+.
Significantly, their skeletal structures are based on
the same series of polyhedra (Figure 9) as the <u>closo</u>
species, but with one vertex of the polyhedron left
vacant. Thus, for both <u>closo</u> and <u>nido</u> species, the
polyhedron on which the structure is based has one
vertex fewer than the number of skeletal bond pairs.

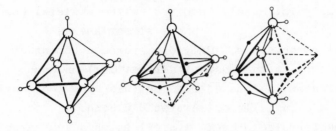

○ Boron   o Terminal hydrogen   ● Bridging hydrogen

Figure 11

The closo- ($B_6H_6^{2-}$), nido- (B_5H_9), and arachno- (B_4H_{10})
structural relationship illustrated for systems with 7 skeletal
bond pairs, formally based on an octahedron.

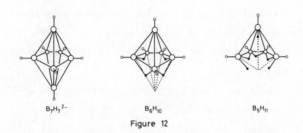

$B_7H_7^{2-}$ B_6H_{10} B_5H_{11}

Figure 12

The closo- ($B_7H_7^{2-}$), nido- (B_6H_{10}), arachno- (B_5H_{11}) relationship
illustrated for systems with 8 skeletal bond pairs, formally
based on a pentagonal bipyramid.

Similar generalisations can be made about the
remaining two categories of borane cluster, the so-
called arachno (web-like) and hypho (net-like) series,
which belong to the formula-types B_nH_{n+6} and B_nH_{n+8}
respectively (39). They may be regarded as clusters
of n BH units, held together by (n + 3) and (n + 4)
skeletal bond pairs respectively, and have structures
based on the polyhedra shown in Figure 9 but with two
(arachno) or three (hypho) vertices unoccupied by
skeletal boron atoms. Their extra hydrogen atoms
adopt positions that may be bridging or terminal, but

are clearly intimately involved in the skeletal bonding
as they lie on the same spherical surface as the
skeletal boron atoms. Only one hydrogen (or other
substituent, in the case of derivatives) on each
skeletal atom is located outside this spherical surface,
held by the radially orientated B-H bond.

The formulae of some typical boranes and carboranes
are listed in Table 1, which also indicates the polyhedra on which their structures are based, and the
skeletal electron count. Two families of closo-, nido-
and arachno-boranes, with structures based on the
octahedron and pentagonal bipyramid, are illustrated
in Figures 11 and 12.

The skeletal structures of boranes and carboranes
can thus be deduced from their formulae. The type
(closo, nido, arachno or hypho) will be apparent from
the formula type ($B_nH_n^{2-}$, B_nH_{n+4}, B_nH_{n+6} or B_nH_{n+8}
respectively, or an isoelectronic carborane). The
number of skeletal bond pairs, s, can also be deduced
from the formula. Expressing the formula as B_nH_{n+c},
then $s = n + \frac{1}{2}c$; expressing the formula as B_xH_y, then
$s = \frac{1}{2}(x + y)$. Knowing s, one can deduce the polyhedron
on which the structure will be based - it will have
(s - 1) vertices. The skeletal atoms are then
allocated to some (nido, arachno or hypho species) or
all (closo species) of the vertices of the polyhedron
in question. If the polyhedron has more than one type
of vertex, or if more than one vertex is left vacant,
then more than one arrangement of the skeletal atoms
is in principle possible, though known precedents
indicate which is most likely. It is commonly a more
highly coordinated vertex that is left vacant, and the
structures of boranes and carboranes are generally the
more compact of the available possibilities, e.g.
the arachno boranes B_4H_{10} and B_5H_{11} have the 3-dimensional

TABLE 1.

Some Typical Boranes and Carboranes

closo species $B_nH_n^{2-}$	nido species B_nH_{n+4}	arachno species B_nH_{n+6}	fundamental polyhedron	v*	s†
$C_2B_3H_5$		$B_3H_8^-$	trigonal bipyramid	5	6
$B_6H_6^{2-}$ CB_5H_7 $C_2B_4H_6$	B_5H_9 $C_2B_3H_7$	B_4H_{10}	octahedron	6	7
$B_7H_7^{2-}$ $C_2B_5H_7$	B_6H_{10} $C_xB_{6-x}H_{10-x}$	B_5H_{11}	pentagonal bipyramid	7	8
$B_8H_8^{2-}$ $C_2B_6H_8$	–	B_6H_{12}	dodecahedron (D_{2d})	8	9
$B_9H_9^{2-}$ $C_2B_7H_9$	B_8H_{12} $C_2B_6H_{10}$	–	tricapped trigonal prism	9	10
$B_{10}H_{10}^{2-}$ $C_2B_8H_{10}$	$B_9H_{12}^-$ $C_2B_7H_{11}$	B_8H_{14}	bicapped square antiprism	10	11
$B_{11}H_{11}^{2-}$ $C_2B_9H_{11}$	$B_{10}H_{14}$ $C_2B_8H_{12}$	B_9H_{15} $C_2B_7H_{13}$	octadecahedron (C_{2v})	11	12
$B_{12}H_{12}^{2-}$ $CB_{11}H_{12}^-$ $C_2B_{10}H_{12}$	$CB_{10}H_{13}^-$ $C_2B_9H_{11}^{2-}$ $C_4B_7H_{11}$	$B_{10}H_{15}^-$ $B_{10}H_{14}^{2-}$	icosahedron	12	13

*v = no. of vertices on polyhedron   †s = no. of skeletal bond pairs

boron arrangements shown in Figs. 11 and 12 rather than
the planar alternatives (axial vertices vacant) that
are in principle available. A detailed discussion of
the shapes possible for boranes and carboranes and
illustrations of further series are given in (39).

This structural pattern shows that the shapes of
the polyhedral fragments found among nido, arachno and
hypho species are appropriate to generate the same
number of skeletal bonding MO's as the complete (closo)
polyhedral cluster. For example, the complete octahedron of $B_6H_6^{2-}$, the square pyramid of B_5H_9, and the
butterfly shape of B_4H_{10} are all appropriate skeletal
shapes to generate 7 skeletal bonding MO's particularly
when, in the nido and arachno cases, the positions of
the extra hydrogen atoms (four bridging H atoms in B_5H_9,
four bridging H atoms and two extra (endo) terminal H
atoms in B_4H_{10} - the atoms shown as filled circles in
Fig. 11), by matching the skeletal symmetry reduce the
energies of (and so stabilize) some of the skeletal
bonding MO's without adding to their number.

3.2.2.   Isolobal Relationships

Large numbers of derivatives (40,41) of borane and
carborane clusters are now known that show how their BH
and CH units can be replaced by an extensive range of
other units, both main group and transition metal,
capable of participating in cluster bonding by using
similar sets of orbitals to those shown in Fig. 10.
In particular, transition metal carbonyl or organometallic units such as $Fe(CO)_3$ or $Co(\eta^5-C_5H_5)$ or related
species can replace the BH units of boranes and carboranes.
For example, whole series of metallocarboranes are
known with general formulae $C_2B_{n-2}H_nM(CO)_3$ (M = Fe, Ru

or Os) or $C_2B_{n-2}H_nM'(\eta^5\text{-}C_5H_5)$ $(M' = $ Co, Rh or Ir)
which adopt <u>closo</u> structures in which the metal atom,
two carbon atoms and the (n - 2) boron atoms together
define the (n + 1) vertices of the appropriate polyhedron. Like a BH unit, these metal units $M(CO)_3$ or
$M'(\eta^5\text{-}C_5H_5)$ can supply a pair of electrons and a set
of three AO's of suitable symmetry for use in cluster
bonding. Since the orbitals in question may well have
considerable <u>d</u> character (42), these metal units are
not strictly isoelectronic with a BH unit, but <u>isolobal</u>
with it, a term used to imply that the number, symmetry
properties, extent in space and energies of the
frontier orbitals of these units are similar (43,44).

The orbitals that a conical metal carbonyl unit
$M(CO)_3$ or metal cyclopentadienyl unit $M(\eta^5\text{-}C_5H_5)$ can
use for cluster bonding are essentially the same as
those already discussed in connection with the bonding
between metals and carbonyl ligands (Fig. 2). A
conical $M(CO)_3$ unit, for example, has a $\underline{d}_{z^2}\text{-}\underline{s}\text{-}p_z$
hybrid orbital pointing along the 3-fold axis, away
from the carbonyl ligands, as the counterpart of the
radially orientated <u>sp</u> hybrid AO of a BH unit, and two
<u>d</u>-<u>p</u> hybrid orbitals as the counterparts of the tangen-
tially orientated <u>p</u> AO's of a BH unit (see Fig. 13).

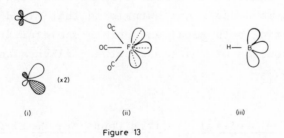

(i)   (ii)   (iii)

Figure 13

The orbitals that a conical $Fe(CO)_3$ unit can use for skeletal bonding,
(i) a radial and pair of tangential hybrid orbitals, or
(ii) an equivalent set of three hybrid orbitals; cf
(iii) the three hybrid sp^3 AO's of a BH unit

An alternative description of these frontier orbitals can be made in terms of three equivalent hybrid orbitals, directed away from the three carbonyl ligands (c.f. a BH unit described in terms of an sp^3-hybridised boron atom), a description that is useful for discussing the orientation of the $M(CO)_3$ unit with respect to the rest of the molecule in certain metal clusters (33). The conical $M(CO)_3$ unit behaves as if it were a fragment of an octahedral $M(CO)_6$ species, with orbitals available in the directions in which the missing carbonyl ligands would have been found (42) (see Fig. 13).

Further comparison of Fig. 13 with Fig. 2 is worthwhile to show how the frontier orbitals of carbon monoxide itself make it a further member of the same isolobal series. Its 'lone pair' HOMO is capable of playing the role of radially orientated cluster orbital, while its pair of LUMO's, the π^* CO orbitals, constitute the tangentially orientated pair of orbitals, suggesting that CO might occupy a polyhedron vertex in a cluster, quite apart from its capacity to act as an external ligand. This feature may contribute to the mobility of carbon monoxide as a ligand in cluster systems. It is also worth noting that an alkyne $RC\equiv CR$ can act as an external ligand or as a source of two skeletal units CR, so it is perhaps not surprising that mixed clusters incorporating both metal and carbon skeletal atoms are common products of reactions between alkynes and metal carbonyls.

3.2.3. The Skeletal Electron Count for Metal Clusters

The method by which one can deduce how many electrons a particular potential cluster fragment can

contribute for skeletal bonding is as follows. Three
AO's in the valence shell of the skeletal atom are
required for cluster use: therefore the number of
electrons available for cluster bonding will be the
number that have to be accommodated in these three
AO's after the remaining valence shell electrons,
whether originating on the skeletal atom or on its
ligands, have been allocated to the other valence shell
orbitals (of which there will be six in the case of a
transition metal, but only one in the case of a main
group element - assuming all valence shell AO's are
used). Tables 2 and 3 show the numbers of electrons

TABLE 2

The Skeletal Electrons (v+x-2) that Main Group
Cluster Units Can Contribute

v*	Main Group Element E	Typical Cluster Unit		
		(x=0)	EH or EX (x=1)	EH$_2$ or EL[†] (x=2)
1	Li	[-1]	0	1
2	Be,Mg,Zn,Cd,Hg	0	1	2
3	B,Al,Ga,In,Tl	1	2	3
4	C,Si,Ge,Sn,Pb	2	3	4
5	N,P,As,Sb,Bi	3	4	5
6	O,S,Se,Te	4	5	[6]
7	F,Cl,Br,I	5	[6]	[7]

*v = no. of valence shell electrons (periodic
group number) of main group element E

x = no. of electrons from ligand(s)

X = a 1-electron ligand

L = a 2-electron ligand

[†]also capable of acting as a divalent unit
supplying 2 fewer electrons

TABLE 3

The Skeletal Electrons (v+x-12) that Transition
Metal Cluster Units Can Contribute*

v	Transition Metal M	Typical Cluster Unit			
		$M(CO)_2$ (x=4)	$M(\eta^5-C_5H_5)$ (x=5)	$M(CO)_3$ (x=6)	$M(CO)_4$† (x=8)
6	Cr,Mo,W	[-2]	-1	0	2
7	Mn,Tc,Re	-1	0	1	3
8	Fe,Ru,Os	0	1	2	4
9	Co,Rh,Ir	1	2	3	[5]
10	Ni,Pd,Pt	2	3	4	

*v = no. of valence shell electrons of M

x = no. of electrons from ligands

† also capable of acting as a divalent unit supplying
2 fewer electrons

that typical cluster bonding units can contribute,
calculated on this basis. Thus units such as BH,
$Fe(CO)_3$, $Co(\eta^5-C_5H_5)$ or a bare tin or lead atom can
each supply 2 electrons, while units such as CH, $Co(CO)_3$,
$Ni(\eta^5-C_5H_5)$, BH_2, or a bare phosphorus or arsenic atom
can supply 3 electrons.

To illustrate the method, a conical $M(CO)_3$ unit
will contain (v + 6) valence shell electrons (v = the
number of valence shell electrons on an isolated metal
atom, i.e. 6 for Cr, 7 for Mn, 8 for Fe etc., while
six electrons come from the three carbonyl ligands).
Allocating twelve of these (v + 6) electrons to the
six AO's not involved in skeletal bonding (3 pairs are
assigned a C→M dative σ-bonding role, 3 pairs a metal→
carbon back π-bonding role) leaves (v - 6) electrons
for skeletal bonding use. Similarly, a cyclopenta-
dienyl-metal unit $M(\eta^5-C_5H_5)$ can supply (v - 7)

electrons, and a metal tetracarbonyl residue $M(CO)_4$,
by using four AO's for C→M bonding and two for back
π-bonding, can in principle supply (v - 4) electrons
for skeletal bonding, thoughly commonly it is found to
function as a divalent unit - a source of two AO's -
supplying (v - 6) electrons. For example, an $Os(CO)_4$
unit, like a CH_2 unit, may be regarded as a divalent
source of two electrons, better suited to an edge-
bridging role on a cluster polyhedron than to
occupancy of a vertex site.

3.2.4. Application of PSEPT to Metal Carbonyl Clusters:
       Systems with Seven Skeletal Bond Pairs

The manner in which PSEPT provides a rationale
for the structures of clusters containing only
transition metal skeletal atoms can be illustrated
by the octahedral rhodium cluster $Rh_6(CO)_{16}$ (Fig. 8),
which appeared anomalous when treated in terms of 2-
centre electron pair bonds. It contains six $Rh(CO)_2$
units (each formally able to supply one electron)
together with an extra four μ_3 carbonyl ligands, each
contributing two electrons. With 6 + 8 i.e. 14
electrons altogether to hold its six skeletal atoms
together, this cluster is formally analogous to the
closo octahedral $B_6H_6^{2-}$ or $C_2B_4H_6$. Its regular octahedral structure is that which is appropriate for a
6-atom, 7-bond pair system. It does not contain a
pair of electrons too many, as the 2-centre bond
approach suggests. The important difference between
the two approaches in this and related octahedral
systems is that PSEPT assumes each skeletal atom uses
3 AO's for skeletal bonding, while the 2-centre bond
treatment requires each atom to use 4 AO's, since each

vertex is connected to four others. For an 86-electron octahedral cluster like $Rh_6(CO)_{16}$, the 2-centre bond approach leads one to expect the octahedral edge bonds to be of average bond order 11/12, i.e. 0.92 (the number of bond pairs divided by the number of edges), whereas PSEPT suggests an edge bond order of 7/12, i.e. 0.58. It has been argued (45,46) that a low bond order for the metal-metal bonds in $Rh_6(CO)_{16}$ is indeed indicated by their length (2.78 Å; cf. 2.69 Å in metallic rhodium, in which the bond order, limited by the total number of valence shell electrons available (9), cannot exceed 9/12, i.e. 0.75). However, since the metal atoms in a metal carbonyl cluster are more positively charged than those in the bulk metal, because of the electron-withdrawing effect of the ligands, the longer bonds in the cluster do not necessarily have a lower bond order. The more highly positively charged each atom becomes, the higher the bond order that will be needed to keep it a fixed distance from its skeletal, similarly charged neighbours. Moreover, it must be stressed that the PSEPT treatment of the 86 valence shell electrons in $Rh_6(CO)_{16}$ and related transition metal clusters in terms of 14 that are metal-metal bonding and 72 that are not is an electron-counting device that oversimplifies the picture (47,48) - the 72 so-called 'non-bonding' electrons may well make a small net bonding contribution to the skeletal bonding, raising the effective metal-metal bond order above 0.58.

In addition to $Rh_6(CO)_{16}$, many other 86-electron octahedral clusters are known. They include the anionic cobalt cluster $[Co_6(CO)_{14}]^{4-}$ which contains twice as many μ_3-carbonyl ligands as does $Rh_6(CO)_{16}$, (48,49) one over each octahedral face. The remaining six ligands occupy terminal sites, one on each metal atom. This difference is attributable to two factors: the tendency of the carbonyl ligands to arrange themselves as far apart as possible, and the capacity of bridging carbonyl groups to withdraw electronic charge from a metal cluster, as already discussed. Increasing the negative charge on a cluster tends to increase the proportion of bridging as opposed to terminal carbonyl groups.

The clusters $Ru_6(CO)_{18}H_2$ (50), $[Ru_6(CO)_{18}H]^-$ (51), $[Os_6(CO)_{18}H]^-$ (52), $[Os_6(CO)_{18}]^{2-}$ (52), $[Co_4Ni_2(CO)_{14}]^{2-}$ (53) and $[Ni_6(CO)_{12}]^{2-}$ (54) provide further examples of 6-atom, 86-electron clusters containing an essentially octahedral arrangement of metal atoms. Other interesting members of the same family include the carbide complexes $Ru_6(CO)_{17}C$ (55), (mesitylene)$Ru_6(CO)_{14}C$ (56), and $[Fe_6(CO)_{16}C]^{2-}$ (57), with core carbon atoms at their centres (Fig. 14). For these clusters, the skeletal electron count includes a contribution of four electrons from the core carbon atom, which clearly makes an important contribution to the skeletal bonding. Of the seven skeletal bonding MO's of an octahedral cluster (58) (Fig. 15), four have symmetries that match those

(i) The <u>closo</u> skeleton of $Ru_6(CO)_{17}C$(54)(mesitylene)$Ru_6(CO)_{17}C$ (55) and $[Fe_6(CO)_{16}C]^{2-}$ (56)

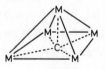

(ii) The <u>nido</u> skeleton of $Fe_5(CO)_{15}C$ (62)

Figure 14

Species with core carbon atoms

230.

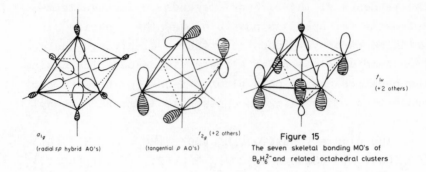

a_{1g}
(radial sp hybrid AO's)

t_{2g} (+2 others)
(tangential p AO's)

t_{1u}
(+2 others)

Figure 15
The seven skeletal bonding MO's of
$B_6H_6^{2-}$ and related octahedral clusters

of the carbon $\underline{s}$ and $\underline{p}$ AO's - the A_{1g} MO generated by an
in-phase combination of the radial AO's matches the
symmetry of the carbon 2s AO, while the set of carbon
2p AO's has symmetry T_{1u}, matching the 'π-bonding'
combination of tangential AO's. Consequently, these
four skeletal MO's will be stabilised by the core
carbon atom. Such an octahedral arrangement of six
metal atoms about a central carbon is an exceedingly
common structural feature of metal carbides (59,60),
and these clusters may be taken as molecular models
of the type of bonding that occurs in the bulk carbide,
fragments of metal carbide surrounded by carbonyl
ligands, in the same way that metal carbonyl clusters
themselves are commonly regarded as fragments of bulk
metal surrounded by carbonyl ligands (9,61,62), though
only in the case of the larger clusters is the metal
atom arrangement really like that in the bulk metal.

These metal clusters incorporating core carbon
atoms incidentally have no counterparts in boron
cluster chemistry. The boranes and carboranes prepared
to date have empty polyhedral skeletons. This is not
surprising, because the smaller closo boranes like
$B_6H_6^{2-}$ are simply not big enough to accommodate a

carbon atom (the octahedral hole has a diameter 0.414d, where d is the octahedral edge length - the hole in $B_6H_6^{2-}$ has a diameter of ca. 0.7 Å). Only for clusters like the icosahedral $B_{12}H_{12}^{2-}$ is there room for a core atom of comparable size to the polyhedral atoms, but for these larger clusters a first row element like carbon or boron, having only s and p AO's in its valence shell, can stabilise only a small proportion of the skeletal bonding MO's (transition metal atoms, by use of their d AO's, can stabilize a much higher proportion, and it is probably significant that 13-atom metal clusters incorporating core metal atoms within 12-vertex (close packed) polyhedra, such as the rhodium carbonyl hydrides $\left[Rh_{13}(CO)_{24}H_{5-n}\right]^{n-}$ (63), are known).

Core carbon atoms, again contributing all their valence shell electrons to the skeletal bonding, feature also in the pentanuclear cluster $Fe_5(CO)_{15}C$ (64) and its ruthenium and osmium analogues (65), which may be regarded as nido clusters containing five skeletal metal atoms held together by seven skeletal bond pairs (see Fig. 14). They are thus counterparts of B_5H_9, and provide further examples of structures that defy description in terms of localized 2-centre electron pair bonds, though intelligible in terms of PSEPT.

Many mixed clusters incorporating both metal and non-metal atoms in their polyhedral skeletons are also recognizable as members of the cluster family with structures clearly related to an octahedron. Representative examples, including the closo $Os_5(CO)_{15}POMe$ (66) and nido $Fe_3(CO)_9S_2$ (67) and $Fe_3(CO)_9Se_2$ (68) are listed in Table 4, which also indicates that metal carbonyl clusters provide examples of yet another class of cluster, that

TABLE 4

Some Clusters that Formally Contain 7 Skeletal Bond Pairs

No. of Skeletal atoms	Cluster Type	Shape	Examples
7	Capped closo	Capped Octahedron	$[Rh_7(CO)_{16}]^{3-}$ (69); $[Rh_7(CO)_{16}I]^{2-}$ (70); $Os_7(CO)_{21}$ (71,72)
6	Closo	Octahedron	$Rh_6(CO)_{16}$ (34); $H_2Ru_6(CO)_{18}$ (50); $[HM_6(CO)_{18}]^-$ (Ru,Os) (51,52); $[Os_6(CO)_{18}]^{2-}$ (52); $[Co_6(CO)_{14}]^{4-}$ (47,49); $[Co_4Ni_2(CO)_{14}]^{2-}$ (53); $[Ni_6(CO)_{12}]^{2-}$ (54); $Ru_6(CO)_{17}C*$ (55); $[Fe_6(CO)_{16}C]^{2-}*$ (57); $Os_5(CO)_{15}POMe$ (66); $Co_4(CO)_{10}C_2Et_2$ (75); $Ru_4(CO)_{12}C_2Ph_2$ (76); $Ir_4(CO)_5(C_8H_{12})_2(C_8H_{10})$ (77); $B_6H_6^{2-}$; $C_2B_4H_6$; CB_5H_7
6	Capped nido	Capped Square Pyramid	$H_2Os_6(CO)_{18}$ (52); $Os_6(CO)_{16}(CMe)_2^+$ (73)
5	Nido	Square Pyramid	$M_5(CO)_{15}C$(Fe,Ru,Os)* (64,65); $Fe_3(CO)_9X_2$(S,Se) (67,68); $Os_3(CO)_{10}C_2Ph_2$ (78); $Os_3(CO)_7(PPh_2)_2$ (benzyne) (79); (cyclobutadiene)$Fe(CO)_3$ (25); $B_4H_8Fe(CO)_3$ (80); $B_4H_8Co(C_5H_5)$ (81); B_5H_9; $C_2B_3H_7$; $C_5H_5^+$ (38,84-88)
4	Arachno	Square	S_4^{2+}; Se_4^{2+} (89); Te_4^{2+} (90); S_2N_2 (91); Bi_4^{2-} (92); $[C_4H_4^{2-}]$
4	Arachno	Butterfly	$Os_4(CO)_{12}H_3I$ (93); B_4H_{10}; C_4H_6(bicyclobutane) (94); $[Fe_4(CO)_{13}H]^-$ (19)

*These clusters contain core carbon atoms, not counted as a skeletal atom

†In this cluster the carbyne units CMe have been treated as external 3-electron ligands, not as skeletal units

containing a 'capped' polyhedral structure in which
one skeletal atom caps a triangular face of the
fundamental polyhedral structure. The main group of
clusters with such structures is that containing n
skeletal atoms held together by n skeletal bond pairs,
which adopt capped <u>closo</u> structures based on the
expected polyhedron (that with (n - 1) vertices) but
with the extra skeletal atom located over one of the
triangular faces where its proximity to three other
metal atoms allows it to make effective use of its
three available AO's without modifying the number of
polyhedral skeletal bonding MO's (35,69-72). The
clusters $[Rh_7(CO)_{16}]^{3-}$ (69), $[Rh_7(CO)_{16}I]^{2-}$ (70), and
$Os_7(CO)_{21}$ (71,72) thus all have the capped octahedral
skeletal structure shown in Figure 16. The capped

(i)   (ii)

Figure 16

(i) The 'capped <u>closo</u>' skeleton of $[Rh_7(CO)_{16}]^{3-}$ (64) and $Os_7(CO)_{21}$ (67)

(ii) The 'capped <u>nido</u>' skeleton of $H_2Os_6(CO)_{18}$ (51) and $Os_6(CO)_{16}(CMe)_2$ (68)

trigonal bipyramidal (<u>i.e.</u> bicapped tetrahedral)
structure of $Os_6(CO)_{18}$ (35), a six skeletal pair
cluster, has already been illustrated (Figure 8).

Table 4 also lists two compounds, $H_2Os_6(CO)_{18}$ (52)
and $Os_6(CO)_{16}(CMe)_2$ (73), for which <u>closo</u> octahedral
metal skeletons were expected, but which actually have
capped <u>nido</u> structures (Figure 16) in which one atom
caps a triangular face of the square pyramid defined
by the remainder. Another unexpected skeletal structure

has been found in the case of another osmium cluster,
$Os_5(CO)_{16}H_2$, in which an $Os(CO)_4$ unit bridges one edge
of an Os_4 tetrahedron (74). (The possibility that an
$Os(CO)_4$ unit, like an isolobal CH_2 unit (42), might
function as a divalent group supplying two electrons
and two AO's and so be more suited to an edge-bridging
than to a vertex-filling role, was noted above). The
$Os_4(CO)_{12}H_2$ residue bridged by the 2-electron donor
$Os(CO)_4$ thereby acquires the six skeletal bond pairs
needed for a tetrahedron.

Despite these few unexpected structures, there is
a wide range of substances, illustrated by the examples
in Table 4, whose shapes represent variations on the
common theme, that groups of trivalent units held
together by seven bond pairs tend to adopt structures
derivable from an octahedron. Note that, in addition
to the examples already discussed, there are several
species incorporating both metal and carbon atoms in
their polyhedral skeletons. For example, in the
acetylene complexes $Co_4(CO)_{10}C_2Et_2$ (75), $Ru_4(CO)_{12}C_2Ph_2$
(76) and $Ir_4(CO)_5(C_8H_{12})_2(C_8H_{10})$ (77), the orientation
of the acetylene with respect to the butterfly-shaped
M_4 residue is such as to complete an octahedron, albeit
severely distorted because of the different sizes of
the metal and carbon atoms (Figure 17), while the M_3C_2
skeletons of the acetylene complex $Os_3(CO)_{10}C_2Ph_2$ (78)
and benzyne complex $Os_3(CO)_7(PPh_2)_2(benzyne)$ (79)
(Figure 17) may be regarded as distorted square
pyramids in which the carbon atoms occupy basal
positions. In mixed clusters containing both metal and
carbon atoms, as in carboranes (39), the carbon atoms
tend to occupy the sites of lower coordination number.
The $\underline{nido}$, square-pyramidal MC_4 skeleton is a well-
established cluster unit in the form of η^4-cyclobuta-
diene complexes such as $C_4R_4Fe(CO)_3$ or $C_4R_4Co(C_5H_5)$ (25),

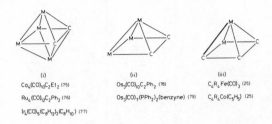

Figure 17

The skeletons of some 7 bond pair clusters with both
metal and carbon skeletal atoms

which have boron counterparts in the form of $B_4H_8Fe(CO)_3$
(80) or $B_4H_8Co(C_5H_5)$ (81), metallo derivatives of
pentaborane (the last compound has a basal metal atom
(81)). Studies of the photoelectron spectra of mixed
metal-boron clusters like $B_4H_8Fe(CO)_3$, in which
$Fe(CO)_3$ groups replace BH units of the parent boranes,
incidentally have lent support to the view that these
isolobal units do indeed have similar capacities for
cluster bonding (82,83).

   Among other entries in Table 4 worth noting are
the <u>nido</u>, non-classical structure of the carbocation
$C_5H_5^+$ (38,84-88), the <u>arachno</u> square planar structures
of species such as Se_4^{2+} (89), Te_4^{2+} (90), S_2N_2 (91)
and Bi_4^{2-} (92) and the <u>arachno</u> butterfly-shaped
structures of $Os_4(CO)_{12}H_3I$ (93) and bicyclobutane,
C_4H_6 (94) (<u>cf</u>. B_4H_{10}, Figure 11).

3.2.5. Systems Formally Containing Six Skeletal Bond Pairs

   Table 5 lists some systems that formally contain
six skeletal bond pairs, for which structures related

TABLE 5

Some Clusters that Formally Contain 6 Skeletal Bond Pairs

No of Skeletal Atoms	Cluster Type	Shape	Examples
6	Capped closo	Capped Trigonal Bipyramid	$Os_6(CO)_{18}$ (35)
5	Closo	Trigonal Bipyramid	$Os_5(CO)_{16}$ (95); $[HOs_5(CO)_{15}]^-$ (96,97); $(RSn)_2Fe_3(CO)_9$ (98); Sn_5^{2-}, Pb_5^{2-} (99); Bi_5^{3+} (100); $Fe_3(CO)_9C_2Ph_2$ (104); $C_2B_3H_5$ (105,106)
4	Nido	Tetrahedral (8)	$M_4(CO)_{12}$; $[M_4(CO)_{10}H_2]^{2-}$ (107); $M'_4(CO)_{12}H_4$ (108); $(C_5H_5)_2M_2M'_2(CO)_8$; $[Re_4(CO)_{12}H_6]^{2-}$ (109); $Co_3(CO)_9CR$ (110); $Co_2(CO)_6C_2R_2$; $(C_5H_5)_2W_2(CO)_4C_2R_2$ (111); $Fe_2(CO)_6B_2H_6$ (112); P_4; $C_4Bu^t_4$ (113)

M = Co, Rh or Ir   M' = Fe, Ru or Os

to the trigonal bipyramid are expected. The capped
closo structure (35) of $Os_6(CO)_{18}$ (Figure 8) has already
been referred to. The expected closo trigonal bipyramidal
structures are adopted by $Os_5(CO)_{16}$ (95), $[Os_5(CO)_{15}H]^-$,
(96,97), $(RSn)_2Fe_3(CO)_9$ (98) (R = $(\eta^5-C_5H_5)Fe(CO)_2$) and
$Pb_5{}^{2-}$ (99), though it is worth noting that in the first
of these, $Os_5(CO)_{16}$, which contains four $Os(CO)_3$ units
and one $Os(CO)_4$ unit (in an equatorial site), the bonds
to the $Os(CO)_4$ unit are significantly longer than those
between the $Os(CO)_3$ units, providing a further
illustration of the artificiality of regarding $Os(CO)_4$
as trivalent. The Group IV anionic clusters $Sn_5{}^{2-}$ and
$Pb_5{}^{2-}$ (99) incidentally are clearly members of a series
of closo clusters $M_n{}^{2-}$ (where M is a Group IV element)
that in principle may prove as extensive as the borane
anion series $B_nH_n{}^{2-}$, though few have yet been
characterised (99-103). Nido structures are expected
for related Group IV anions, $M_n{}^{4-}$.

The orientation of the acetylenic ligand with
respect to the triangular metal carbonyl residue in
the complex $Fe_3(CO)_9C_2Ph_2$ (104) can be rationalised as
placing one carbon atom in an equatorial position, and
one in an axial position, of a distorted trigonal
bipyramidal closo Fe_3C_2 cluster. In the carborane
$C_2B_3H_5$, which may be regarded as the parent member of
this cluster family in more senses than one - it is
the parent trigonal bipyramidal borane cluster ($B_5H_5{}^{2-}$
is unknown), and was the first carborane described in
the literature (105) - both carbon atoms occupy the
lower coordination axial sites.

By far the largest group of metal carbonyl clusters
formally held together by six skeletal bond pairs,
indeed by any number of electrons, is the group of
tetrahedral metal carbonyl clusters (8), of which there
is room to cite only a few representative formula types

and some specific examples in Table 5 (35,104-113).
These have no direct analogues in boron cluster
chemistry. The only tetrahedral boron cluster known
is B_4Cl_4, which has four skeletal bond pairs.
Tetrahedral 6 bond pair clusters can anyway be
classified in PSEPT only by the artificial device of
regarding their structures as nido fragments of
trigonal bipyramids. It is for the tetrahedron that
the '(n + 1)-rule' linking vertices (n) with skeletal
bond pairs (n + 1) breaks down (114). Unlike the
polyhedra in Figure 9, the tetrahedron requires either
n bond pairs (in A_1 and T_2 MO's) or (n + 2) bond pairs
(in A_1 + T_2 + E MO's) for a closed shell electronic
configuration. The difference from the other polyhedra arises because only for the tetrahedron does the
number of AO's used by each skeletal atom match its
number of neighbours. This is why the bonding can be
described quite satisfactorily in terms of either six
edge bonds or four face bonds, depending on the
orientation of the three valence orbitals (Fig. 13(ii))
of each cluster unit (33). For example, in the anionic
rhenium carbonyl hydride cluster $[Re_4(CO)_{12}H_6]^{2-}$, the
carbonyl ligands of each $Re(CO)_3$ unit are staggered
with respect to the edges of the tetrahedron, implying
that the three orbitals available for metal-metal
bonding, which point away from the carbonyl ligands,
are orientated over the tetrahedral edges and can
therefore form six Re-Re edge bonds which in this
cluster are believed to be protonated, i.e. they are
3-centre ReHRe edge bonds. In the related compound
$Re_4(CO)_{12}H_4$, by contrast, the $Re(CO)_3$ units are
orientated so that the carbonyl groups eclipse the
tetrahedral edges, leaving the metal-metal bonding
orbitals pointing over the faces of the Re_4 tetrahedron.
The four 3-centre Re_3 bonds so formed (this cluster has

two fewer electron pairs than $[Re_4(CO)_{12}H_6]^{2-}$ are believed to be protonated, so forming four 4-centre Re_3H face bonds (see Figure 18).

(i)   (ii)

Figure 18

Orientation of the carbonyl ligands, and implied orientation
of the skeletal AO's, in (i) $Re_4(CO)_{12}H_4$ and (ii) $[Re_4(CO)_{12}H_6]^{2-}$
(cf Fig 13(ii))

The family of clusters with six skeletal bond pairs in principle also includes triangular metal clusters, such as the trinuclear dodecacarbonyls $M_3(CO)_{12}$ of the iron subgroup metals, which might be classed as <u>arachno</u> species. Although their triangular shapes show that the structural pattern still holds for such compounds, it is unhelpful and misleading to treat them as systems containing six bond pairs, as really they contain only three. Three of the six pairs formally available for skeletal bonding are actually nonbonding. This is yet another example of the artificiality of assuming that $M(CO)_4$ is a source of 3 AO's and 4 electrons, rather than a source of 2 AO's and 2 electrons.

An orbital and electron count can be used to illustrate the problem in a more general way, and show that trinuclear clusters that formally, by PSEPT, contain six skeletal bond pairs, actually can make effective skeletal bonding use of only three of these pairs. Suppose we have a set of three cluster-forming units MX_n, each of which contains 16 electrons and so can function either as a source of 2 AO's and 2 electrons or as a source of 3 AO's and 4 electrons.

(i) MX_n as a source of 2 AO's and 2 electrons:
the basis set for skeletal bonding will consist of 6
AO's, which will therefore generate 6 MO's, in three
of which six electrons, (3 pairs), need to be accommodated.
Clearly all three pairs can be bond pairs in the case
of a triangular array of the three MX_n units.

(ii) MX_n as a source of 3 AO's and 4 electrons:
the basis set will consist of nine AO's and generate
nine MO's, in six of which accommodation will be needed
for the twelve electrons present. Clearly it is
unrealistic to expect to generate six bonding MO's from
a basis set of only 9 AO's, whatever the grouping. For
a triangular group of three MX_n units, it is realistic
to expect 3 bonding, 3 nonbonding and 3 antibonding
MO's to be generated. Filling the first six of these
leads to the same bonding description as in (i) above.

A similar cautionary note is applicable to other
arachno and hypho systems involving relatively few
skeletal atoms. Although the skeletal electron count
of Tables 2 and 3 provides an extremely useful basis
for structural classification, the number of electrons
involved in bonding the skeletal atoms together may be
significantly fewer than the number formally counted
as 'skeletal bonding electrons'. The remainder,
however, will normally be bonding the skeletal atoms
to ligands located in the general directions where
skeletal atoms would be found in more complete cluster
systems (see, e.g., the endo hydrogen atoms of B_4H_{10}
and B_5H_{11} - filled circles, Figs. 11 and 12).

For these smaller metal carbonyl clusters,
therefore, although the nido and arachno classifications
remain useful pointers to molecular shape, the bonding
between the skeletal atoms is probably best described
in terms of localised 2- or 3-centre bonds where the

symmetry allows this, or in terms of qualitative
localised MO treatments that use as their basis sets
the number of AO's each cluster unit can realistically
contribute (48,114,115).

3.3. Higher Nuclearity Metal Carbonyl Clusters

A few examples could be cited to suggest that the
analogy between boron clusters and metal carbonyl
clusters persists for higher nuclearity metal clusters.
However, most of the larger metal carbonyl clusters
known adopt structures that are not analogous to
boranes or carboranes, and no simple bonding approach
has been developed yet to rationalise existing
structures, let alone to predict structures. The
structures have been illustrated in detail in other
sections of this book, (see, e.g., Chapter 2) and will
not be reproduced again here. Many contain fragments
of close-packed metal lattices, incorporating core
metal or carbon atoms, unlike the empty polyhedra of
boranes and carboranes, and bonding rationales that
use the metal itself as a basis, and take account of
the way the site symmetry influences the bonding
potential of a skeletal atom (115), appear likely to
be helpful.

3.4. Clusters Incorporating Skeletal Carbon Atoms

One area in which PSEPT and the borane analogy

242.

remains of considerable use has already been referred
to, and this concerns mixed clusters incorporating
both metal and carbon atoms in their molecular
skeletons. Table 6 lists some systems that have
already been mentioned, and many others (6,116-120),
classified in a manner that shows how extensive this
particular family of clusters has become. Several
systems that contain exclusively carbon skeletal
atoms are included, e.g. aromatic ring systems such as
$C_4H_4^{2-}$, $C_5H_5^{-}$, C_6H_6, $C_7H_7^{+}$ are arachno members, and
their metal complexes are nido members, of this family.
A few non-aromatic hydrocarbons and carbo-cations also
have a close structural and bonding relationship to
metal clusters and to boranes and carboranes.
Figure 17 has already illustrated some organometallic
systems that formally contain seven skeletal bond
pairs. Figure 19 shows a family of systems that
formally contain eight skeletal bond pairs, including
cyclobutane as a hypho member (in which four of the
eight 'skeletal' bond pairs clearly bond endo
hydrogen atoms) (117).

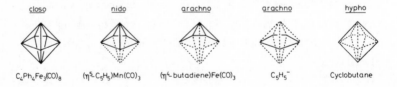

| closo | nido | arachno | arachno | hypho |

$C_4Ph_4Fe_3(CO)_8$ $(\eta^5\text{-}C_5H_5)Mn(CO)_3$ $(\eta^4\text{-butadiene})Fe(CO)_3$ $C_5H_5^{-}$ Cyclobutane

Figure 19

Some 8 skeletal bond pair systems incorporating
carbon atoms

TABLE 6

Hydrocarbon and Metal-Hydrocarbon Systems that Show Closo-, Nido- and Arachno-Relationships Like Those of Boranes (6,116)*

Fundamental Polyhedron	s	Closo Species ($n = s - 1$)	Nido species ($n = s - 2$)	Arachno species ($n = s - 3$)
trigonal bipyramid	6	$C_2Ph_2Fe_3(CO)_9$ (104)	$C_2Ph_2Co_2(CO)_6$; $C_2Ph_2Ni_2Cp_2$ $C_2R_2W_2(Cp)_2(CO)_4$ (111); $C_4Bu^t_4$ (113)	$[C_2H_4Pt(PPh_3)_3; C_3H_7^+]$
octahedron	7	$C_2Et_2Co_4(CO)_{10}$ (75) $C_2Ph_2Ru_4(CO)_{12}$ (76) $Ir_4(CO)_5(C_8H_{12})(C_8H_{10})$ (77)	(η^4-cyclobutadiene)$CoCp$ (25) (η^4-cyclobutadiene)$Fe(CO)_3$ (25); $C_2Ph_2Os_3(CO)_{10}$ (78); $C_5H_5^+$ (38,84-88)	(η^3-allyl)$Co(CO)_3$ (25); bicyclo[1.1.0]butane, C_4H_6 (94); (cyclobutadiene)$^{2-}$
pentagonal bipyramid	9	$C_4R_4Fe_3(CO)_8$ (25)	$C_5H_5Mn(CO)_3$ (25); $[C_6Me_6]^{2+}$ (120); $C_4R_4Fe_2(CO)_6$ (25)	(η^4-$CH_2:CHCH:CH_2)Fe(CO)_3$ (25); (cyclopentadienyl)$^-$
hexagonal bipyramid**	9	–	$(C_6H_6)_2Cr,Mo,W$; $(C_6H_6)_2Cr(CO)_3$ (25)	benzene
dodecahedron	9	$(EtCCH)_2Fe_4(CO)_{11}$ (118)	–	(η^3-$C_5H_5)AlMe_2$; benzvalene; hexamethylbicyclo[2.1.1]-hexenyl cation, $C_6Me_6H^+$ (119)
heptagonal bipyramid**	10	–	$C_7H_7V(CO)_3$	cycloheptatrienyl cation

*n = No. of skeletal atoms, s = No. of skeletal electron pairs

**The hexagonal bipyramid and heptagonal bipyramid are alternative shapes to the dodecahedron and tricapped trigonal prism compatible with 9 and 10 skeletal bond pairs respectively

4. TRANSITION METAL HALIDE CLUSTERS

The octahedral metal clusters formed by niobium, tantalum, molybdenum and tungsten in their lower halide chemistry differ from the octahedral metal carbonyls already discussed in that for these systems it is more realistic to treat the skeletal bonding as involving four AO's per metal atom.

The two main structural types (Figure 20) (121-125)

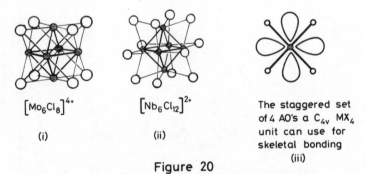

$\left[Mo_6Cl_8\right]^{4+}$

(i)

$\left[Nb_6Cl_{12}\right]^{2+}$

(ii)

The staggered set of 4 AO's a C_{4v} MX_4 unit can use for skeletal bonding

(iii)

Figure 20

Structures of octahedral metal-halide clusters, and AO's available for skeletal bonding.

contain M_6X_8 and M_6X_{12} units. The former, exemplified
by $Mo_6Cl_8^{4+}$, has a μ_3-halogen atom X located over the
centre of each of the eight octahedral faces. The
latter, exemplified by $Nb_6Cl_{12}^{2+}$, has twelve μ_2 halogen
atoms located over the centres of the twelve octahedral
edges. Both types have a flattened square pyramidal
arrangement of halogen atoms about each metal atom,
which contributes four appropriate orbitals (e.g. $\underline{dsp}^2$
hybrid AO's - the metal atom is not far out of the
plane of the four attached halogen atoms). One further
metal AO (e.g. a $\underline{pd}$ hybrid AO) points radially outward,
away from the centre of the cluster, and is available
for bonding the $\underline{exo}$ ligand normally associated with
these clusters, commonly a further halogen or pseudo-
halogen. There remain four AO's pointing in the
general direction of the other cluster atoms. They
can be represented by a set of hybrid AO's, staggered
with respect to the four halogen atoms of the MX_4 unit
(Fig. 20(iii)). The number of skeletal bonding MO's
they generate (twelve for $Mo_6Cl_8^{4+}$, eight for $Nb_6Cl_{12}^{2+}$)
reflects their orientation which in turn reflects the
orientation of the C_{4v} MX_4 units with respect to the
4-fold axes of the octahedra. An electron count for
the units Mo_6^{12+} and Nb_6^{14+}, which formally constitute
the metal cores of these clusters, shows them to
contain 12 and 8 skeletal bond pairs, respectively, as
appropriate for closed shell conformations.

   Since an octahedron has twelve edges and eight
faces, it is possible to describe the metal-metal
bonding in these species in terms of localised 2- or
3-centre bonds, using twelve 2-centre octahedral edge
bonds for $Mo_6Cl_8^{4+}$ and eight 3-centre octahedral face
bonds for $Nb_6Cl_{12}^{2+}$. This does not mean that the
molybdenum cluster is held together more strongly than
the niobium cluster, because if one takes account of

the contribution to skeletal bonding made by the edge-
or face-bridging chloride ions, the two types of
cluster are seen to be effectively isoelectronic.
Each contains 20 electron pairs associated with the
metal-metal and metal-halogen bonding, occupying MO's
of symmetries $A_{1g}(2)$, A_{2u}, E_g, $T_{1u}(2)$, $T_{2g}(2)$ and T_{2u}
(122).

An important difference between the chloride
ligands in these clusters and the carbonyl ligands in
metal carbonyl clusters is worth noting. Whereas
carbonyl groups that coordinate only through their
carbon atoms act invariably as 2-electron ligands,
regardless of whether they occupy bridging or terminal
sites, the numbers of electrons contributed by halide
ligands vary with the mode of coordination (2 for a
terminal X^-, 4 for a μ_2-X^-, 6 for a μ_3-X^-).

The correspondence between the number of AO's
available for skeletal bonding on each skeletal atom
and the number of neighbouring skeletal atoms (4) in
these halide clusters, allowing 2-centre edge bond or
3-centre face bond descriptions, is reminiscent of the
tetrahedral metal carbonyl clusters $[Re_4(CO)_{12}H_6]^{2-}$
and $Re_4(CO)_{12}H_2$ discussed above (33,109). Just as the
tetrahedron is the only deltahedron in which all
vertices have three neighbours, so is the octahedron
the only deltahedron in which all vertices have 4
neighbours.

5. CLUSTERS FORMED BY THE COPPER GROUP METALS

Two other categories of cluster compound that are
formed by transition metals and are worthy of brief
discussion are those formed by copper and gold. In
both cases, the type of cluster formed reflects the
tendency of these metals not to use all the nine
orbitals available in their valence shells, because
the relatively large energy gap between the valence
shell s and p orbitals makes the latter less accessible
for bonding, particularly in low oxidation state
derivatives. A reduced tendency to use all nine
valence shell orbitals is a familiar feature of the
metals towards the right of the transition series,
particularly the heavier ones, where '16-electron'
species are more usual than '18-electron' species
among their mononuclear organometallic derivatives, for
example (25), and where nido-type structures may be
found for metallo-boranes or carboranes where the
electron count (assuming all nine orbitals to be used)
might have suggested a closo structure (126).

5.1. Copper Clusters

The few copper clusters that have been structurally
characterised are organo derivatives in which the
metal is formally in the +1 oxidation state, and in
which the skeleton is held together by 'electron
deficient' (i.e. 3 centre 2 electron) bridge bonding
by alkyl or aryl groups rather than by direct
unassisted metal-metal bonding.

For example, the trimethylsilylmethyl derivative
$(CuCH_2SiMe_3)_4$ (127) crystallises as the tetramer in
which a square planar arrangement of metal atoms is
held together by bridging silylmethyl groups (Figure
21). The bond angles (CuCCu 74°, CCuC 164°) are
consistent with the bonding interpretation shown -
linearly (sp) hybridised copper atoms involved in 3-
centre Cu_2C bonds (the valence shell d AO's in such a
copper(I) derivative are expected to be completely
filled). The Cu$\cdots$Cu distance of 2.42 Å is short
enough to allow significant overlap of the metal
orbitals with the carbon sp^3 hybrid AO, but is not

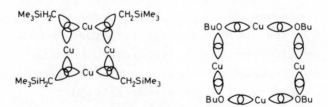

Figure 21

The tetrameric copper(I) derivatives $(CuCH_2SiMe_3)_4$ and
$(CuOBu^t)_4$

regarded as indicative of direct metal-metal bonding. If the metal atoms had been able to make use of more AO's for cluster bonding, a more compact structure, like the tetrahedral Li_4 or folded chair-shaped Li_6 arrangements present in lithium alkyls (128) might have been expected. A similar roughly square planar arrangement of four copper atoms is to be found in the t-butoxide, $(CuOBu^t)_4$ (129) (Fig. 21), though with a greater $Cu\cdots\cdots Cu$ distance (2.71 Å), as expected for an electron precise (as opposed to electron deficient) bridge (in an electron deficient bridge, both metal orbitals overlap the same bridge atom orbital, so requiring them to be closer together than in an electron precise bridge, where the bridging atom supplies two AO's and where there is no net metal$\cdots\cdots$ metal bonding (36,130).

Incorporating a Lewis base unit in the bridging organic group of an organocopper(I) tetramer induces trigonal hybridisation ($\underline{sp}^2$) of the metal, and consequent puckering of the Cu_4 ring, as has been found in the p-tolylcopper(I) derivative with a dimethylaminomethyl substituent in the $\underline{ortho}$ position, $[CuC_6H_3(Me)(CH_2NMe_2)]_4$ (Fig. 22) (131). The changed

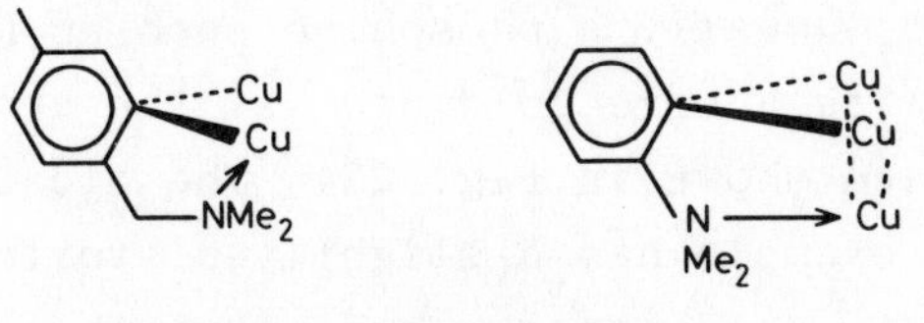

Figure 22

(i) The edge-bridging dimethylaminomethylphenyl ligand of the tetramer $[CuC_6H_3Me(CH_2NMe_2)]_4$ (131) which contains a butterfly-shaped Cu_4 skeleton.

(ii) The face-bridging dimethylaminophenyl ligand of the Cu_6 clusters $(CuC_6H_4NMe_2)_4(CuX)_2$ (132,133) which contain distorted octahedral Cu_6 skeletons.

metal hybridisation in this compound is reflected in
its smaller Cu····Cu distance (2.38 Å). Use of the
related 2-dimethylaminophenyl ligand (Fig. 22) has
allowed the isolation and structural characterisation
of two distorted octahedral Cu_6 clusters, of formulae
$[Cu(2-Me_2NC_6H_4)]_4 \cdot (CuBr)_2$ (132) and $[Cu(2-Me_2NC_6H_4)]_4 \cdot$
$[(CuC\equiv C(p-tolyl)]_2$ (133). In these, the phenyl rings
form electron deficient bridges between pairs of
copper atoms, while the dimethylamino substituent
coordinates to a third copper atom. The bromine atoms
(132) or acetylide residues (133) perform edge-bridging
roles. Since these clusters are held together entirely
by the bridging capacity of their ligands, they clearly
differ markedly from the octahedral metal halide and
metal carbonyl clusters already discussed.

5.2. Gold Clusters

Gold clusters differ from copper clusters in that,
at least for the species isolated so far, bridging of
the metal atoms by ligands does not account for their
existence. They consist of cationic or neutral
aggregates of gold atoms, each with one terminally
attached ligand, usually a phosphine such as PPh_3, in
several cases with a core gold atom at the centre.
Some examples are shown in Fig. 23. The dication
$[AuPR_3]_6^{2+}$ for example has a slightly distorted octahedral set of six gold atoms, each of which bears a
phosphine ligand held by a bond pointing radially
outwards from the centre of the cluster (134).
Similar radially-orientated bonds to one ligand on
each peripheral atom also characterise the structure
of the triply-charged cation $[Au(AuPR_3)_8]^{3+}$ (135) and

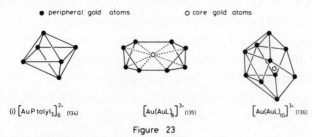

(i) $\left[\text{AuPtolyl}_3\right]_6^{2+}$ (134) $\left[\text{Au(AuL)}_8\right]^{3+}$ (135) $\left[\text{Au(AuL)}_{10}\right]^{3+}$ (136)

Figure 23

The metal atom arrangements in some cationic
gold clusters

neutral $\left[\text{Au(AuPR}_3)_7(\text{AuI})_3\right]$ (136), although these
latter clusters have core gold atoms at their centres.
The arrangements of the peripheral gold atoms appear
reminiscent of the polyhedral fragment structures of
many neutral boranes - both centred clusters bear
close relationships to centred icosahedra, with four
or two vertices unoccupied respectively. However, the
structures are not those that might have been inferred
by PSEPT, and it is evidently unrealistic to expect the
gold atoms in these clusters to make use of all their
valence shell AO's because of the large s-p energy gap.
A useful basis for understanding their bonding is to
regard each AuL unit (where L is a 2-electron donor,
e.g. PR_3) as a source of one electron and one sp hybrid
orbital for skeletal bonding (137). Consistent
with this interpretation, the distance between the
central gold atom and the peripheral atoms is
significantly less than the distance between peripheral
atoms, though the higher formal oxidation state of the
central gold atom (138) should not be ignored in this
connection. Mössbauer spectroscopic studies (139)
support the suggestion of essentially linear hybridisation
at the gold atoms, though the manner in which they

group together is suggestive of a significant bonding interaction between the peripheral atoms.

Closed-shell electronic configurations are indicated by extended Hückel MO treatments of the centred clusters (137), but the electronic configuration of the distorted octahedral $(AuPR_3)_6^{2+}$ (it is flattened along one 3-fold axis) remains something of a puzzle. Under octahedral symmetry, six $AuPR_3$ units each furnishing one radially orientated sp hybrid AO would generate only one bonding MO, the fully in phase A_{1g} combination of all six AO's (Fig. 15). The remaining five MO's consist of a set of three T_{1u} MO's (c.f. Fig. 15), which in the absence of a stabilising contribution from tangential p AO's would be feebly antibonding, and a pair of E_g antibonding MO's. There is thus neither obvious accommodation, in a closed shell, for the four available skeletal bonding electrons, nor is the distortion (compression along a 3-fold axis) obviously that required by the partial occupancy of the T_{1u} MO's. It is possible that not all the atoms present were located. It has been pointed out that a regular octahedral cluster $[Au_6(PR_3)_6C]^{2+}$, with a core carbon supplying four more electrons and stabilising the occupied A_{1g} and T_{1u} orbitals, would have a closed shell electronic configuration (137).

6.   CONCLUSION

This survey has been concerned with simple ways
of describing the bonding in transition metal clusters.
In it, the important and varied role played by the
ligands has been noted, the scope and limitations of
2-centre bonding descriptions of the skeletal bonds
have been discussed, and the advantage has been argued
of looking at clusters as built up from components
within which the bonding is relatively clear cut, so
making clear the number of AO's and electrons such
units can supply for skeletal bonding use. Transition
metal carbonyl clusters include a large number of
systems in which the component units supply three AO's
apiece, and in this connection there is a clear
relationship between many intermediate-sized metal
carbonyl clusters and borane clusters. Indeed, there
appears to be a continuum of such clusters that
includes also many mixed clusters, organometallic and
organic systems, as well as various other main group
metal and metalloid systems. At the left of the
transition series, the metal halide clusters formed
by the niobium and molybdenum subgroup metals, in
contrast to metal carbonyl clusters, provide examples
of systems where the building units supply four AO's

apiece, while at the right of the transition series, clustering entails the use of fewer AO's, essentially only one or two per cluster unit. These changes across the transition metals arise in large part because the valence shell, and particularly the d orbitals, are becoming progressively more completely filled, though the changing relative energy levels of the $\underline{d}$, $\underline{s}$ and $\underline{p}$ AO's (which are in turn sensitive to the ligands present) also play their part.

Ultimately, a satisfactory understanding of the structures and bonding of transition metal clusters can only come from detailed structural and theoretical studies. Elegant examples of the ways that thorough and systematic studies of series of closely related systems, differing only in the numbers of electrons they contain, can provide a qualitative picture of the MO ordering in cluster systems are provided by the studies of Dahl and his co-workers on 3- (140-143) or 4- (144-147) metal atom systems. Elsewhere, extended Hückel (115,148), topological (149), graph theoretical (150) and SCF-X$_\alpha$ (151) theoretical approaches have been used to consider known and hypothetical cluster systems and to explore the extent to which metal clusters reproduce the properties of metal surfaces (9,151-153). Although such studies have shown that simplistic bonding descriptions should only be used with caution, simple ways of correlating molecular shapes and electron numbers appear likely to remain useful.

7. REFERENCES

1. E. W. Abel and F. G. A. Stone, Quart. Rev. Chem. Soc. (London), 23, 325 (1969).
2. R. D. Johnston, Advan. Inorg. Chem. Radiochem., 13, 471 (1970).
3. R. B. King, Progr. Inorg. Chem., 15, 287 (1972).
4. D. L. Kepert and K. Vrieze, Comprehensive Inorg. Chem. (ed. A. F. Trotman-Dickenson) 4, 197, Pergamon, Oxford (1973).
5. K. Wade, Advan. Inorg. Chem. Radiochem., 18, 1 (1976).
6. P. Chini, G. Longoni and V. G. Albano, Advan. Organometal. Chem., 14, 285 (1976).
7. P. Chini and B. T. Heaton, Topics in Current Chem., 71, 1 (1977).
8. E. L. Muetterties, T. N. Rhodin, E. Band, C. F. Brucker and W. R. Pretzer, Chem. Rev., 79, 91 (1979).
9. M. J. Bennett and R. Mason, Proc. Chem. Soc., 273 (1963).
10. I. H. Hillier, M. F. Guest, B. R. Higginson and D. R. Lloyd, Molecular Physics, 27, 215 (1973).

256.

11.   E. J. Baerends and P. Ros, Molecular Physics, <u>30</u>,
          1735 (1975).

12.   J. B. Johnson and W. G. Klemperer, J. Amer. Chem.
          Soc., <u>99</u>, 7132 (1977).

13.   J. C. Green, E. A. Seddon and D. M. P. Mingos,
          J. Chem. Soc. Chem. Comm., 94 (1979).

14.   P. S. Braterman, Structure and Bonding, <u>10</u>, 57
          (1972); <u>26</u>, 1 (1976).

15.   J. Evans, Advan. Organometal. Chem., <u>16</u>, 319 (1977).

16.   J. A. Connor, Topics in Current Chem., <u>71</u>, 71 (1977).

17.   C. E. Housecroft, K. Wade and B. C. Smith,
          J. Organometal. Chem., <u>170</u>, C1 (1979).

18.   C. J. Commons and B. F. Hoskins, Austral. J. Chem.,
          <u>28</u>, 1663 (1975).

19.   M. Manassero, M. Sansoni and G. Longoni, J. Chem.
          Soc. Chem. Comm., 919 (1976).

20.   S. W. Ulmer, P. M. Skarstad, J. M. Burlitch and
          R. E. Hughes, J. Amer. Chem. Soc., <u>95</u>, 4469
          (1973).

21.   N. E. Kim, N. J. Nelson and D. F. Shriver, Inorg.
          Chim. Acta, <u>7</u>, 392 (1973).

22.   D. F. Shriver, J. Organometal. Chem., <u>94</u>, 259 (1975).

23.   J. S. Kristoff and D. F. Shriver, Inorg. Chem., <u>13</u>,
          499 (1974).

24.   H. A. Hodali, D. F. Shriver and C. A. Ammlung,
          J. Amer. Chem. Soc., <u>100</u>, 5239 (1978).

25.   G. E. Coates, M. L. H. Green and K. Wade,
          "Organometallic Compounds", 3rd Edn. Vol. 2,
          "The Transition Metals", Methuen, (1968).

26.   C. A. Tolman, Chem. Soc. Rev., <u>1</u>, 337 (1972).

27.   F. A. Cotton and J. M. Troup, J. Amer. Chem. Soc.,
          <u>96</u>, 4155 (1974).

28.   R. Mason and A. I. M. Rae, J. Chem. Soc. (A), 778
          (1968).

29.   E. R. Corey and L. F. Dahl, Inorg. Chem., <u>1</u>, 521 (1962)

30. A. G. Orpen, A. V. Rivera, E. G. Bryan, D. Pippard,
    G. M. Sheldrick and K. D. Rouse, J. Chem.
    Soc. Chem. Comm., 723 (1978).

31. R. D. Wilson and R. Bau, J. Amer. Chem. Soc., 98,
    4687 (1976).

32. G. Huttner and H. Lorenz, Chem. Ber., 107, 996
    (1974).

33. R. Hoffmann, B. E. R. Schilling, R. Bau, H. D. Kaesz
    and D. M. P. Mingos, J. Amer. Chem. Soc.,
    100, 6088 (1978).

34. E. R. Corey, L. F. Dahl and W. Beck, J. Amer.
    Chem. Soc., 85, 1202 (1963).

35. R. Mason, K. M. Thomas and D. M. P. Mingos, J. Amer.
    Chem. Soc., 95, 3802 (1973).

36. K. Wade, J. Chem. Soc. Chem. Comm., 792 (1971);
    Electron Deficient Compounds, Nelson, London
    (1971).

37. D. M. P. Mingos, Nature (London) Phys. Sci., 236,
    99 (1972).

38. R. E. Williams, Inorg. Chem., 10, 210 (1971).

39. R. E. Williams, Advan. Inorg. Chem. Radiochem.,
    18, 67 (1976).

40. R. Snaith and K. Wade, MTP Internat. Rev. Sci.,
    Inorg. Chem. Ser. 1, 1, 139 (1972), Ser. 2, 1,
    95 (1975).

41. K. P. Callahan and M. F. Hawthorne, Advan. Organo-
    metal. Chem., 14, 145 (1976).

42. M. Elian and R. Hoffmann, Inorg. Chem., 14, 1058
    (1975).

43. M. Elian, M. M. L. Chen, D. M. P. Mingos and
    R. Hoffmann, Inorg. Chem., 15, 1148 (1976).

44. D. M. P. Mingos, Advan. Organometal. Chem., 15, 1
    (1977).

45. K. Wade, Inorg. Nucl. Chem. Letters, 14, 71 (1978).

258.

46. C. E. Housecroft, K. Wade and B. C. Smith, J. Chem.
    Soc. Chem. Comm., 765 (1978).

47. D. M. P. Mingos, J. Chem. Soc. Dalton, 133 (1974).

48. J. W. Lauher, J. Amer. Chem. Soc., 100, 5305 (1978).

49. V. Albano, P. L. Bellon, P. Chini and V. Scatturin,
    J. Organometal. Chem., 16, 461 (1969)

50. M. R. Churchill and J. Wormald, J. Amer. Chem. Soc.,
    93, 5670 (1971).

51. C. R. Eady, B. F. G. Johnson, J. Lewis, M. C. Malatesta,
    P. Machin and M. McPartlin, J. Chem. Soc.
    Chem. Comm., 945 (1976).

52. M. McPartlin, C. R. Eady, B. F. G. Johnson and
    J. Lewis, J. Chem. Soc. Chem. Comm., 883
    (1976).

53. V. G. Albano, G. Ciani and P. Chini, J. Chem. Soc.
    Dalton, 432 (1974).

54. J. C. Calabrese, L. F. Dahl, A. Cavalieri, P. Chini,
    G. Longoni and S. Martinengo, J. Amer. Chem.
    Soc., 96, 2616 (1974).

55. A. Sirigu, M. Bianchi and E. Benedetti, Chem.
    Comm., 596 (1969).

56. R. Mason and W. R. Robinson, Chem. Comm., 468
    (1968).

57. M. R. Churchill and J. Wormald, J. Chem. Soc.
    Dalton, 2410 (1974).

58. H. C. Longuet-Higgins, Quart. Rev. Chem. Soc.
    (London), 11, 121 (1957).

59. R. E. Rundle, Acta Crystallogr., 1, 180 (1948).

60. W. A. Frad, Advan. Inorg. Chem. Radiochem., 11,
    153 (1968).

61. E. L. Muetterties, Bull. Soc. Chim. Belg., 84,
    959 (1975); 85, 451 (1976).

62. R. Ugo, Catal. Rev., 11, 225 (1975).

63. S. Martinengo, B. T. Heaton, R. J. Goodfellow and
    P. Chini, J. Chem. Soc. Chem. Comm., 39 (1977).

64. E. H. Braye, L. F. Dahl, W. Hübel and
     D. L. Wampler, J. Amer. Chem. Soc., 84,
     4633 (1962).

65. C. R. Eady, B. F. G. Johnson, J. Lewis and
     T. Matheson, J. Organometal. Chem., 57,
     C82 (1973).

66. J. M. Fernandez, B. F. G. Johnson, J. Lewis and
     P. R. Raithby, J. Chem. Soc. Chem. Comm.,
     1015 (1978).

67. L. F. Dahl and P. W. Sutton, Inorg. Chem., 2,
     1067 (1963).

68. C. H. Wei and L. F. Dahl, Inorg. Chem., 4, 493
     (1965).

69. V. G. Albano, P. L. Bellon and G. F. Ciani, Chem.
     Comm., 1024 (1969).

70. V. G. Albano, G. Ciani, S. Martinengo, P. Chini
     and G. Giordano, J. Organometal. Chem., 88,
     381 (1975).

71. C. R. Eady, B. F. G. Johnson and J. Lewis, J. Chem.
     Soc. Dalton, 2606 (1975).

72. C. R. Eady, B. F. G. Johnson, J. Lewis, R. Mason,
     P. B. Hitchcock and K. M. Thomas, J. Chem.
     Soc. Chem. Comm.. 385 (1977).

73. C. R. Eady, J. M. Fernandez, B. F. G. Johnson,
     J. Lewis, P. R. Raithby and G. M. Sheldrick,
     J. Chem. Soc. Chem. Comm., 421 (1978).

74. J. J. Guy and G. M. Sheldrick, Acta Cryst., B34,
     1725 (1978).

75. L. F. Dahl and D. L. Smith, J. Amer. Chem. Soc.,
     84, 2450 (1962).

76. B. F. G. Johnson, J. Lewis, B. E. Reichert,
     K. T. Schorpp and G. M. Sheldrick, J. Chem.
     Soc. Dalton, 1417 (1977).

77. G. F. Stuntz, J. R. Shapley and C. G. Pierpoint,
     Inorg. Chem., 17, 2596 (1978).

78. C. G. Pierpoint, Inorg. Chem., 16, 636 (1977).

79. C. W. Bradford, R. S. Nyholm, G. J. Gainsford,
    J. M. Guss, P. R. Ireland and R. Mason,
    J. Chem. Soc. Chem. Comm., 87 (1972).

80. N. N. Greenwood, C. G. Savory, R. N. Grimes,
    L. G. Sneddon, A. Davison and S. S. Wreford,
    J. Chem. Soc. Chem. Comm., 718 (1974).

81. L. G. Sneddon and D. Voet, J. Chem. Soc. Chem.
    Comm., 118 (1976).

82. J. A. Ulman and T. P. Fehlner, J. Amer. Chem.
    Soc., 100, 449 (1978).

83. J. A. Ulman, E. L. Andersen and T. P. Fehlner,
    J. Amer. Chem. Soc., 100, 456 (1978).

84. W.-D. Stohrer and R. Hoffmann, J. Amer. Chem. Soc.,
    94, 1660 (1972).

85. S. Masamune, M. Sakai and H. Ona, J. Amer. Chem.
    Soc., 94, 8955 (1972).

86. S. Masamune, M. Sakai, H. Ona and A. J. Jones,
    J. Amer. Chem. Soc., 94, 8956 (1972).

87. M. J. S. Dewar and R. C. Haddon, J. Amer. Chem.
    Soc., 95, 5836 (1973).

88. W. J. Hehre and P. von R. Schleyer, J. Amer. Chem.
    Soc., 95, 5837 (1973).

89. I. D. Brown, D. B. Crump, R. J. Gillespie and
    D. P. Santry, Chem. Comm., 853 (1968).

90. T. W. Couch, D. A. Lokken and J. D. Corbett,
    Inorg. Chem., 11, 357 (1972).

91. C. M. Mikulski, P. J. Russo, M. S. Saran,
    A. G. MacDiarmid, A. F. Garito and
    H. J. Heeges, J. Amer. Chem. Soc., 97, 6358
    (1975).

92. A. Cisar and J. D. Corbett, Inorg. Chem., 16,
    2482 (1977).

93. B. F. G. Johnson, J. Lewis, P. R. Raithby,
    G. M. Sheldrick, K. Wong and M. McPartlin,
    J. Chem. Soc. Dalton, 673 (1978).

94. K. W. Cox and M. D. Harmony, J. Chem. Phys., $\underline{50}$, 1976 (1969).

95. B. F. G. Johnson, C. R. Eady, J. Lewis,
    B. E. Reichert and G. M. Sheldrick, J. Chem.
    Soc. Chem. Comm., 271 (1976).

96. C. R. Eady, J. J. Guy, B. F. G. Johnson, J. Lewis,
    M. C. Malatesta and G. M. Sheldrick, J. Chem.
    Soc. Chem. Comm., 807 (1976).

97. J. J. Guy and G. M. Sheldrick, Acta Cryst., $\underline{B34}$, 1722 (1978).

98. T. J. McNeese, S. S. Wreford, D. L. Tipton and
    R. Bau, J. Chem. Soc. Chem. Comm., 390 (1977).

99. J. D. Corbett and P. A. Edwards, J. Chem. Soc.
    Chem. Comm., 984 (1975); Inorg. Chem., $\underline{16}$, 903 (1977).

100. J. D. Corbett, Inorg. Chem., $\underline{7}$, 198 (1968).

101. J. D. Corbett and P. A. Edwards, J. Amer. Chem.
    Soc., $\underline{99}$, 3313 (1977).

102. C. H. E. Belin, J. D. Corbett and A. Cisar, J.
    Amer. Chem. Soc., $\underline{99}$, 7163 (1977).

103. R. W. Rudolph, W. L. Wilson, F. Parker, R. C. Taylor
    and D. C. Young, J. Amer. Chem. Soc., $\underline{100}$, 4629 (1978).

104. J. F. Blount, L. F. Dahl, C. Hoogzand and
    W. Hübel, J. Amer. Chem. Soc., $\underline{88}$, 292 (1966).

105. I. Shapiro, C. D. Good and R. E. Williams, J.
    Amer. Chem. Soc., $\underline{84}$, 3837 (1962).

106. E. A. McNeill, K. L. Gallaher, F. R. Scholer and
    S. H. Bauer, Inorg. Chem., $\underline{12}$, 2108 (1973).

107. V. G. Albano, G. Ciani, M. Manassero, F. Canziani,
    G. Giordano, S. Martinengo and P. Chini,
    J. Organometal. Chem., $\underline{150}$. C17 (1978).

108. R. D. Wilson, S. M. Wu, R. A. Love and R. Bau,
    Inorg. Chem., $\underline{17}$, 1271 (1978).

109. H. D. Kaesz, B. Fontal, R. Bau, S. W. Kirtley and
     M. R. Churchill, J. Amer. Chem. Soc., 91,
     1021 (1969).

110. B. R. Penfold and B. H. Robinson, Accounts Chem.
     Res., 6, 73 (1973)

111. D. S. Ginley, C. R. Bock, M. S. Wrighton,
     B. Fischer, D. L. Tipton and R. Bau,
     J. Organometal. Chem., 157, 41 (1978).

112. E. L. Andersen and T. P. Fehlner, J. Amer. Chem.
     Soc., 100, 4606 (1978).

113. G. Maier, S. Pfriem, U. Schäfer and R. Matusch,
     Angew. Chem. Internat. Edn., 17 520 (1978).

114. C. Glidewell, Inorg. Nucl. Chem. Letters, 11,
     761 (1975).

115. J. W. Lauher, J. Amer. Chem. Soc., 101, 2604 (1979).

116. K. Wade, Inorg. Nucl. Chem. Letters, 8, 563 (1972).

117. C. E. Housecroft and K. Wade, Tetrahedron Letters,
     3175 (1979).

118. E. Sappa, A. Tiripicchio and M. T. Camellini,
     J. Chem. Soc. Dalton, 419 (1978).

119. H. Hogeveen and P. W. Kwant, J. Amer. Chem. Soc.,
     95, 7315 (1973).

120. H. Hogeveen and P. W. Kwant, J. Amer. Chem. Soc.,
     96, 2208 (1974).

121. F. A. Cotton and T. E. Haas, Inorg. Chem., 3, 10
     (1964).

122. S. F. A. Kettle, Theoret. Chim. Acta, 3, 211 (1965).

123. W. F. Libby, J. Chem. Phys., 46, 399 (1967).

124. H. Müller, J. Chem. Phys., 49, 475 (1968).

125. R. F. Schneider and R. A. Mackay, J. Chem. Phys.,
     48, 843 (1968).

126. F. G. A. Stone, J. Organometal. Chem., 100, 257
     (1975).

127. J. A. J. Jarvis, B. T. Kilbourn, R. Pearce and
     M. F. Lappert, J. Chem. Soc. Chem. Comm.,
     475 (1973).

128. B. J. Wakefield, "Organolithium Compounds",
Pergamon, Oxford (1974).

129. T. Greiser and E. Weiss, Chem. Ber., 109, 3142
(1976).

130. R. Mason and D. M. P. Mingos, J. Organometal.
Chem., 50, 53 (1973).

131. G. van Koten and J. G. Noltes, J. Organometal.
Chem., 84, 129 (1975).

132. J. M. Guss, R. Mason, K. M. Thomas, G. van Koten
and J. G. Noltes, J. Organometal. Chem., 40,
C79 (1972).

133. R. W. M. ten Hoedt, J. G. Noltes, G. van Koten
and A. L. Spek, J. Chem. Soc. Dalton, 1800
(1978).

134. P. Bellon, M. Manassero and M. Sansoni, J. Chem.
Soc. Dalton, 2423 (1973).

135. P. Bellon, F. Cariati, M. Manassero, L. Naldini
and M. Sansoni, Chem. Comm., 1423 (1971).

136. P. Bellon, M. Manassero and M. Sansoni, J. Chem.
Soc. Dalton, 1481 (1972).

137. D. M. P. Mingos, J. Chem. Soc. Dalton, 1163 (1976).

138. C. Battistoni, G. Mattogno, F. Cariati, L. Naldini
and A. Sgamellotti, Inorg. Chim. Acta., 24,
207 (1977).

139. F. A. Vollenbroek, P. C. P. Bouten, J. M. Trooster,
J. P. van den Berg and J. J. Bour, Inorg.
Chem., 17, 1345 (1978).

140. C. E. Strouse and L. F. Dahl, Discuss. Farad. Soc.,
47, 93 (1969).

141. D. L. Stevenson, C. H. Wei and L. F. Dahl, J. Amer.
Chem. Soc., 93, 6027 (1971).

142. L. F. Dahl and C. E. Strouse, J. Amer. Chem. Soc.,
93, 6032 (1971).

143. P. D. Frisch and L. F. Dahl, J. Amer. Chem. Soc.,
94, 5082 (1972).

264.

144. Trinh-Toan, W. P. Fehlhammer and L. F. Dahl,
     J. Amer. Chem. Soc., $\underline{94}$, 3389 (1972);
     $\underline{99}$, 402 (1977).

145. G. L. Simon and L. F. Dahl, J. Amer. Chem. Soc.,
     $\underline{95}$, 2164, 2175 (1973).

146. R. S. Gall, C. T. Chu and L. F. Dahl, J. Amer.
     Chem. Soc., $\underline{96}$, 4019 (1974).

147. Trinh-Toan, B. K. Teo, J. A. Ferguson, T. J. Meyer
     and L. F. Dahl, J. Amer. Chem. Soc., $\underline{99}$,
     408 (1977).

148. R. C. Baetzold and R. E. Mack, J. Chem. Phys.,
     $\underline{62}$, 1513 (1975).

149. R. B. King, J. Amer. Chem. Soc., $\underline{94}$, 95 (1972).

150. R. B. King and D. H. Rouvray, J. Amer. Chem.
     Soc., $\underline{99}$, 7834 (1977).

151. R. P. Mesmer, S. K. Knudson, K. H. Johnson,
     J. B. Diamond and C. Y. Yang, Phys. Rev.
     Sect. B., $\underline{13}$, 1396 (1976).

152. A. B. Anderson and R. Hoffmann, J. Chem. Phys.,
     $\underline{61}$, 4545 (1974).

153. L. W. Anders, R. S. Hanson and L. S. Bartell,
     J. Chem. Phys., $\underline{62}$, 1641 (1975).

Transition Metal Clusters
Edited by Brian F.G. Johnson
© 1980 John Wiley & Sons Ltd.

CHAPTER 4

Cubane Clusters

C. D. Garner

Department of Chemistry, The University of Manchester, Manchester, U.K.

1. INTRODUCTION

The parent molecule from which this class of complexes derive their title is the hydrocarbon cubane, C_8H_8, the molecules of which, Figure 1, consist of a cube of carbon atoms enclosed within a cube of hydrogen atoms (1). A convenient synthesis of this

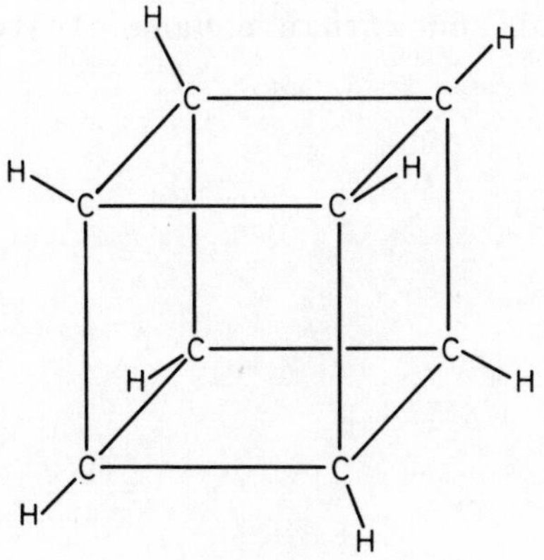

Figure 1

Molecular structure of cubane [2]

molecule from $[Fe(\eta^4-C_4H_4)(CO)_3]$ has been described
(2). Compounds which contain a cubane 'cluster' as
an integral part of their molecular architecture,
constitute an identifiable and interesting if diverse,
class of substances. Some of these compounds contain
a sub-unit which meets the specifications (3) often
used to define a metal cluster, however, many of them
do not and some of them, including cubane itself,
involve no metal atoms at all. Furthermore, the
cubane frameworks of the members of this class lack
the regularity and/or high symmetry of the parent
molecule. Nevertheless, the structural relationships
which exist between cubane 'clusters' make it
attractive to discuss all of these systems together.

Subsequent to a consideration of certain
structural aspects of the cubane framework, this
review presents a general survey of compounds of this
class, organised according to the group of the
Periodic Table to which the metal of the cluster
belongs. The concluding section discusses some
bonding aspects of and within cubane clusters.

2.   STRUCTURAL ASPECTS OF THE CUBANES

The polyhedron of cubane (Fig. 1) has full O_h
point symmetry but, as virtually all of the cubane
'clusters' have atoms of two different elements (A
and B) at alternate corners of the cube, Figure 2(a),
the highest molecular symmetry which can be obtained
is $\underline{T}_d$. The A_4 and B_4 arrangements each constitute a
tetrahedron and thus the cubane framework can be
considered as deriving from two interpenetrating
tetrahedra, Figure 2(b). These tetrahedra need not
be of the same size to preserve $\underline{T}_d$ symmetry but only
when they have this equivalence will A-$\hat{B}$-A = B-$\hat{A}$-B = 90°
and the polyhedron have planar faces. X-ray
crystallographic studies of certain cubane clusters,
for example $\left[Fe_4(NO)_4(\mu_3\text{-}S)_4\right]$ (4), have identified
intramolecular intermetallic separations that are
within the distance normally considered indicative of
a direct metal-metal bonding interaction. For such
complexes, it is possible to view the cubane moiety as
derived from a central tetrahedral core (A_4), above
each face of which is centred a triply bridging atom
B, Figure 2(c). If B were to be situated at the
centre of each face then A-$\hat{B}$-A and B-$\hat{A}$-B would be 120°
and 33.6^o respectively. The regular cube is obtained

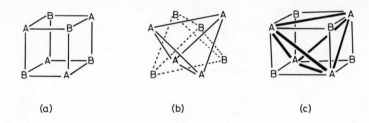

(a)   (b)   (c)

Figure 2

Alternative perspectives of an A_4B_4

cubane framework

when each B is situated at a distance of $0.409x(A-A)$
from the centre of an A_3 face, along the normal to that
face (5). When B is at a greater distance than this
along the normal, $A-\hat{B}-A < B-\hat{A}-B$. Thus the relative
magnitude of these angles gives a clear indication of
the departure from a regular cubic structure due to
the difference in size of the A_4 and B_4 tetrahedra.
This may arise because of the different atomic volumes
of the atoms A and B and/or because of particular
attractive or repulsive interactions within the
structure. For a cubane cluster of $\underline{T}_d$ symmetry, the
relationship between the A---A separation, a, the B---B
separation, b, and the A-B bond length, c, is

$$3(a^2 + b^2) - 2ab = 8c^2$$

Restricting the discussion to a consideration of the
dominant interactions being within one homonuclear
tetrahedron, two examples may be cited to illustrate
the general point. (1) For a given A-B distance, any
attractive interactions between the atoms A will, if
the cubane framework is to be preserved, lead
to a concomitant increase in the

B–B separations and a reduction in A–$\hat{\text{B}}$–A versus B–$\hat{\text{A}}$–B.
For example, $[Co_4(\eta^5-C_5H_5)_4(\mu_3-S)_4]$, which has Co–S
bonds of length 2.23(1) Å and involves no cobalt-
cobalt bonding over the separations of 3.30(5) Å, has
S---S separations of 2.99(3) Å and average Co–$\hat{\text{S}}$–Co and
S–$\hat{\text{Co}}$–S interbond angles of 84.3 and 95.3°, respectively.
Whereas $[Fe_4(NO)_4(\mu_3-S)_4]$, which has Fe–S bonds of
length 2.217(2) Å and is considered to involve a single
iron-iron bond over each edge (length 2.634(1) Å) of
the Fe_4 tetrahedron, has S----S distances of 3.503(2) Å
and Fe–$\hat{\text{S}}$–Fe and S–$\hat{\text{Fe}}$–S interbond angles of 73.4 and
104.4°, respectively (4). (2) Similarly, an increase
in the B----B distances leads to a concomitant
decrease in A-----A and a reduction in the magnitude of
A–$\hat{\text{B}}$–A versus B–$\hat{\text{A}}$–B. As an example consider
$[Cu_4(Et_3P)_4(\mu_3-X)_4]$ (where X = Cl, Br or I) and for which
the Cu–X bonds are of length 2.438(1), 2.544(1) and
2.684(1) Å, respectively; the Cu----Cu distances decrease
systematically from 3.211(2) Å, to 3.184(2) Å, to
2.927(2) Å as the X----X separations increase from
3.657(2) Å, to 3.932(1) Å, to 4.380(1) Å. The
Cu–$\hat{\text{X}}$–Cu and X–$\hat{\text{Cu}}$–X interbond angles vary as 82.4, 77.5
and 66.1 and 97.2, 101.3 and 109.4° respectively (7).

Distortions of an A_4B_4 cubane framework may occur
for one or more reasons which include: (a) crystal
packing forces, e.g. $[Os_4(CO)_{12}(\mu_3-O)_4]$ which has $\underline{D_{2d}}$
symmetry (8), (b) particular non-bonded repulsive
interactions within the cubane 'cluster', e.g.
$[Cu_4(MePh_2P)_4(\mu_3-I)_4]$, in which the Cu_4I_4 unit has $\underline{C_2}$
symmetry (9), (c) asymmetric attractive forces within
one of the tetrahedral sub-units, e.g.
$[Fe_4(\eta^5-C_5H_5)_4(\mu_3-S)_4]$, the Fe_4S_4 framework of which
has $\underline{D_{2d}}$ symmetry with two Fe–Fe bonds of length
2.650(6) Å and four Fe---Fe distances of 3.363(10) Å
(10), and (d) Jahn-Teller distortions to remove orbital

degeneracy within the molecular orbitals of the cluster, as suggested (11) for $[Fe_4(PhCH_2S)_4(\mu_3-S)_4]^{2-}$, which contains an Fe_4S_4 fragment of $\underline{D}_{2d}$ symmetry with two Fe-Fe distances of 2.776(11) Å and four of 2.732(3) Å.

A number of distortions of the cubane framework can, in principle, occur and preserve elements of symmetry present for $\underline{T}_d$ symmetry. The principal symmetries encountered for cubane complexes are presented in Table 1, together with the sets of structural elements obtained for each point group. The two important axes of reference for these distortions are a $\underline{C}_3$, A-B body diagonal of the cube, and an $\underline{S}_4(\underline{C}_2)$, perpendicular to two parallel A_2B_2 faces. For an elongation or compression of the A_4 and/or B_4 tetrahedron along an A-B body diagonal, the highest symmetry which can be obtained is $\underline{C}_{3v}$ and, failing that $\underline{C}_3$. If the A_4 and/or B_4 tetrahedron is elongated or compressed down an $\underline{S}_4$ axis, the highest molecular symmetry which can be obtained is $\underline{D}_{2d}$. This is a popular distortion for cubane structures and two representations of this are illustrated in Figure 3, for an elongation of the A_4 array and a flattening of the B_4 array. Fig. 3(a) shows the displacements from a regular cubic geometry and the location of the symmetry elements and Fig. 3(b) the projection down the $\underline{S}_4$ axis. The lower symmetry point groups retain some of the symmetry elements of Fig. 3(a): $\underline{S}_4-\underline{S}_4,\underline{C}_2$; $\underline{D}_2-\underline{C}_2,2\underline{C}_2'$; $\underline{C}_{2v}-\underline{C}_2,2\sigma_v$; $\underline{C}_s-\sigma_v$; $\underline{C}_2-\underline{C}_2$.

TABLE 1

Point Groups Encountered for the Cubane 'Clusters' and
Sub-divisions* of Structural Elements Therein.

Element	T_d	C_{3v}	C_3	D_{2d}	S_4	D_2	C_{2v}	C_s	C_2
A–A; B–B	6	2(3)	2(3)	4+2	4+2	3+2	4+2(1)	2(2)+2(1)	2(2)+2(1)
A–B̂	12	6+2(3)	4(3)	8+4	3(4)	3(4)	2(4)+2(2)	6(2)	6(2)
A–B̂–A; B–Â–B	12	6+2(3)	4(3)	8+4	3(4)	3(4)	2(4)+2(2)	6(2)	6(2)

* 2(3)≡3+3 etc.

274.

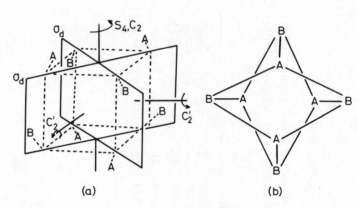

(a)   (b)

Figure 3

Perspectives of a D_{2d} A_4B_4 arrangement

3. COMPOUNDS CONTAINING CUBANE 'CLUSTERS'

3.1. Survey

The first examples of complexes which contain eight metal atoms of the same type arranged in a cubic manner derived from copper(I) compounds involving dianionic sulphur chelates, e.g. $[Cu_8\{S_2C_2(CN)_2\}_6]^{4-}$ (12a). More recently, $[Ni_8(CO)_8(\mu_4\text{-PPh})_6]$ has been characterised (12b). $[Me_4\{\mu_3\text{-HgMo(CO)}_3(\eta^5\text{-C}_5H_5)\}_4]$, Figure 4 contains an

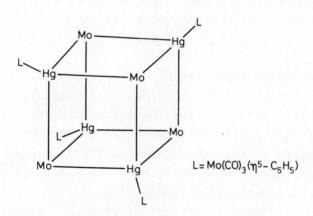

$$L = Mo(CO)_3(\eta^5\text{-}C_5H_5)$$

Figure 4

Structure of $[Mo_4\{\mu_3\text{-HgMo(CO)}_3(\eta^5\text{-C}_5H_5)\}_4]$ [13]

A_4B_4 cubane framework also composed only of metal atoms (13). Typically, cubane 'clusters' involve a set of four metal atoms (A), plus their accompanying ligands located external to the cube and a set of four non-metal atoms (B), which may also be bonded to other atoms or groups outside the cube. Although it is usually the case, there is no requirement that each of the A_4 and B_4 subgroups are homogenous. A wide range of species have been shown capable of fulfilling the triply bridging role of the B atoms, including CO, OH^-, OR^-, SR^-, NR^{2-}, AsR^{2-}, halide, chalcogenide and pnictide ions. The cubane skeleton readily accommodates both the tetrahedral and octahedral co-ordination geometries so popular for complexes about a single metal centre. Furthermore, cubane frameworks have been characterised with the metal atoms possessing a trigonal arrangement of ligand atoms, e.g. $[Mo_4\{\mu_3-HgMo(CO)_3(\eta^5-C_5H_5)\}_4]$ (Fig. 4) (13) and $[Tl_4(\mu_3-OMe)_4]$ (14), and a five co-ordinate geometry, e.g. $[Fe_4\{S_2C_2(CF_3)_2\}_4(\mu_3-S)_4]^{2-}$ (15). Also an unsaturated organic molecule can form a π-complex to each of the metal atoms in a cubane cluster, as in $[Co_4(\eta^5-C_5H_5)_4(\mu_3-P)_4]$ (16) and $[Ru_4(\eta^6-C_6H_6)_4(\mu_3-OH)_4]^{4+}$ (17).

The cubane framework has a basic geometry and sufficient flexibility which can accommodate the constraints, imposed by bonding and repulsive interactions about and between the atoms located at the vertices of the cube, for a very wide range of compounds. Also, with respect to the solid state for which most structural determinations are accomplished, the cubane 'cluster' arrangement generally appears as a compact unit with exact or approximate $\underline{T}_d$ symmetry, thus producing efficient crystal packing. There appear to be no general synthetic procedures which favour the

formation of cubane 'clusters', however, several of
these complexes appear to have been obtained by a
condensation reaction, subsequent to or in conjunction
with, the loss of ligands from a pair of dimeric
species, each possessing a central framework which
corresponds to the A_2B_2 face of the cube.

3.2. s- and p-Block Elements

Given the large number of cubane 'cluster'
compounds which have been characterised for d-transition
metals, it is surprising that more have not been
identified for the s- and p-block elements. However,
this may reflect the current popularity of these two
branches of inorganic chemistry, rather than any
inherent structural preferences.

Lithium methyl, $[Li_4(\mu_3-Me)_4]$ (18) is an
interesting example of an electron deficient cubane
cluster, each methanide ion having only one pair of
electrons available for cluster bonding. The bonding
within the cubane skeleton of the alkali metal
t-butoxides and trimethylsiloxides, $[M_4(\mu_3-OYMe_3)_4]$
(where M = K, Rb, or Cs; Y = C or Si) (19) may be
described in more conventional terms since each
oxygen atom has three lone pairs of electrons
available for this framework.

Beryllium dimethyl reacts with trimethylsilanol
to form $[Be_4Me_4(\mu_3-OSiMe_3)_4]$, which involves an almost
perfectly cubic Be_4O_4 core with Be-O = 1.73(2)
O--O = 2.47(2), Be---Be = 2.45(2) $\hat{A}$, O-Be-O = 91(1)$^\circ$
and Be-Ô-Be = 90(1)$^\circ$. This structure would also
appear to be obtained for $[BeMe(OR)_4]$ (where R = Ph, or
$PhCH_2$) (20) compounds.

278.

The adducts from Ph_3Al and primary amines undergo decomposition in boiling benzene or toluene, eliminating benzene which, in the case of aromatic amines having no <u>ortho</u>-substituent yield the $[PhAlNAr]_4$ tetramers. These units clearly possess the same number of valence electrons as the hypothetical cubane C_8Ph_8 and thus the adoption of an Al_4N_4 cubane skeleton may be readily rationalised. The crystal structure of $[Al_4Ph_4(\mu_3-NPh)_4]$ has been determined; the molecule has crystallographic $\underline{S}_4$ symmetry and a regular cube as the central feature with Al-N = 1.91(2), Al---Al = 2.70(2), N---N = 2.71(2) Å, Al-$\hat{N}$-Al = 89.8(4) and N-$\hat{Al}$-N = 90.2(4)^O (21).

Thallium(I) alkoxides have been known for over a century and may be prepared by the reaction of the metal with the alcohol under reflux. The majority of these compounds are liquids but the methoxide is an exception and may be obtained in a crystalline state. X-ray diffraction studies showed that the constituent molecules of this latter compound have a cubane structure, with the thallium(I) atoms forming a regular tetrahedron of edge 3.8 Å (14). A study of the Raman spectroscopic properties of $[Tl_4(\mu_3-OEt)_4]$ and $[Pb_4(\mu_3-OH)_4]^{4+}$ led to the conclusion that only very weak metal-metal bonding could be present in these complexes (22). A conclusion consistent with the structural data obtained for the thallium(I) methoxide tetramer and also the results of an X-ray scattering experiment for solutions containing the lead(II) hydroxide tetramer, which indicated intra-molecular Pb----Pb separations of <u>ca</u>. 3.83 Å (23).

Low-polymer organosilosesquioxanes, $(RSiO_{3/2})_{\underline{n}}$, prepared by alkali-catalyzed siloxane rearrangement of the products of hydrolysis of $RSiCl_3$ compounds, have been obtained as crystalline cubic octamers for a wide range of aliphatic R groups. The structure illustrated in Figure 5 has been suggested for these octamers on the basis of X-ray powder diffraction data. $[Sn_4(Ph)_4(\mu_3-P)_4]$ has been characterised as one of the oxidation products formed

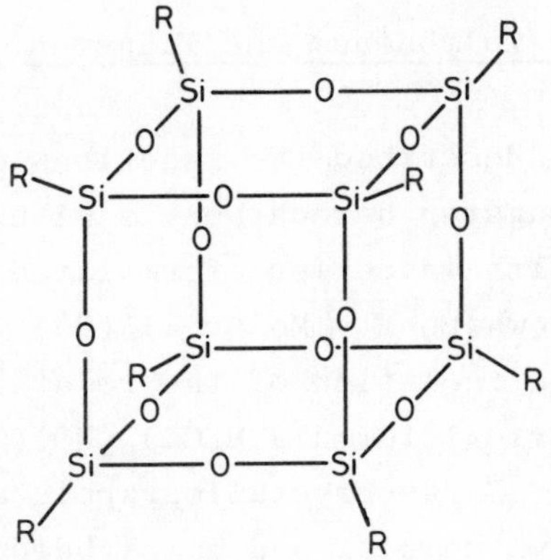

Figure 5

Structure postulated for $(RSiO_{3/2})_8$ oligomers [24]

in the reaction between $Sn(Ph)_4$ and P_4 in a sealed tube at $300^{\circ}C$. A more efficient synthesis of this cubane compound is achieved by reacting $PhSnCl_3$ with PH_3 in the presence of Et_3N (1:1:3) (25).

3.3. d-Transition Metals

3.3.1. Chromium, Molybdenum and Tungsten

Hieber et al. described the reactions of
molybdenum and tungsten hexacarbonyls with ethanolic
KOH that yield salts which were formulated as
$K_3[M_2(CO)_6(OH)_3]$ (where M = Mo or W) (26). Subsequent
studies led to the isolation of the related nitrosyl
complexes of empirical formula $M(CO)_2(NO)(OH)L$ (where
L = H_2O or R_3PO). X-ray crystallographic studies of
a tungsten carbonyl complex and a molybdenum nitrosyl
carbonyl adduct established their tetrameric nature
and the respective formulations as $[W_4(CO)_{12}(\mu_3-OH)_4]^{4-}$
and $[Mo_4(CO)_8(NO)_4(\mu_3-OH)_4]$,4L (27). The metal-metal
separations within these cubane 'clusters' are in the
range 3.4-3.5 Å, distances which preclude any direct
metal-metal bonding interactions. $[(\eta^5-C_5H_5)MS]_4$ (where
M = Cr (27) or Mo (28)) have been reported and the
formation of a cubane structure by these molecules can
be rationalised since their central framework is
isoelectronic with that of $[Fe_4(NO)_4(\mu_3-S)_4]$.
$[Mo_4\{\mu_3-HgMo(CO)_3(\eta^5-C_5H_5)\}_4]$ has been isolated as a
bi-product in the synthesis of $[(\eta^5-C_5H_5)Mo(CO)_3(C_4H_7)]$
from $[Mo_2(\eta^5-C_5H_5)_2(CO)_6]$, 2-butenyl chloride and Na/Hg.
The molybdenum atoms in this cubane skeleton are
unusual in that they appear to be bonded only to three
mercury atoms (Fig. 4); the possibility of hydride
atoms on these molybdenums was excluded on the basis
of [1]H n.m.r. investigations. The Mo_4Hg_4 core involves

three different sets of Mo-Hg distances of length
2.559(3), 3.162(3) and 3.192(3) Å which, when compared
to the length (2.692(3) Å) of the Hg-Mo(CO)$_3$(η^5-C$_5$H$_5$)
bond, suggests that single Mo-Hg bonds occur over the
four shortest edges of the cubane skeleton but only
weakly attractive forces obtain over the 3.2 Å
separations (13).

D. T. Sawyer et al. (30) have suggested that
[Mo$_4$O$_4$(S$_2$CNEt$_2$)$_4$(μ_3-S)$_4$]$^{2-,4-}$ may be the products of
the one- and two- electron reduction of
[Mo$_2$O$_2$S$_2$(S$_2$CNEt$_2$)$_2$]. The dimeric anions
[Fe$_6$Mo$_2$S$_8$(SR)$_9$]$^{3-}$ (where R = Et, CH$_2$CH$_2$OH, Ph or
p-Cl-C$_6$H$_4$) have been prepared by reacting FeCl$_3$ with
NaSR and [NH$_4$]$_2$[MoS$_4$] in methanol. These species have
a structure comprised of two cubane 'clusters', each
containing three tetrahedrally co-ordinated iron atoms
and one octahedrally co-ordinated molybdenum atom,
linked by three μ_2-thiolato-groups over a Mo---Mo
separation of 3.66(2) Å (Figure 6). ^{57}Fe Mössbauer

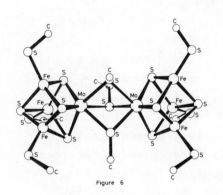

Representation of the atomic arrangements in [Fe$_6$Mo$_2$S$_8$(SR)$_9$]$^{3-}$
complexes (for clarity only those carbons bonded to
sulphur have been included) [31]

Figure 6

studies indicate that the iron atoms in these anions
are electronically equivalent with an oxidation state
of ca. +2.5 (32). The anions undergo two reversible

one-electron reductions (31), implying the existence
of the series $[Fe_6Mo_2S_8(SR)_9]^{n-}$ (where $\underline{n}$ = 3, 4 or 5).
$[Fe_6Mo_2S_9(SEt)_8]^{3-}$, for which the metal atoms have a
net oxidation state one greater than for the
$[Fe_6Mo_2S_8(SR)_9]^{3-}$ ions has also been characterised
(33). This anion possesses a central core, the
structure of which is very similar to that shown in
Figure 6, except that one of the μ_2-thiolato-groups is
replaced by a μ_2-sulphido-group, resulting in a
contraction of the Mo---Mo separation to 3.306(3) Å.
The $[Et_4N]^+$ salts of $[Fe_6Mo_2S_8(SR)_9]^{3-}$ (where R = Et
or CH_2CH_2OH (32)) and $[Fe_6Mo_2S_9(SEt)_8]^{3-}$ (33),
crystallise in the hexagonal space group $\underline{P6}_{3/\underline{m}}$, with
the anions having crystallographically
imposed $\underline{C}_{3h}$ symmetry (with disorder of some of the
carbon atoms for the former anions and the μ_2-bridging
groups for the latter) and this symmetry is closely
approximated by the Fe-Mo-S framework of the anion in
$[Bu_4^nN]_3[Fe_6Mo_2S_8(SPh)_9]$ (31). The $\{Fe_3MoS_4\}$ cubane
skeleton in these structures has $\underline{C}_{3v}$ symmetry and the
intercubane dimensions of $[Fe_6Mo_2S_8(SEt)_9]^{3-}$ are:
S---S = 3.668(5) and 3.577(5), Mo---Fe = 2.726(2),
Fe---Fe = 2.699(3), Mo-S = 2.347(3), Fe-S = 2.259(4)
and 2.276(4) Å; S-M̂o-S = 102.7(1), S-F̂e-S = 108.5(1)
and 104.1(3), Mo-Ŝ-Fe = 72.5(1), and Fe-Ŝ-Fe = 73.4(1)
and 72.7(1)° (32). The molybdenum environment in
these iron-molybdenum-sulphur cubane 'clusters' is very
similar to that suggested for this element in the
nitrogenase enzymes and their FeMo-cofactor, on the
basis of extended X-ray absorption fine structure
measurements for the $\underline{K}$-edge of the molybdenum in
these biological systems (34).
$[Et_3N]_3[Fe_6W_2S_8(\mu_2-OMe)_3(SPh)_6]$ has also been prepared.
The constituent anions have crystallographic $\underline{C}_{3h}$
symmetry, with two $\{Fe_3WS_4\}$ cubane clusters linked by

the bridging methoxide groups over a W---W separation of 3.174(1) Å (35). Müller et al have reported the structure of the novel cage compound $[W_2S_8Ag_4(PPh_3)_4]$, the skeleton of which (Figure 7) consists of two interconnected six-membered rings, the arrangement of which resembles a fusion of two $\{Ag_3WS_4\}$ cubanes linked over a common Ag_2S_2 face.

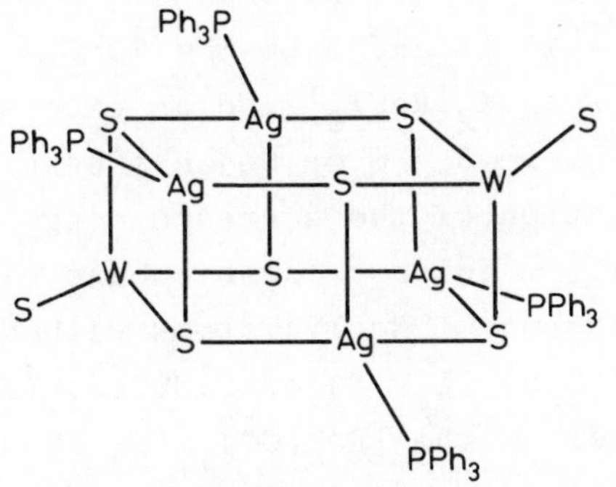

Figure 7

A diagrammatic representation of the structure

of $[W_2S_8Ag_4(PPh_3)_4]$ [36]

3.3.2. Manganese and Rhenium

The reaction of $M(CO)_5Br$ (where M = Mn or Re) with a thiol (RSH) leads to the formation of the corresponding $[M_2(CO)_8(\mu_2\text{-SR})_2]$ compound, the

molecules of which lose a further CO to yield
$[M_4(CO)_{12}(\mu_3-SR)_4]$. I.r. spectroscopic studies
indicated that these tetramers have a regular
tetrahedral structure and do not contain direct metal-
metal bonding interactions (37). These conclusions
were confirmed by an X-ray crystallographic study of
$[Re_4(CO)_{12}(\mu_3-SMe)_4]$ (38). The molecules have
crystallographically imposed $\underline{C}_2$ symmetry and
dimensions within the following ranges: Re---Re = 3.85 -
3.96, S----S = 3.05 - 3.17, Re-S 2.48 - 2.52 Å;
Re-S-Re = 100.7 - 104.1 and S-Re-S = 75.8 - 77.9°.
The reaction between $K_2[ReCl_6]$ and an excess of KCN in
a melt of KXCN (where X = S or Se) has been shown to
lead to the formation of the corresponding
$[Re_4(CN)_{12}(\mu_3-X)_4]^{4-}$ anion. These cubane clusters
contain a tetrahedron of Re(IV) atoms with mean Re-Re
distances of 2.755(5) (X = S) or 2.805(5) (X = Se)
which (in contrast to the $[Re_4(CO)_{12}(\mu_3-SMe)_4]$
structure) correspond to a single Re-Re bond over each
edge of the tetrahedron (39).

3.3.3. Iron

The first tetranuclear cyclopentadienylmetal
complex to be reported was $[Fe_4(\eta^5-C_5H_5)_4(CO)_4]$, which
was isolated by King (40), following the prolonged
refluxing of $[(\eta^5-C_5H_5)Fe(CO)_2]_2$ in xylene under
nitrogen. Molecular weight, [1]H n.m.r., and i.r.
measurements led King to suggest that the constituent
molecules contained a tetrahedral arrangement of the
iron atoms, with a triply bridging carbonyl group
situated above each face, and a $(\eta^5-C_5H_5)$ ligand
attached to each iron atom. These conclusions were

confirmed using X-ray crystallography by Dahl et al.
(41). Although the $[Fe_4(\eta^5-C_5H_5)_4(\mu_3-CO)_4]$ molecules
have no crystallographically imposed symmetry,
assuming each cyclopentadienyl ring to have cylindrical
symmetry, the atomic arrangement closely corresponds
to $\underline{T}_d$ symmetry. The length of each edge of the Fe_4
tetrahedron is 2.52(1) Å, a distance which
corresponds to that expected for an iron-iron single
bond; the other dimensions of the cubane are listed in
Table 9 (p.286). Each carbonyl group in these molecules
is capable of acting as a Lewis base and
$[Fe_4(\eta^5-C_5H_5)_4(\mu_3-COAlEt_3)_4]$ has prepared (42). The
series of complexes $[Fe_4(\eta^5-C_5H_5)_4(\mu_3-CO)_4]^{\underline{x}}$ (where
$\underline{x}$ = -1, 0, +1 or +2) has been established (40, 43) and
the monocation has been characterised as the $[PF_6]^-$ salt
(44). The cation has crystallographic $\underline{C}_2$ symmetry and
a structure which closely resembles that of the neutral
homologue. The $\{Fe_4C_4\}$ cubane framework approximates
to $\underline{D}_{2d}$ symmetry with two Fe-Fe bonds of 2.495(11) Å
and four of 2.473(7) Å; other dimensions are given in
Table 9 (p.286).

$\quad [Fe_4(\eta^5-C_5H_5)_4(\mu_3-S)_4]$ has been prepared by
refluxing a solution of $[Fe(\eta^5-C_5H_5)(CO)_2]_2$ with S_8
(45) or cyclohexene sulphide (46), in toluene or benzene,
respectively. Recrystallisation of the crude product
from bromobenzene yielded an orthorhomic (45) crystal
modification, whereas from chloroform a monoclinic
(46) form wasobtained. X-ray crystallographic
determinations have been accomplished for both forms.
There are no significant differences between the
molecular parameters in the two phases and, assuming
the cyclopentadienyl fragments to have cylindrical
symmetry, the molecular structure corresponds to $\underline{D}_{2d}$
symmetry. The distortions of the $\{Fe_4S_4\}$ core from
$\underline{T}_d$ symmetry are pronounced in that there are two

Fe-Fe separations of 2.650(6) Å and four of 3.363(10) Å, two S---S distances of 3.334(9) and four of 2.880(12) Å (46). Cyclic voltammetric studies (43) of $[Fe_4(\eta^5-C_5H_5)_4(\mu_3-S)_4]$ have established the existence of the corresponding +1, +2 and +3 cationic and -1 anionic species. The electrochemical oxidations and reductions which relate these complexes are all reversible and the cyclic voltammogram for the reduction of $[Fe_4(\eta^5-C_5H_5)_4(\mu_3-S)_4]^{2+}$ shows four reversible, one-electron waves, Figure 8 (47). Infra-red and electron spin resonance data indicate

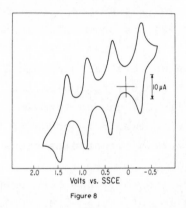

2.0   1.5   1.0   0.5   0   -0.5
Volts vs. SSCE

Figure 8

Cyclic voltammogram of $[Fe_4(\eta_5-C_5H_5)_4 \, (\mu_3-S)_4] \, [PF_6]_2$
in 0.1M $[Bu_4^tN][PF_6]$ MeCN at a platinum bead electrode [47]

that this $\{Fe_4S_4\}$ cubane cluster remains intact in solution throughout these five oxidation states and this has been confirmed, for two of the cations, by X-ray crystallographic studies of $[Fe_4(\eta^5-C_5H_5)_4(\mu_3-S)_4]Br$ (48) and $[Fe_4(\eta^5-C_5H_5)_4(\mu_3-S)_4][PF_6]_2$ (47). $[Fe_4(\eta^5-C_5H_5)_4(\mu_3-S)_4]^+$ may be obtained in almost quantitative yield from oxidation of the neutral compound by $AgBF_4$, Br_2 or I_2. The lattice of this bromide salt contains cations with crystallographically imposed $\underline{C}_2$ symmetry. The $\{Fe_4S_4\}$ cubane core closely approximates to $\underline{D}_2$ symmetry, with three pairs of Fe---Fe and S---S distances of lengths, 2.652(9), 3.188(3), and 3.319(3) Å and 2.879(6), 3.062(6) and 3.389(9) Å, respectively. $[Fe_4(\eta^5-C_5H_5)_4(\mu_3-S)_4]^{2+}$ has been obtained by the controlled potential electrolysis of the neutral parent molecule. The $[PF_6]^-$ salt crystallises with this cation on a site of $\underline{S}_4$ symmetry; the Fe---Fe distances group as a set of four shorter (2.834(3) Å and two longer (3.254(3) Å) and conversely the S---S distances group as a set of two shorter (2.820(6) Å) and four longer (3.304(5) Å). Full structural details of the cubane clusters within $[Fe_4(\eta^5-C_5H_5)_4(\mu_3-S)_4]^{x+}$ (where x = 0, 1 or 2) are included in Table 9 (p.327).

$[Fe_4(NO)_4(\mu_3-S)_4]$ may be prepared by reacting $[Fe(CO)_3(NO)]_2Hg$ and elemental sulphur in refluxing toluene for 16 hr. under nitrogen. $[Fe_4(NO)_4(\mu_3-S)_2(\mu_3-NCMe_3)_2]$ is obtained in a corresponding reaction which employs $(Me_3CN)_2S$ in place of the sulphur. The former complex exhibits one $\nu(N-O)$ infra-red absorption at 1780 cm^{-1} and the latter three such bands at 1745(s), 1760(vs) and 1792(w) cm^{-1}. Structural details are available for both compounds. Although $[Fe_4(NO)_4(\mu_3-S)_4]$ molecules strictly have no symmetry, the atomic arrangement (Fig. 9) corresponds closely to $\underline{T}_d$ symmetry, consistent with the single $\nu(N-O)$ infra-red absorption. This Fe$_4$ tetrahedron appears to involve a single Fe-Fe bond over each edge of length 2.634(1) Å and the S---S distances are 3.503(2) Å (4). $[Fe_4(NO)_4(\mu_3-S)_2(\mu_3-NCMe_3)_2]$ molecules have a structure analogous to that depicted in Fig. 9, with two of the (μ_3-S) ligands replaced by (μ_3-NCMe_3) groups. These

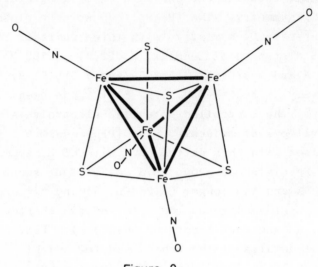

Figure 9

Structure of $[Fe_4(NO)_4(\mu_3\text{-}S)_4]$ molecules [4]

molecules have crystallographic $\underline{C}_2$ symmetry and, apart from the methyl groups, approximate $\underline{C}_{2v}$ symmetry for which, respectively, four and two ν(N-O) stretching modes should be infra-red active. The Fe-Fe bonds vary in length from 2.496(1) to 2.642(1) Å and further details for this and its more symmetry analogue are included in Table 9 (p.327). Cyclic voltammetric studies have indicated that a one-electron oxidation and a one electron-reduction product exist for $[Fe_4(NO)_4(\mu_3\text{-}S)_4]$ and four successive one-electron reduction occur for $[Fe_4(NO)_4(\mu_3\text{-}S)_2(\mu_3\text{-}NCMe_3)_2]$ (4).

Balch prepared $[Fe_4\{S_2C_2(CF_3)_2\}_4(\mu_3\text{-}S)_4]$ by refluxing $[Fe_2(CO)_6\{S_2C_2(CF_3)_2\}]$ and S_8 together in xylene. Schrauzer $\underline{et}$ $\underline{al}$.had previously reported the analogous complex with the $S_2C_2(Ph)_2$ dithiolene ligand, obtained by the slow addition of $FeSO_4, 7H_2O$ in $H_2O/MeOH$ to a refluxing solution of benzoin and P_4S_{10} in xylene. The former complex is readily reduced by dissolution in basic solvents; from the burgundy coloured solution formed by adding this compound to DMSO, $[Fe_4\{S_2C_2(CF_3)_2\}_4(\mu_3\text{-}S)_4]^{2-}$ has been isolated as $[Ph_4As]^+$ and $[Bu_4^nN]^+$ salts (49). The crystal structure of $[Bu_4^nN]_2[Fe_4\{S_2C_2(CF_3)_2\}_2(\mu_3\text{-}S)_4]$ has been determined (15) but the disorder of the cations prevented the accuracy of the determination being as high as normally expected. The $\{Fe_4S_4\}$ cubane framework approximates to $\underline{D_{2d}}$ symmetry, with four shorter Fe---Fe distances of 2.73(10) Å and two longer ones of 3.23(5) Å and conversely four longer S---S distances of 3.47(9) Å and two shorter ones of 2.87(12) Å. The pattern and values of these distances resemble those of $[Fe_4(\eta^5\text{-}C_5H_5)_4(\mu_3\text{-}S)_4]^{2+}$ and this similarity would appear to arise because these $\{Fe_4S_4\}$ cubane frameworks are formally isoelectronic.

Röttinger and Vahrenkamp (50) attempted the preparation of $\{Fe_4(RY)_4\}$ (where Y = P or As) clusters, on the bases that an extensive range of the corresponding sulphido- derivatives have been characterised and phosphinidine and arsinidine groups can reasonably be expected to function as triply bridging ligands. This idea has been confirmed through the preparation of $[Fe_4(CO)_{12}(\mu_3\text{-}AsMe)_4]$, which may be obtained by the controlled decomposition of $[Fe(CO)_4(AsH_2Me)]$. The molecular structure of this cubane moiety approximates to $\underline{T_d}$ symmetry with As---As separations of 2.97 Å and involves no direct iron-iron bonding over the separations of 3.76 Å.

3.3.4. {FeS$_4$} Cubane Clusters in Biological Systems and Their Synthetic Analogues

Iron-sulphur proteins have attracted much attention during the past two decades during which time there has been a considerable improvement in the understanding of the structure and function of these systems (51-54). The iron-sulphur proteins constitute a large and important group of biological materials which are probably present in and basic to all forms of life. They are involved in reactions in bacteria such as hydrogen uptake and evolution, ATP formation, pyruvate metabolism, nitrogen fixation, and photosynthetic electron transport, that are generally considered to have been a prerequisite for the development and maintenance of all living processes. Certain of these key reactions are also vital for the function of higher organisms, as are nitrate, nitrite and sulphite reduction, xanthine and aldehyde oxidation and mitochondrial electron transport conversions in which iron-sulphur proteins participate. The iron-sulphur proteins range from small units, of molecular weight ca. 6,000 containing a small number of iron atoms, to complex aggregates containing a large number of iron atoms and other centres, e.g. the molybdenum-iron protein of nitrogenase has a molecular weight of ca. 220,000 and contains 28-32 Fe and 2 Mo atoms. The presently well-defined iron-sulphur proteins of molecular weight ca. 6,000-20,000 are classified as rubredoxins (Rd) if they contain one or two iron atoms per molecule and no sulphide, or a ferredoxin (Fd) if they contain two, four, or eight

iron atoms <u>and</u> an equivalent amount of sulphide per
molecule. Three different types of iron-sulphur
centre have been characterised in these proteins
(Fig. 10) and in each case the iron atom is surrounded

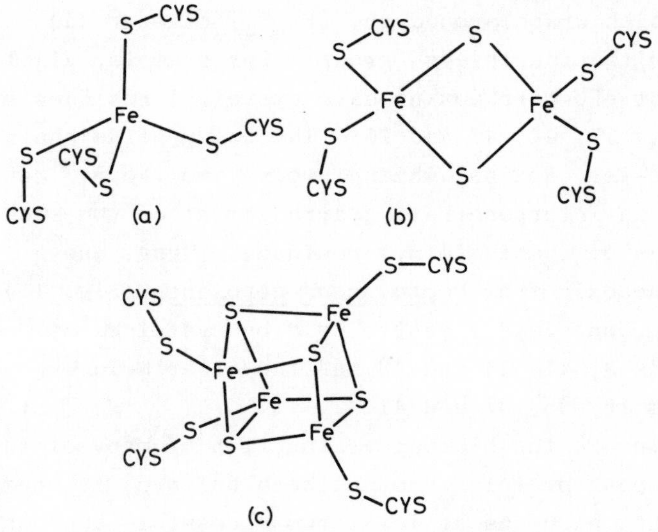

Figure 10

Fe-S centres present in iron sulphur proteins,
(a) Rubredoxin, (b) 2Fe-Ferredoxin, (c) 4Fe- and 8Fe-Ferredoxin

by an approximately tetrahedral array of four sulphur
atoms. Therefore, all the iron-sulphur proteins
contain a number of cysteinyl residues which is equal
to or greater than the number of iron atoms therein.
The amino acid sequences of the simple iron-sulphur

proteins are remarkably similar and, within a given
protein class, the relative positions of the cysteines
are maintained. This implies that special
folding properties, involving some or all of the
conserved residues, dictate a particular stereochemistry for the iron-sulphur site(s). 4Fe-ferredoxins
have the pattern cys-A-A-cys-A-A-cys in common, followed
by a comparatively long portion of the polypeptide
chain which wraps around the $\{Fe_4S_4\}$ cluster and
includes the fourth-cys- group. For example, algal
and plant 4Fe-ferredoxins have cysteinyl residues at
positions 37, 44, 47 and 75. The polypeptide chain of
some 8Fe-ferredoxins, which involve two $\{Fe_4S_4\}$ centres,
contain an interpenetrating arrangement of two such
groupings of the cysteinyl residues. Thus, the
8Fe-ferredoxin from Peptococcus aerogenes (Fig. 11)
involves one $\{Fe_4S_4\}$ centre held by cysteines at
positions 9, 11, 14 and 43 and the other held by
residues 18, 35, 37 and 41.

Although the biological function of many of the
iron-sulphur proteins has not been defined, it seems
clear that each has at least two accessible adjacent
redox states per active site and that they function
primarily in electron transport sequences, rather
than acting as a site for substrate binding and
conversion. Discussions have been presented, based
upon the biochemical roles of the various types of
iron-sulphur proteins, suggesting that the earliest
electron-transport systems to evolve involved $\{Fe_4S_4\}$
cubane centres at 8Fe- or 4Fe-ferredoxins. Both of
these types of centre have been structurally
characterised in a protein environment. Chromatium
vinosum high potential iron protein (HiPIP) contains
a single polypeptide of 85 amino acid residues and
encloses an $\{Fe_4S_4\}$ cubane cluster, the basic framework

of which remains intact in the oxidised (HiPIP$_{ox}$) and
reduced (HiPIP$_{red}$) forms of the protein (57).
Peptococcus aerogenes ferredoxin contains two separate
{Fe$_4$S$_4$} centres, Fig. 11, the centres of which are
some 12 Å apart, enveloped by a 'linear' polypeptide

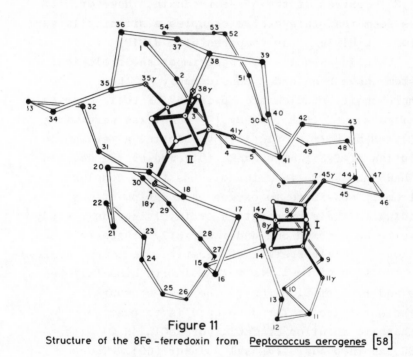

Figure 11

Structure of the 8Fe -ferredoxin from Peptococcus aerogenes [58]

of 54 amino acids. The {Fe$_4$S$_4$} centres in both of
these proteins are surrounded principally by non-
polar amino acid side chains, although (vide infra),
it does appear that the peptide is also linked to the

cluster by hydrogen bonding from amide groups to some
of the sulphido- and cysteinyl-sulphur atoms (58).
The average dimensions determined for the iron-sulphur
centres of these proteins are summarised in Table 2.
These results establish that the same basic
$[Fe_4(S-cys-)_4(\mu_3-S)_4]$ unit exists for the HiPIP
oxidised and reduced systems and for both of the
$\{Fe_4S_4\}$ centres of the 8Fe-ferredoxin. However, it
does seem that this centre occupies a slightly larger
volume in $HiPIP_{red}$ as compared with $HiPIP_{ox}$ (57).

   Detailed studies of these important biological
systems have been aided considerably by the
comprehensive studies accomplished for their chemical
analogues. $[Fe_4(SR)_4(\mu_3-S)_4]^{2-}$ complexes were first
isolated by Holm et al (11) by reacting a methanolic
solution of $FeCl_3$ with NaSR, followed by the addition
of NaHS and NaOMe in methanol. As these complexes are
extremely sensitive to oxygen, all manipulations
leading to their preparation or involving their study
have to be carried out under strictly oxygen-free conditions. An alternative preparative procedure involves
the reaction of $FeCl_3$ in methanol with LiSR followed by
the addition of Li_2S (59). The easiest route to
these complexes so far developed is to react a
methanolic solution of $FeCl_3$ (or $FeCl_2,4H_2O$) with
NaSR and then add elemental sulphur (60). The
$[Fe_4(SR)_4(\mu_3-S)_4]^{2-}$ ions are obtained from all of these
routes as nice crystalline materials by precipitation
with a quaternary ammonium cation, followed by
recrystallisation from a suitable solvent. The
formation of these complexes from the reactants
requires some change in the oxidation state of the iron
during the reaction. Since iron(III) or iron(II) lead
to the formation of the dianion, it seems clear that
this represents the thermodynamically stable state of

TABLE 2

Average Dimensions (and their e.s.d's) of $[Fe_4(S-cys)_4(\mu_3-S)_4]$ centres in certain proteins.

	$HiPIP_{ox}$ [a]	$HiPIP_{red}$ [a]	$Fd_{ox}I$ [b]	$Fd_{ox}II$ [b]
Distance (Å)				
Fe----Fe	2.72(4)	2.81(4)	2.73(6)	2.67(7)
S----S	3.55(6)	3.65(11)	3.52(7)	3.49(12)
Fe–S	2.26(8)	2.32(9)	2.23(12)	2.22(14)
Fe–S(cys)	2.20(2)	2.22(3)	2.22(16)	2.25(17)
Angle (°)				
Fe–Ŝ–Fe	74(1.3)	76(2.4)	75(2)	75(2)
S–F̂e–S	104(2.4)	104(2.6)	103	104
S–Fe–S(cys)	115(4.9)	116(5.3)	115(3)	115(3)

a   C. W. Carter, Jr., J. Kraut, S. T. Freer, R. A. Alden, L. C. Sieker, E. T. Adman and L. H. Jensen, Proc. Natl. Acad. Sci. U.S.A., 69, 3526 (1972).

b   see C. W. Carter, Jr., 'Iron-Sulfur Proteins', W. Lovenberg (ed.), Academic Press, New York, Vol. III, p167 (1977).

these cubane clusters.

The structure of several of these synthetic analogues have been determined and relevant dimensions are listed in Table 3. These data, when compared with those in Table 2, clearly show that the dimensions of the $\{Fe_4S_4\}$ synthetic analogues have a very close correspondence to those of the natural systems, the magnitude of the structural parameters being relatively insensitive to the nature of the terminal thiol. Furthermore, the cubane cage undergoes a slight but significant expansion following the one-electron reduction of $[Fe_4(SPh)_4(\mu_3-S)_4]^{2-}$ (61), in a manner similar to that observed for the $\{Fe_4S_4\}$ centre of <u>Chromatium</u> HiPIP (57). None of these cubane clusters has any crystallographically imposed symmetry. However, all of the dianions (11, 61-64) exhibit a very slight collective distortion from <u>ca</u>. T_d to <u>ca</u>. D_{2d} symmetry, in the sense of having a compressed Fe_4 and/or S_4 tetrahedron (two longer and four shorter edges) which may also be true for the natural systems. $[Fe_4(SPh)_4(\mu_3-S)_4]^{3-}$ also has a $\{Fe_4S_4\}$ centre which approximates to D_{2d} symmetry but with an elongated $\{Fe_4S_4\}$ core. These distortions have been suggested to occur as a consequence of a first-order Jahn-Teller effect which removes the orbital degeneracy expected for the T_d geometry (p.271) (11).

The thiol exchange reaction
$$[Fe_4(SR)_4(\mu_3-S)_4]^{2-} + \underline{n}R'SH \rightleftharpoons [Fe_4(SR)_{4-\underline{n}}(\mu_3-S)_4]^{2-} + \underline{n}RSH$$
proceeds readily at room temperature in non-aqueous solvents. The position of this equilibrium lies well to the right if $pK_a(R'SH) << pK_a(RSH)$, e.g. for $R = Bu^t$ and $R' = aryl$, the substitution is complete for an R'SH:Fe ration of <u>ca</u>. 1.3:1 (62). This thiol exchange appears to proceed in a stepwise manner, each substitution being initiated by protonation of the

TABLE 3

Average Dimensions[a] (and their e.s.d.'s)[b] of $[Fe_4(SR)_4(\mu_3-S)_4]^{2-,3-}$ complexes

	[c] $[FeS(S(CH_2)_2COO)]_4^{6-}$	[d] $[FeS(SCH_2CH_2OH)]_4^{2-}$	[e] $[FeS(SCH_2Ph)]_4^{2-}$	[f] $[FeS(SPh)]_4^{2-}$	[g] $[FeS(SPh)]_4^{3-}$
Distance (Å)					
Fe --- Fe	2.778(3) 2.743(3)	2.729(16)	2.776(11) 2.732(3)	2.730(3) 2.739(4)	2.744(16)
S ---- S	3.613(3) 3.596(3)	3.69(3) 3.57(3)	3.645(2) 3.586(4)	3.650(10) 3.592(6)	3.605(15) 3.685(13)
Fe - S	2.261(3) 2.300(3)	2.24(1) 2.31(1)	2.239(4) 2.310(3)	2.267(5) 2.296(4)	2.352(10) 2.288(10)
Fe - S(R)	2.250(3)	2.26(2)	2.251(3)	2.263(3)	2.295(7)
Angle (°)					
Fe - Ŝ - Fe	74.1(2)	73.4(10)	73.8(2)	73.5(5)	72.9(6)
S - F̂e - S	103.9(2)	106.2(9) 103.5(5)	104.1(2)	104.3(9)	104.8(8)
S - F̂e - S(R)	109.9 - 119.3(2)	110.0 - 119.0(4)	110.2 1 117.3(2)	100.2 - 135.7(1)	98.3 - 123.6(2)

a Å, degrees and grouped for idealised D_{2d} symmetry in sets of (2+4) or (4+8) if values sufficiently distinct

b Probably underestimated

c H. L. Carrell, J. P. Glusker, R. Job and T. C. Bruice, J. Amer. Chem. Soc., 99, 3683 (1977).

d G. Christou, C. D. Garner and M. G. B. Drew, submitted for publication.

e B. A. Averill, T. Herskovitz, R. H. Holm and J. A. Ibers, J. Amer. Chem. Soc., 95, 3523 (1973).

f L. Que, Jr., M. A. Bobrik, J. A. Ibers and R. H. Holm, J. Amer. Chem. Soc., 96, 4168 (1974).

g E. J. Laskowski, R. B. Frankel, W. O. Gillum, G. C. Papaefthymiou, J. Renaud, J. A. Ibers and R. H. Holm, J. Amer. Chem. Soc., 100, 5322, (1978).

coordinated RS$^-$ group by R'SH, followed by the rapid dissociation of RSH and co-ordination of R'S$^-$ (65). This facile thiol exchange, together with the observation that the low molecular weight iron-sulphur proteins can be reversibly unfolded by solvents such as DMSO, led to the idea (66) that it should be possible to displace iron-sulphur cores from proteins using an appropriate thiol (e.g. PhSH). The spectroscopic properties of the complexes thus formed could be compared directly with those of the synthetic analogues to confirm or establish their identity. This approach has been applied successfully to distinguish 2Fe- and 4Fe-ferredoxin centres in proteins (66, 67). Orme-Johnson et al. (54) have developed a procedure of 'thiol' exchange whereby the iron-sulphur core of one holo-protein may be extracted and bound to another apoprotein; the iron-sulphur core of the latter having been removed by acidification. The reconstitution of active Clostridium pasteurianum from the apoprotein and $[NMe_4]_2[Fe_4(SCH_2CH_2OH)_4(\mu_3-S)_4]^{2-}$ (68) demonstrates the complete interchange between natural and synthetic clusters. (Reconstitution can also be achieved from the apoprotein and iron (II, III) and sulphide salts (69)).

The oxidation states which are attainable for the $[Fe_4(SR)_4(\mu_3-S)_4]$ complexes synthesised so far are summarised in Table 4. These have been established by an extensive series of polarographic and other electrochemical measurements (70). $[Fe_4(SR)_4(\mu_3-S)_4]^{3-}$ complexes may be prepared by the reduction of the corresponding dianion by sodium acenaphthylenide (70) and the R=Ph and CH$_2$Ph derivatives have been fully characterised (61, 71). However, no mono- or tetra-anionic species have yet been isolated. The $[Fe_4(SR)_4(\mu_3-S)_4]^{2-,3-}$ complexes are well suited to

TABLE 4

Oxidation States Attained in $\left[Fe_4(SR)_4(\mu-S_3)_4\right]^a$
Synthetic Analogues and 4Fe- and 8Fe-ferredoxins.

$$\left[Fe_4(SR)_4(\mu_3-S)_4\right]^{4-} \rightleftharpoons \left[Fe_4(SR)_4(\mu_3-S)_4\right]^{3-} \rightleftharpoons \left[Fe_4(SR)_4(\mu_3-S)_4\right]^{2-} \rightleftharpoons \left[Fe_4(SR)_4(\mu_3-S)_4\right]^{-}$$

Specifications	$\underline{b}_C{}^{2-}$	$\underline{b}_C{}^{-}$	$\underline{b}_C$	$\underline{b}_C{}^{+}$
	$Fd_{\underline{s-red}}$	$Fd_{\underline{red}}$	$Fd_{\underline{ox}}$	$Fd_{\underline{s-ox}}$
		HiPIP $\underline{s-red}$	HiPIP $\underline{red}$	HiPIP $\underline{ox}$
Formal oxidation states of Fe	4Fe II	3Fe II + Fe III	2Fe II + 2Fe III	Fe II + 3Fe III

$\underline{a}$ the charge on the complex assumes conventional RS$^-$ ligands: this charge cannot be specified for the proteins

$\underline{b}$ see C. W. Carter et al., Proc. Natl. Acad. Sci. U.S.A., 68, 3526 (1972).

spectroscopic studies and a number of detailed
investigations have been accomplished. These have
been of value and interest in their own right and
have established the correspondence, between the
oxidation state levels operative in the 4Fe- and 8Fe-
ferredoxins and HiPIP proteins and those characterised
for the synthetic analogues. This correspondence is
detailed in Table 4. A confusing aspect of the early
studies of iron-sulphur proteins was the fact that
proteins with apparently the same $\{Fe_4S_4\}$ centres
exhibited considerably different redox properties.
Thus $HiPIP_{red}$ is <u>oxidised</u> at +350 mV to a paramagnetic
species ($\overline{72}$) whereas the paramagnetic form of the iron-
sulphur centre(s) in the 4Fe- and 8Fe-ferredoxins is
produced upon <u>reduction</u> at a potential between -380
and -480 mV (73). The confirmation (70) that the
$\{Fe_4S_4\}$ centre of $HiPIP_{red}$ is isoelectronic with that
of Fd_{ox} removed this $\overline{\text{confusion}}$. Evidence in support
of this conclusion has been obtained from a number of
studies. (i) Variable temperature magnetic
susceptibility studies of Fd_{ox} (74), $HiPIP_{red}$ (75),
and $\left[Fe_4(SR)_4(\mu_3\text{-S})_4\right]^{2-}$ complexes (76) have established
that all of these systems have a magnetic moment
equivalent to <u>ca</u>. 1.0 BM per iron atom at room
temperature. As the temperature approaches absolute
zero, so the magnetic susceptibility tends to zero.
This behaviour is consistent with a description of the
electronic structure of the $\{Fe_4S_4\}$ centres as having
a spin singlet ground state and thermally accessible
excited states for which $S \neq 0$. (ii) Electron spin
resonance spectroscopy has been widely applied to the
study of iron-sulphur proteins and their synthetic
analogues (77). The product of one-electron reduction
of Fd_{ox} in aqueous solution and $\left[Fe_4(SR)_4(\mu_3\text{-S})_4\right]^{2-}$
complexes is an e.s.r. active complex with $g_{av} < 2.0$;

for the former a spectrum indicative of rhombic symmetry is usually observed, with g-values of _ca_. 1.89, 1.93 and 2.07, g_{av} = 1.96 (78) (see Fig. 12) and $[Fe_4(SCH_2CH_2OH)_4(\mu_3-S)_4]^{3-}$ in water exhibits an e.s.r. spectrum typical of axial symmetry with $\underline{g}_\perp$ = 1.93 and $\underline{g}_{//}$ = 2.05 ($\underline{g}_{av.}$ = 1.97) (79). The e.s.r. spectrum of the fully reduced form of an 8Fe-ferredoxin shows additional features consistent with a weak spin-coupling between the two Fd_{red} centres over the separation (e.g. Fig. 11) of <12 Å (80). The reduction of $HiPIP_{red}$ has not been achieved in aqueous solution. However, when dissolved in a solution of $DMSO/H_2O$ (_ca_. 4:1 by volume) in which the tertiary structure of the protein unfolds, $HiPIP_{s-red}$ can be obtained by the reduction of $HiPIP_{red}$ by dithionite. $HiPIP_{s-red}$ has an axially symmetric e.s.r. spectrum with $\underline{g}_\perp$ = 1.93 and $\underline{g}_{//}$ = 2.04 ($\underline{g}_{av.}$ = 1.97) (81). The $\{Fe_4S_4\}$ centres of the ferredoxin proteins can be oxidised by ferricyanide to produce Fd_{s-ox} which exhibits an e.s.r. signal for which g_{av} >2.0 (see Fig. 12) (82). The oxidised form of _Chromatium_ HiPIP has an e.s.r. spectrum which is probably axially symmetric with $\underline{g}_{//}$ = 2.12 and $\underline{g}_\perp$ = 2.04 ($\underline{g}_{av}$ = 2.07) (77). The e.s.r. spectrum of $[Fe_4(SCH_3CH_2OH)_4(\mu_3-S)_4]^-$ in aqueous solution is axially symmetric with g-values at 1.963 and 2.006 (79). (iii) Detailed [57]Fe Mössbauer studies of the iron-sulphur proteins and their synthetic analogues have been accomplished (77, 84). The results obtained are consistent with the concepts deduced from the magnetic susceptibility and e.s.r. studies of these systems, in addition they provide information concerning the oxidation state of the iron atoms and the magnetic interactions within the iron-sulphur cores. The Mössbauer chemical shift is a measure of the electron density at the nucleus studied and, for

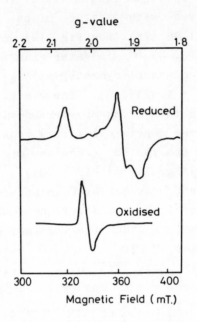

g-value

2·2   2·1   2·0   1·9   1·8

Reduced

Oxidised

300   320   360   400

Magnetic Field (mT.)

Figure 12

E.s.r. spectra of the 4Fe-ferredoxin

from <u>Bacillus stearothermophilus</u> [83]

the iron-sulphur proteins and their synthetic analogues, an approximately linear correlation is obtained between the chemical shift and the oxidation state of the iron atom. The limits of the range are neatly delineated by the iron(II) and (III) centres in rubredoxin, which have chemical shifts of 0.65 and 0.25 mm.s^{-1}, respectively (at 77°K, relative to pure iron metal at room temperature) (86). The $\{Fe_4S_4\}$ centres of Fd_{ox} and $HiPIP_{red}$ have a non-magnetic ground state with total spin of zero: the iron atoms are almost equivalent and the chemical shift of 0.42 mm.s^{-1} is consistent with an average oxidation state for these atoms of +2.5. This compares with the value of ca. 0.44 mm.s^{-1} obtained for $[Fe_4(SR)_4(\mu_3-S)_4]^{2-}$ complexes (87). Mössbauer studies of the oxidised $\{Fe_4S_4\}$ (HiPIP$_{ox}$) centre of Chromatium vinosum indicate (77, 85) that the ground state has a net spin of $\frac{1}{2}$, the iron atoms are nearly equivalent, and the chemical shift of 0.31 mm.s^{-1} corresponds to an average oxidation state of +2.75. The corresponding HiPIP$_{s-red}$ centre also appears to have a ground state with a net spin of $\frac{1}{2}$, four equivalent iron atoms and a chemical shift of 0.59 mm.s^{-1} which corresponds to an average oxidation state of +2.25 (77). Similar conclusions apply to Fd$_{red}$ centres and $[Fe_4(SR)_4(\mu_3-S)_4]^{3-}$ complexes. (iv) ^1H and ^{13}C nuclear magnetic resonance spectroscopy has been applied extensively to monitor the environment of the redox active sites of the iron-sulphur proteins and the observation of similar spectra from the synthetic analogues has clarified the interpretation of these data (84). Proteins which contain $\{Fe_4S_4\}$ cores typically exhibit ^1H resonances downfield of the normal features characteristic of the polypeptide. These distinct resonances are due to the $\beta-CH_2$ protons of the cysteinyl residues ligated to the $\{Fe_4S_4\}$ cluster, shifted because of contact interactions with these paramagnetic cores. $[Fe_4(SR)_4(\mu_3-S)_4]^{2-}$ complexes containing alkylthiolato-groups have the ^1H resonances of the $\beta-CH_2$ groups shifted in the same direction and to the same extent (10-15 ppm) as those (8-20 ppm) for the proteins containing the corresponding centres. The magnitude of this downfield shift increases with increasing temperature and directly parallels the increase in the magnitude susceptibility

over the same temperature range (88). (v) The
visible and ultra violet absorption spectra of
$[Fe_4(SR)_4(\mu_3-S)_4]$ chromophores are dominated by
intense sulphur to iron charge-transfer transitions.
The spectral profiles of the synthetic analogues vary
with R (70) and no complete equivalence with a
spectrum of the corresponding centres in the proteins
has been obtained. However, reduction of
both the analogue and natural centres results in a
significant reduction in the intensity of the
absorptions in the visible spectral region (84).
Recent studies (89) have demonstrated the potential
value of magnetic circular dichroism as a technique
for detailing the electronic transitions of iron-
sulphur clusters.

The principal natural functions of the iron-
sulphur proteins appear to derive from their ability
to transfer one (Rd, 2Fe- and 4Fe-ferredoxins) or two
(8Fe-ferredoxins) electrons, usually at a potential
not far removed from that of the hydrogen electrode.
Typically 4Fe- and 8Fe- centres shuttle between the
C^- and C states (Table 4) (90) at a potential between
-280 and -430 mV (Table 5). The corresponding potential
for the synthetic analogues in organic media is
typically ca. -1,000 mV, the potential being more
negative for electron releasing R groups. However,
studies of the proteins and the analogues under
identical conditions, including measurements as a
function of solvent composition, have established
that for many systems the effect of the protein
environment is relatively small and amounts to
<-120 mV (91). Nevertheless, the facts that the
$\{Fe_4S_4\}$ clusters of HiPIP and Fd normally shuttle
between the redox levels C/C^+ and C^-/C, respectively,
and that reduction of the former centre can be

TABLE 5

Redox Potentials[a] for {Fe$_4$S$_4$} Centres in Proteins and Synthetic Analogues

Species	Nature and Potential (mV) of Redox Couple			Medium
	c^{2-}/c^-	c^-/c	c/c^+	
Bacillus stearothermophilus[b]		−280		aq.
Desulfovibrio desulfuricans[b]		−330		aq.
Chromatium HiPIP[b]			+350	aq.
Chromatium Fd[b]		<−640		DMSO–H$_2$O (4:1)
Chromatium Fd[c]		−490		aq.
Azotobacter vinelandii FdI[c]			−420, +350	aq.
[Fe$_4$(S−(RS)−Cys(Ac)NHMe)$_4$(μ_3−S)$_4$]$^-$ [b]		−490 (−730)		aq.
[Fe$_4$(SCH$_2$CH$_2$OH)$_4$(μ_3−S)$_4$]$^-$ [b]		−510 (−750)		aq.
[Fe$_4$(SCH$_2$CH$_2$OH)$_4$(μ_3−S)$_4$]$^-$ [b]		−1,170		DMSO
[Fe$_4$(S−Bu^t)$_4$(μ_3−S)$_4$]$^-$ [b]	−2,160	−1,420	−120	DMF
[Fe$_4$(S−Bu^t)$_4$(μ_3−S)$_4$]$^-$ [b]	−2,020	−1,330		DMF
[Fe$_4$(SPh)$_4$(μ_3−S)$_4$]$^-$ [b]	−1,750	−1,040		DMF

a Italicised values are vs. standard hydrogen electrode, ph ∿7 25°C, other values are vs. saturated calomel electrode, 25°C; sources of data are given in Refs. (70, 84, 90).

b 4 Fe-Fd centres and their analogues

c 8 Fe-Fd centres

achieved only in a denaturing solvent (81), clearly
indicate that the protein tertiary structure <u>can</u>
control the redox properties of an $\{Fe_4S_4\}$ centre.
Furthermore, <u>Azotobacter</u> <u>vinelandii</u> FdI, an 8Fe-
ferredoxin, would be remarkable if it really contains two
$\{Fe_4S_4\}$ centres, both of which shuttle between C and
C^+ states, one at a potential of +350 mV and the other
at -420 mV (82). A number of interdependent ideas
have been advanced to explain how a protein could
affect $\{Fe_4S_4\}$ redox potentials. These include:
(i) hydrogen bonding. The X-ray crystallographic
studies of <u>Chromatium</u> HiPIP (57) and <u>P</u>. <u>aeogenes</u> 8Fe-
ferredoxin (58) have identified specific hydrogen
bonding interactions involving hydrogens of protein
amide groups and the sulphur atoms of the redox
centres. The number and strength of such interactions
could affect the redox potentials of the $\{Fe_4S_4\}$ cores,
thus the existence of more NH---S bonds in the
ferredoxin than in the HiPIP protein could explain why
the former operates between the C^- and C states,
whereas the latter operates between C and C^+. Also,
the more negative redox potentials observed for both
protein and analogue complexes in DMSO-H_2O solutions,
as compared to aqueous media, could be due to the
disruption of hydrogen bonds by the denaturing media
(82). (ii) The particular deformation imposed on the
cubane cluster by the bonding and non-bonding interactions with the protein. This cannot be ascertained
until more accurate structural data become available
for these proteins, however, resonance Raman
spectroscopic studies have indicated (93) that the
iron-sulphur frameworks of the proteins have a lower
symmetry than those of the synthetic analogues.
(iii) Variation in the thickness and dielectric
constant of the environment of the $\{Fe_4S_4\}$ core (94).

(iv) The actual charge on the cluster and/or the close proximity of charge groups to the $\{Fe_4S_4\}$ cores; although this is not evident in the redox potentials for the C^-/C conversions of the analogues for which R is $(CH_2)_2CO_2^-$ or CH_2CHMe_2 (92). The relative importance of these various effects, or whether others are operative, remains to be established. Also, the mechanism(s) for electron influx to and egress from these $\{Fe_4S_4\}$ cubane clusters within the protein have yet to be characterised. Although all of these proteins have some amino acids with aromatic residues, the electron transfer does not obligatorily involve these groups, as an 8Fe-ferredoxin lacking aromatic substituents has been prepared and shown to be fully active (95).

The demonstration of the validity of the $\left[Fe_4(SR)_4(\mu_3-S_4)\right]^{2-}$ complexes as analogues of the natural systems has been extended to include their chemical function. Thus those water soluble complexes for which $R = (CH_2)_2OH$ or the octapeptide derivative $Ac-(gly)_2-cys-(gly)_2-(cys)-(gly)_2-NH_2$ can replace Spirulina maxima ferredoxin as the electron mediator in the in vitro hydrogen evolving system involving dithionite and Clostriduim pasteurianum hydrogenase (96).

Other investigations of the chemical reactivity of $\left[Fe_4(SR)_4(\mu_3-S)_4\right]^{2-}$ complexes have been concerned principally with the replacement of the RS^- ligands by other nucleophiles. Reaction with benzoyl halides in acetonitrile effects conversion to the corresponding $\left[Fe_4X_4(\mu_3-S)_4\right]^{2-}$ (where X = Cl or Br) complex. Treatment of the chloride complex with sodium iodide in acetonitrile then yields the corresponding iodo-derivative (97). The structure of the cubane core of $\left[Fe_4Cl_4(\mu_3-S)_4\right]^{2-}$ is almost identical with that for the corresponding thiolato-complexes (Table 3).

Although there is no crystallography imposed symmetry, the structure again closely approximates to $\underline{D}_{2d}$ symmetry with a slightly elongated Fe_4 tetrahedron (two Fe---Fe distances of average length 2.755(9) Å and four of 2.771(3) Å and a significantly compressed S_4 tetrahedron (two S---S distances of average length 3.637(5) Å and four of 3.550(2) Å (98). Similar substitution reactions have been effected using MeCOX (where X = Cl or $MeCO_2$), for which the complete series of complexes $\left[Fe_4(X)_n(SR)_{4-n}(\mu_3-S)_4\right]^{2-}$ (n = 1-4) has been detected, and $(CF_3\overline{C}O)_2O$, $\overline{(CF_3SO_2)}_2O$ and CF_3SO_3Me, which yield the corresponding $\underline{n}$ = 1 and X = CF_3CO_2 or CF_3SO_3 complexes (99). The magnetic properties of the $\{Fe_4S_4\}$ cluster are not greatly affected by substitution of these various nucleophiles for the thiolate ligands but the redox and absorption spectroscopic properties are. Electrochemical studies have shown that the redox potentials are much less negative than those of the tetrathiolato-complexes (Table 5), presumably reflecting an increased electron affinity of the central core when bound to more electronegative ligands. The potential for the 2-/3- coupld for a solvated Fe_4S_4 core has been estimated at $\underline{ca}$. -600 mV (vs. the saturated calomel electrode) on the basis of these studies. $\left[Fe_4X_4(\mu_3-S)_4\right]^{2-}$ (where X = Cl or Br) exhibit almost reversible 2-/3- redox couples, with potentials of -790 and -740 mV, respectively, in acetonitrile solution. The corresponding 3-/4- reductions are irreversible as is the 2-/3- reduction for the tetraiodo- complex (at $\underline{ca}$. -730 mV). The exchange of thiolate ligands for any of the above nucleophiles results in a decrease in the intensity of the absorptions in the visible spectral region, and $\left[Fe_4(O_2CMe)_4(\mu_3-S)_4\right]^{2-}$ has no absorption maxima in this region. These data support the conclusion (100) that the lowest lying, sulphur to metal charge-transfer, transitions of the $[Fe_4(SR)_4(\mu_3-S)_4]^{2-}$ complexes involve the thiolato-sulphurs.

A number of observations have been made which indicate that the substitution and reaction chemistry of $[Fe_4(SR)_4(\mu_3-S)_4]^{n-}$ complexes may extend beyond simple ligand exchange reactions. Bruice et al. (101) have studied acid-base equilibria, solvolysis and hydrolysis of these complexes and have shown that the hydrolysis of the $\{Fe_4S_4\}$ cubane core proceeds by protonation which may occur on the core faces. The observation (99) that the rate of the reaction between $[Fe_4(SPh)_4(\mu_3-S)_4]^{3-}$ and $(MeCO)_2O$ is much faster than that involving the corresponding dianion, indicates that reduction activates the cluster towards electro-philic attack. This conclusion is probably relevant to the ability of $[Fe_4(SR)_4(\mu_3-S)_4]^{2-}$ complexes, in the presence of an excess of the thiol, to catalyse the conversion of R'NC to RSCH=NR', possibly via the involvement of $[Fe_4(SR)_4(\mu_3-S)_4]^{4-}$ (102). The enhanced reactivity of the C^- (and possible C^+) states of the $\{Fe_4S_4\}$ clusters would seem to be important in respect of certain of their biological functions. Thus a preparation of hydrogenase from Clostridium pasteurianum, possessing one $\{Fe_4S_4\}$ centre, can exist in three different oxidation levels, which span a total change of two electrons, and CO has been shown to bind directly to the cluster in both its oxidised and reduced states (103). The binding of CO to $[Fe_4(SPh)_4(\mu_3-S)_4]^{3-}$ has been established (104) by e.s.r. spectroscopy. These data, together with the inhibition of H_2 oxidation by CO, suggest that H_2 also binds directly to $\{Fe_4S_4\}$ centres during the process of hydrogen uptake. An iron-hydride intermediate has been postulated, deriving either from the protonation of the C^- cluster or H_2 addition to the C^+ cluster; a proton base adjacent to this iron centre could bind the (second) proton. The process of H_2 evolution may therefore involve the two protons being 'collected' on the proton base and the C^- cluster.

Despite this large amount of information now available for the iron-sulphur proteins and their synthetic analogues, the field remains open for much more study. The recent developments have indicated that substitution and/or addition reactions may have a considerable importance in the biological functions of these centres. Another intriguing possibility is that $\{Fe_4Se_4\}$ cores may also have a role in certain biological systems, e.g. formate dehydrogenase (105). $[Fe_4(SR)_4(\mu_3\text{-}Se)_4]^{2-}$ complexes have been prepared by Christou et al. (106) and Holm et al. (107) using analogous procedures to those described for the μ_3-sulphido-derivatives. This latter study prepared the complete family of the $[Fe_4(XPh)(\mu_3\text{-}Y)_4]^{2-}$ (where X, Y = S, Se) complexes and structurally characterised $[Me_4N]_2[Fe_4(SPh)_4(\mu_3\text{-}Se)_4]$. Although this anion has no crystallographic symmetry, it closely corresponds to $\underline{D}_{2d}$ symmetry, with two Fe---Fe separations (mean 2.773(7) Å) shorter than the other four (mean 2.788(7) Å), two Se---Se separations (mean 3.901(12) Å) longer than the other four (mean 3.827(9) Å), and four Fe-S bonds (mean 2.385(2) Å) shorter than the outer eight (mean 2.417(5) Å). The average Fe-Se-Fe and Se-Fe-Se interbond angles are 71(1) and 106(1)$^\circ$, respectively. Compared to its μ_3-sulphido- counterpart, $[Fe_4(SPh)_4(\mu_3\text{-}Se)_4]^{2-}$ exhibits a red-shifted absorption spectrum and a small (ca. 50 mV) positive shift in the redox potentials for the 2-/3- and 3-/4- couples.

3.3.5. Ruthenium and Osmium

In a manner similar to that described earlier for its iron analogue, prolonged boiling of $[(\eta^5-C_5H_5)Ru(CO)_2]_2$ in xylene under nitrogen, results in the loss of half of the carbon monoxide and the formation of the deep purple coloured $[Ru_4(\eta^5-C_5H_5)_4(\mu_3-CO)_4]$. The i.r., mass, and ^1H n.m.r. spectroscopic properties of this complex are consistent with it having the structure of the iron derivative (108a).

$[Ru_4(\eta^6-C_6H_6)_4(\mu_3-OH)_4](SO_4)_2$, $12H_2O$ has been isolated as a product of the reaction between $[(\eta^6-C_6H_6)RuCl_2]$ and aqueous Na_2CO_3 (in the molar ratio 1:2) in the presence of an excess of Na_2SO_4. The cation has crystallographic $\underline{T}_d$ symmetry with Ru---Ru = 3.29, O---O = 2.62, Ru-O = 2.12 Å, O-$\hat{\text{Ru}}$-O = 76.3 and Ru-$\hat{\text{O}}$-Ru = 102.2° (17). Small quantities of $[Os_4(CO)_{12}(\mu_3-O)_4]$ have been obtained as a bi-product of the reaction between OsO_4 and CO in xylene, which leads to $[Os_3(CO)_{12}]$ as the major product. This cubane 'cluster' has crystallographic $\underline{D}_{2d}$ symmetry with no direct metal-metal bonding over the Os---Os separations of 3.253(4) and 3.190(4) Å (8).

3.3.6. Cobalt and Nickel

The reaction of $Co(O_2CMe)_2$, $4H_2O$ with $SbCl_3$ in MeOH at $150^\circ C$ under 200 atm. of CO and 100 atm. of H_2, serendipitously leads to the formation of $[Co_4(CO)_{12}(\mu_3-Sb)_4]$, which is deposited on the walls of the pressure vessel as shiny black insoluble crystals (108). $[Co_4(\eta^5-C_5H_5)_4(\mu_3-P)_4]$ was prepared in a more deliberate manner by refluxing a toluene solution of $[Co(\eta^5-C_5H_5)(CO)_2]$ with white phosphorus (16). The former compound does not contain any Co---Co bonding interactions as the edges of the Co_4 tetrahedron are 4.115(4) $\overset{\circ}{A}$ in length, whereas the latter compound would appear to involve two Co-Co bonds of length 2.504(2) $\overset{\circ}{A}$, with the other four Co---Co distances of 3.630(2) $\overset{\circ}{A}$ precluding any further such interactions. The latter molecule also involves two short P----P contacts of 2.568(2) $\overset{\circ}{A}$, a distance which could be considered indicative of a direct bonding interaction between phosphorus atoms (109). The reaction of $[Co(NO)(CO)_2(PPh_3)]$ with an excess of $(Me_3CN)_2S$ in refluxing toluene, produces the very soluble, brown crystalline product $[Co_4(NO)_4(\mu_3-NCMe_3)_4]$. The infra-red spectrum of this compound contains a single strong $\nu(N-O)$ band at 1722 cm^{-1}, suggesting a structure involving one terminal nitrosyl ligand per cobalt atom. This has been confirmed by an X-ray crystallographic study which identified a cubane cluster, with a triply bridging NR group being located over each face of the Co_4 tetrahedron, the overall molecular structure approximating to $\underline{D}_2$ symmetry. The metal-metal distances range from 2.46 to 2.72 $\overset{\circ}{A}$ and thus some direct bonding interaction is presumed to occur between each pair of cobalt atoms (110). The reaction of elemental sulphur with $[Co(\eta^5-C_5H_5)CO]$ produces $[Co_4(\eta^5-C_5H_5)_4(\mu_3-S)_2(\mu_3-S_2)_2]$, in which a distorted tetrahedral array of $Co(\eta^5-C_5H_5)$ moieties are linked by two bridging sulphido- and two bridging disulphido-ligands situated above the faces of a Co_4

tetrahedron, the edges of which (3.2 - 3.7 Å) are too
long for any direct metal-metal bonding interaction.
These disulphido-groups are oriented such that one of
the atoms is co-ordinated to two of the cobalts and
the other to the third of the triangular face (111).
This compound reacts with an excess of Ph_3P with the
cleavage of the disulphide groups and sulphur abstrac-
tion to form the more regular cubane complex
$[Co_4(\eta^5-C_5H_5)_4(\mu_3-S)_4]$. This latter complex may be
oxidised by $Ag[PF_6]$ to the corresponding monocation.
The dimensions of these clusters are included in
Table 9 (p. 327) and a comparison of the corresponding
bond lengths suggests that the oxidation removes an
electron from an orbital which is primarily metal-
metal antibonding in character (6).

$[Ni_8(CO)_8(\mu_4-PPh)_6]$ has been isolated as black
plate-like crystals following the addition of $PhPCl_2$
to a suspension of $[NMe_4]_2[Ni_3(CO)_3(\mu_2-CO)_3]$ in dry
THF. The molecule can be considered an analogue of
cubane as it contains a cube of nickel atoms with a
single bond over each edge of length 2.636(3) -
2.681(3) Å; each nickel atom is bonded to a terminal
CO group and situated above the centre of each face
of the cube is a PPh group (12b).

The addition of KOH in methanol to a solution of
$M(pd)_2$ (where M = Co or Ni and pd = 2,4-pentane-
dionate) in methanol results in the formation of the
corresponding $[M_4(pd)_4(MeOH)_4(\mu_3-OMe)_4]$ complex (112).
These two complexes are isomorphous and their structure,
as determined for the cobalt derivative is illustrated
in Figure 13. The molecules have $\underline{C}_2$ crystallographic
symmetry with Co---Co separations between 3.10(1) and
3.17(1) Å. The corresponding nickel complex has a
room temperature magnetic moment of 3.3 B.M. per
nickel, a value consistant with the presence of

314.

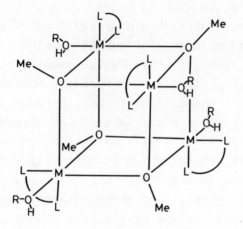

Figure 13

Arrangement of the atoms for

$$[M(LL)_4(ROH)_4(\mu_3-OMe)_4]$$

(where $M = Co$ or Ni, $LL = MeCOCHCOMe$
and $R = Me$ [112] or $M = Ni$, $LL = \underline{o} - O \cdot C_6H_4 \cdot CHO$,
and $R = Me$ or Et [113, 114])

octahedrally co-ordinated nickel(II). A study of the
magnetic susceptibility from 296 to 1.6° K has
demonstrated that the ground molecular spin state has
S = 4 (<u>i.e.</u> eight electron spins parallel) and these
electrons are ferromagnetically coupled. The intramolecular Ni---Ni coupling constant is only 7 cm^{-1}
(<u>ca.</u> 10^{-3} eV) and this is considered to arise as a
result of superexchange over the μ_3-OMe groups,
rather than to any direct Ni-Ni interactions (113).
Schrauzer and Kohnle (114) concluded, on the basis of
molecular weight, spectral and magnetic data that
$\{Ni(O.C_6H_4.CHO)(OMe)(MeOH)\}$ was tetrameric with a
cubic structure. This was confirmed by Andrew and
Blake (115) for the corresponding ethanol solvate,
the structure of which is illustrated in Figure 13;
the molecule has $\underline{S_4}$ crystallographic symmetry but
closely approximates to $\underline{T_d}$ symmetry with Ni---Ni
separations of 3.08(2) $\overset{o}{A}.\overline{\ }$ A weak ferromagnetic
exchange occurs within these cubane frameworks with
coupling constants of 4 and 7 cm^{-1}, for the methanol
and ethanol adducts, respectively. A hydroxonickel(II)
tetramer $\{Ni_4(OH)_4\}^{4+}$ aq. has been postulated as the
major cationic species formed in solutions of $Ni(ClO_4)_2$
(0.1 to 1.0 mol l^{-1}) on the basis of potentiometric,
spectrophotometric and kinetic studies (116).

3.3.7. Platinum

$[Pt_4Me_{12}(\mu_3-Cl)_4]$ was one of the first compounds
shown to have the cubane core, in a study (117) which
(inadvertently (118)) showed this also to be true for
$[Pt_4Me_{12}(\mu_3-OH)_4]$. Four compounds of this type have
been structurally characterised and the average

TABLE 6

Average Dimensions (Å, degrees) of the Cubane Frameworks
of $[Pt_4R_{12}(\mu-_3X)_4]$ compounds

	Pt---Pt (Å)	X---X (Å)	Pt-X (Å)	X-P̂t-X (°)	Pt-X̂-Pt (°)	Ref.
$[Pt_4Me_{12}(\mu_3-OH)_4]$	3.430(2)	2.78(1)	2.22(1)	77.6(2)	101.2(2)	a
$[Pt_4Me_{12}(\mu_3-Cl)_4]$	3.73(3)	3.28	2.48	81	99	b
$[Pt_4Et_{12}(\mu_3-Cl)_4]$	3.903(15)	3.34	2.58(5)	81(2)	98	c
$[Pt_4Me_{12}(\mu_3-T)_4]$	3.9	4.0	2.83(4)	91(2)	89(2)	d

a T. G. Spiro, D. H. Templeton and A. Zalkin, Inorg. Chem., 7, 2165 (1968).

b R. E. Rundle and J. H. Sturdivant, J. Am. Chem. Soc., 69, 1561 (1947).

c R. N. Hargreaves and M. R. Truter, J. Chem. Soc. (A), 90 (1971).

d G. Donnay, L. B. Coleman, N. G. Krieghoff and D. O. Cowan, Acta Cryst., B24, 157 (1968).

dimensions of their cubane frameworks are listed in
Table 6. The ^{1}H n.m.r. spectrum of $[Pt_4Me_{12}(\mu_3\text{-OH})_4]$
contains a seven-line pattern for the hydroxo-proton
consistent with a coupling with three equivalent
platinum atoms (^{195}Pt, I = $\frac{1}{2}$). On the basis of
similar ^{1}H n.m.r. data, $[PtMe_3(SH)]_4$ is suggested (119)
to have the analogous cubane structure. Tipper's
compound, $Pt(C_3H_6)Cl_2$, which is prepared by the
treatment of a solution of chloroplatinic acid in
acetic anhydride with cyclopropane, has been
formulated as the cubane 'cluster' $[Pt_4(C_3H_6)_4Cl_4(\mu_3\text{-Cl})_4]$
on the basis of infra-red, mass spectroscopic and other
data (120).

3.3.8. Copper and Silver

The addition of an aqueous solution of copper(II)
bromide to one of $K_2\{S_2C_2(CN)_2\}$ produces the
$[Cu_8\{S_2C_2(CN)_2\}_6]^{4-}$ complex. This anion consists of
a distorted cubic array of copper atoms, with interatomic distances, which range from 2.783 to 2.865 Å,
embedded within a group of twelve sulphur atoms
located at the vertices of a distorted icosahedron
(12). Other such systems have also been characterised
with dithiosquarato- and 1,1-dicarboethoxy-2-2-ethyl-
enedithiolato- ligands and the Cu_8 cube appears to be
a regular feature for copper(I) complexes involving
this type of ligand (121). A study of the reactivity
of these systems has commenced and, in one instance,
proton addition has been shown to lead to the formation
of a species containing an assembly of ten copper
atoms (122).

One method for the characterisation of tertiary

phosphines and arsines consists of the preparation of
their highly crystalline gold(I) chloride complexes
$Au(YR_3)Cl$ (where Y = P or As). Mann et al. (123)
showed that copper(I) and silver(I) iodides also form
derivatives with the same empirical formula, however,
molecular weight determinations showed that these
gold(I) complexes are monomeric whereas the copper(I)
and silver(I) complexes are tetrameric. The X-ray
crystallographic studies of two of these complexes
represented the first structural characterisation of
cubant 'clusters'. A comprehensive development of
the initial observations has been accomplished by
Churchill et al. and Teo and Calabresi (7, 9, 124-127)
and this series of complexes most clearly manifest
the effects of non-bonded interactions upon the
geometry of the cubane skeleton. Table 7 summarises
one major conclusion to emerge from these studies,
namely that those tetrameric complexes, for which the
non-bonded repulsions are too large to be accommodated
within the cubane framework, adopt the chair
conformation. The atomic arrangements for these two
geometries is illustrated in Figure 14 for
$[Ag_4(PPh_3)_4I_4]$ which can exist in either the cubane
or the chair conformation in the solid state. Slow
crystallisation of the complex from chloroform/ether
yields monoclinic crystals containing molecules with
the cubane framework, whereas slow crystallisation
from methylene chloride/ether yields triclinic crystals
of composition $Ag_4(PPh_3)_4I_4 \cdot 1.5CH_2Cl_2$, in which the
molecules adopt the chair conformation (124).

Factors which affect the relative stabilities of
the cubane and step structures within this group of
molecules include (125): (i) the magnitude of the X---X,
M---M, H---X, H---H, and other non-bonded interactions,
(ii) the relative stabilities of the tetrahedrally and

TABLE 7

Structures adopted by $M_4(Ph_3Y)_4X_4$ (where M = Cu or Ag, Y = P or As, and X = Cl, Br or I) molecules[a].

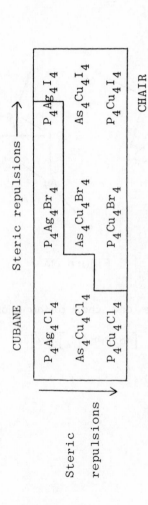

CUBANE   Steric repulsions $\longrightarrow$

	CUBANE		CHAIR
Steric repulsions $\longrightarrow$	$P_4Ag_4Cl_4$	$P_4Ag_4Br_4$	$P_4Ag_4I_4$
	$As_4Cu_4Cl_4$	$As_4Cu_4Br_4$	$As_4Cu_4I_4$
	$P_4Cu_4Cl_4$	$P_4Cu_4Br_4$	$P_4Cu_4I_4$

Steric repulsions $\longrightarrow$

a after B.-K. Teo and J. C. Calabrese, Inorg. Chem., 15, 2474 (1976).

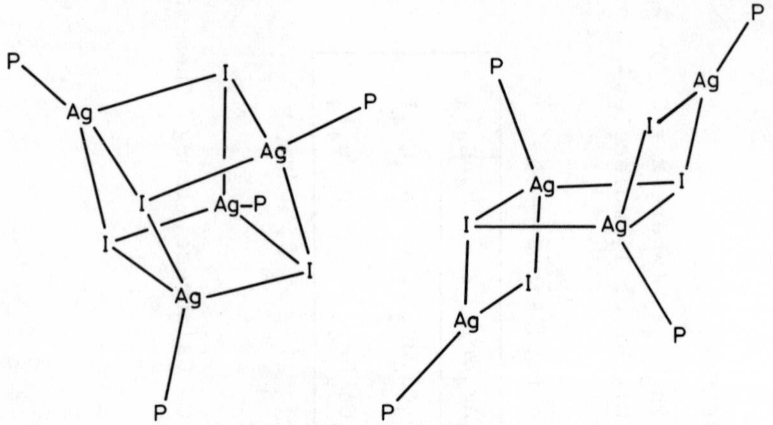

Figure 14

Cubane and chair conformations for

$$\left[Ag_4(PPh_3)_4\ I_4\right]\ \left[124\right]$$

trigonally co-ordinated M atoms, (iii) the relative
stability of μ_3- and μ_2- bonded X atoms, and (iv)
crystal packing interactions. The series of complexes
$[Cu_4(PR_3)_4(\mu_3-I)_4]$ illustrate the sequence of
structural consequences with increasing steric crowding;
the PEt_3 complex has a 'regular' cubane framework (126),
this is distorted in the $PMePh_2$ complex (9), and the
PPh_3 complex adopts the step structure (127). Table 8
lists selected intramolecular dimensions for those
molecules of this $[M_4(YR_3)_4(\mu_3-X)_4]$ series which adopt
a cubane geometry. Certain systematic structural
variations are evident for these molecules. As the
size of the halogen increases then the size of the X_4
tetrahedron also increases, maintaining the inter-
halogen distances at least at their van der Waals
separations; thus to maintain the appropriate M-X bond
lengths, the size of the M_4 tetrahedron is reduced.
(The only exception to this is the $[Ag_4(PPh_3)_4(\mu_3-Br)_4]$
complex, where the crystallography imposed $\underline{C_3}$ symmetry
has been suggested (124) to result in the atypically
large Ag---Ag separations). The contraction of the M_4
tetrahedron causes some increase in the repulsions
between the YR_3 groups and the cubane framework which
is manifest as a slight increase in the length of the
M-Y bonds. As seen clearly for the $[M_4(PPh_3)_4(\mu_3-Cl)_4]$
complexes, the replacement of copper by silver leads
to an expansion of the whole cubane skeleton.

3.3.9. Zinc, Cadmium and Mercury

Dimethyl- and diethylzinc react with phenol and
some aliphatic alcohols to form the corresponding
$[Zn_4(R)_4(\mu_3-OR')_4]$ complex (128). The molecules of

TABLE 8

Average selected intramolecular dimensions (Å, degrees) of $[M_4(YR_3)_4(\mu_3-X)_4]$

(where M = Cu or Ag, Y = P or As, and X = Cl, Br, or I) complexes

Compound	Crystal Symmetry	M---M	X---X	M-X	M-Y	M-X-M	X-M-X	Ref
$[Cu(PEt_3)Cl]_4$	T_d	3.211	3.657	2.438	2.176	82.4	97.2	a
$[Cu(PEt_3)Br]_4$	T_d	3.184	3.932	2.544	2.199	77.5	101.3	a
$[Cu(PEt_3)I]_4$	T_d	2.927	4.380	2.684	2.254	66.1	109.4	b
$[Cu(PPh_3)Cl]_4$	C_2	3.306	3.576	2.444	2.193	85.2	94.1	c
$[Cu(PPh_2Me)I]_4$	C_2	2.930	4.406	2.689	2.250	65.8	109.5	d
$[Cu(AsEt_3)I]_4$	T_d	2.783	4.424	2.677	2.361	62.6	111.5	b
$[Ag(PPh_3)Cl]_4$	C_2	3.633	3.837	2.653	2.382	86.5	92.7	e
$[Ag(PPh_3)Br]_4$	C_3	3.825	4.082	2.800	2.422	86.2	93.7	f
$[Ag(PPh_3)I]_4$	C_1	3.483	4.582	2.910	2.458	73.6	104.1	g
$[Ag(PEt_3)I]_4$	D_{2d}	3.208	4.753	2.919	2.438	66.7	109.0	h

a   M. R. Churchill et al., Inorg. Chem., 14, 2041 (1975).

b   M. R. Churchill and K. L. Kalra, Inorg. Chem., 13, 1899 (1974).

c   M. R. Churchill and K. L. Kalra, Inorg. Chem., 13, 1065 (1974).

d   M. R. Churchill and F. J. Rotella, Inorg. Chem., 16, 3267 (1977).

e   B.-K. Teo and J. C. Calabrese, Inorg. Chem., 15, 2467 (1976).

f   B.-K. Teo and J. C. Calabrese, J. Chem. Soc. Chem. Comm., 185 (1976).

g   B.-K. Teo and J. C. Calabrese, Inorg. Chem., 15, 2474 (1976).

h   M. R. Churchill and B. G. DeBoer, Inorg. Chem., 14, 2502 (1975).

the compound with R=R'=Me have a cubane framework with T_d symmetry and Zn-O = 2.09 $\overset{o}{A}$, Zn-O-Zn = 96, and O-Zn-O = 83° (129). The reaction between Me_2Zn and MeOH yields a further product of composition $Zn_7O_8Me_{14}$, the ^1H n.m.r. spectrum of which was interpreted on the basis of a dicubane structure (130). This novel possibility was subsequently confirmed by an X-ray crystallographic study, the two cubes having a common zinc atom (131).

$[Hg_4Me_4(\mu_3-OSiMe_3)_4]$ has also been characterised by X-ray crystallography; this complex dissociates into monomers in benzene solution whereas the analogous zinc and cadmium complexes maintain their tetrameric structure (132).

4. BONDING CONSIDERATIONS

The symmetrized combinations of the valence orbitals of the constituents of a $[M_4L_4(\mu_3-X)_4]$ cubane cluster with $\underline{T_d}$ symmetry are: M_4(or X_4)$-(n-1)d$, $a_1+2e+2t_1+3t_2$; ns, $\overline{a_1}+t_2$; np, $a_1+e+t_1+2t_2$; $L_4-\sigma$, a_1+t_2, π, $e+t_1+t_2$. The essentials of the bonding interactions are more conveniently discussed using a localised molecular orbital treatment, in which the 36 valence orbitals on the (transition) metal atoms are grouped in 3 discrete $(a_1+e+t_1+2t_2)$ sets: (i) orbitals external to the cluster which are used for π- and/or σ-bonding to the ligands L or are non-bonding; (ii) orbitals used to form the twelve M-X bonds which comprise the cubane skeleton: (iii) orbitals for metal-metal bonding which group as (a_1+e+t_2) bonding and (t_1+t_2) antibonding. (In a similar manner, the ns and np valence orbitals of the atoms X_4 are conveniently subdivided into a group (a_1+t_2) directed away from the cube and a group $(a_1+e+t_1+2t_2)$ for bonding to M_4). The relative energies of these groups of orbitals typically appear to be: M-X bonding, M-L bonding <M-M bonding <(M non-bonding) <M-M antibonding <M-L antibonding <M-X antibonding.

The cubane skeleton of $[Li_4(\mu_3-Me)_4]$ is held

together by 8 electrons which are accommodated in (a_1+t_2) M-X bonding orbitals.

The principal structural constraint which determines the gross shape of a cubane cluster is the extent of the direct metal-metal bonding. This varies from zero to a maximum of a single metal-metal bond over each of the six M---M distances. Cubane clusters so far characterised for the metals of the s and p block elements have no electrons involved in metal-metal bonding. Dahl et <u>al</u>. have synthesised and characterised several iron and cobalt cubane clusters with varying extents of metal-metal bonding. The dimensions of the central moieties of these clusters are listed in Table 9, the complexes being ordered according to the number of electrons in the orbitals used to describe the direct metal-metal interactions. Maximum metal-metal bonding is nominally achieved when there are 12 electrons to be accommodated in these orbitals, since these fill the (a_1+e+t_2) bonding set; electrons additional to these occupy the (t_1+t_2) anti-bonding orbitals. Thus the average metal-metal separation of the complexes in Table 9, generally lengthens as the M_4 electronic configuration increases from 12 to 24. $\left[Fe_4(\eta^5-C_5H_5)_4(\mu_3-CO)_4\right]$ (41) has a 12 electron Fe_4 framework since: the 4 C_5H_5 groups provide $(4 \times 5) = 20$ electrons for the 4 $Fe-C_5H_5$ bonds and the other 4 are provided by the iron atoms; the 4 CO groups provide $(4 \times 2) = 8$ electrons for the 12 Fe-C bonds and the iron atoms provide the remaining 16 electrons; Fe^o has 8 valence electrons and thus $4(8-1-4) = 12$ electrons are accommodated in the metal-metal orbitals. This is also the case for $\left[Fe_4(NO)_4(\mu_3-S)_4\right]$ (4) since: the 4 NO groups provide $(4 \times 3) = 12$ electrons for the Fe-NO $4(\sigma+2\pi)$ bonds, the remaining 12 electrons are provided by the iron atoms;

TABLE 9

Correlation of the Geometry of Cubane Clusters with their M_4 Electronic Configuration

Complex	M_4 Electronic Configuration	Symmetry		M---M (Å)
		Crystal	Ideal	
$[Fe_4(\eta^5-C_5H_5)_4(\mu_3-Co)_4]^+$	11	$\underline{C}_{\underline{s}}$	$\underline{D}_{\underline{2d}}$	[2] 2.495(11) [4] 2.473(7)
$[Fe_4(\eta^5-C_5H_5)_4(\mu_3-Co)_4]$	12	$\underline{C}_1$	$\underline{T}_{\underline{d}}$	[6] 2.520(15)
$[Fe_4(NO)_4(\mu_3-S)_4]$	12	$\underline{C}_1$	$\underline{T}_{\underline{d}}$	[6] 2.634(1)
$[Fe_4(NO)_4(\mu_3-S)_2(\mu_3-NR)_2]$ R = CMe$_3$	12	$\underline{C}_1$	$\underline{C}_{\underline{2v}}$	[1] 2.496(1) [1] 2.642(1) [4] 2.562(1)
$[Co_4(NO)_4(\mu_3-NR)_4]$ R = CMe$_3$	16	$\underline{C}_2$	$\underline{D}_2$	[2] 2.460(2) [2] 2.544(2) [2] 2.717(2)
$[Fe_4(\eta^5-C_5H_5)_4(\mu_3-S)_4]^{2+}$	18	$\underline{S}_4$	$\underline{D}_{\underline{2d}}$	[2] 3.254(3) [4] 2.834(3)
$[Fe_4(\eta^5-C_5H_5)_4(\mu_3-S)_4]^+$	19	$\underline{C}_2$	$\underline{D}_2$	[2] 2.652(4) [2] 3.188(3) [2] 3.319(3)
$[Fe_4(\eta^5-C_5H_5)_4(\mu_3-S)_4]$	20	$\underline{C}_2$	$\underline{D}_{\underline{2d}}$	[2] 2.650(6) [4] 3.363(10)
$[Co_4(\eta^5-C_5H_5)_4(\mu_3-P)_4]$	20	$\underline{C}_5$	$\underline{D}_{\underline{2d}}$	[2] 2.504(2) [4] 3.630(2)
$[Co_4(\eta^5-C_5H_5)_5(\mu_3-S)_4]^+$	23	$\underline{S}_4$	$\underline{D}_{\underline{2d}}$	[2] 3.330(5) [4] 3.172(5)
$[Co_4(\eta^5-C_5H_5)_4(\mu_3-S)_4]$	24	$\underline{C}_2$	$\underline{T}_{\underline{d}}$	[6] 3.30(4)
$[Co_4(Co)_{12}(\mu_3-Sb)_4]$	24	$\underline{D}_{\underline{2d}}$	$\underline{T}_{\underline{d}}$	[6] 4.115(4)

X---X (Å)	M-X (Å)	M-X̂-M (°)	X-M̂-X (°)	Ref.
[2] 2.98(2) [4] 3.11(1)	[4] 2.03(3) [8] 1.96(2)	[4] 79(1) [8] 77(1)	[4] 99(1) [8] 102(11)	a
[6] 3.05	[12] 1.986(9)	[12] 78.8(4)	[12] 100.2(6)	b
[6] 3.503(2)	[12] 2.217(2)	[12] 73.4(1)	[12] 104.4(1)	c
[1] 2.880(7) [1] 3.507(3) [4] 3.208(4)	[2] 1.914(3) [2] 2.222(2) [4] 1.908(3) [4] 2.224(2)	[2] 72.9(1) [2] 81.6(1) [4] 70.4(1) [4] 84.2(1)	[2] 98.0(2) [2] 104.0(1) [4] 101.4(1) [4] 101.6(1)	c
[2] 2.68(1) [2] 2.83(1) [2] 2.89(1)	[4] 1.88 [4] 1.92 [4] 1.92	[4] 80.8(6) [4] 84.0(3) [4] 89.6(3)	[4] 88.3(3) [4] 96.0(4) [4] 99.1(5)	d
[2] 2.820(6) [4] 3.304(5)	[4] 2.156(3) [8] 2.208(4)	[4] 94.9(1) [8] 81.0(1)	[4] 79.3(1) [8] 98.4(1)	e
[2] 2.879(6) [2] 3.062(6) [2] 3.389(8)	[4] 2.185(5) [4] 2.212(5) [4] 2.246(5)	[4] 74.2(2) [4] 92.1(2) [4] 96.3(2)	[4] 80.5(2) [4] 87.5(2) [4] 100.9(2)	e
[2] 3.334(9) [4] 2.880(13)	[4] 2.250(10) [8] 2.204(8)	[4] 73.9(2) [8] 98.0(3)	[4] 98.2(2) [8] 80.5(3)	e
[2] 3.138(2) [4] 2.568(2)	[4] 2.256(1) [8] 2.216(1)	[4] 68.8(3) [8] 108.6(3)	[4] 90.2(4) [8] 70.1(2)	f
[2] 2.930(10) [4] 3.083(8)	[4] 2.215(5) [8] 2.219(5)	[4] 97.2(3) [8] 91.4(2)	[4] 82.6(2) [8] 88.1(2)	g
[6] 2.99(1)	[12] 2.230(3)	[12] 95(1)	[12] 84(1)	g
[6] 3.156(2)	[12] 2.814(2)	[12] 103.8(1)	[12] 74.3(1)	h

a Trinh-Toan, W. P. Fehlhammer and L. F. Dahl, J. Amer. Chem. Soc., 94, 3389 (1972).

b M. A. Neuman, Trinh-Toan and L. F. Dahl, J. Amer. Chem. Soc., 94, 3383 (1972).

c R. S. Gall, C. Ting-Wah Chu and L. F. Dahl, J. Amer. Chem. Soc., 96, 4019 (1974).

d R. S. Gall, N. G. Connelly and L. F. Dahl, J. Amer. Chem. Soc., 96, 4017 (1974).

e Trinh-Toan, B. K. Teo, J. A. Ferguson, T. J. Meyer and L. F. Dahl, J. Amer. Chem. Soc., 99, 408 (1977) and references therein.

f G. L. Simon and L. F. Dahl, J. Amer. Chem. Soc., 95, 2175 (1973).

g G. L. Simon and L. F. Dahl, J. Amer. Chem. Soc., 95, 2164 (1973).

h A. S. Foust and L. F. Dahl, J. Amer. Chem. Soc., 92, 7237 (1970).

the 4 S^{2-} ions provide the 24 electrons for the 12 Fe-S bonds; Fe^{II} has the d^6 configuration and thus $4(6-3) = 12$ electrons are available for metal-metal bonding. $[Co_4(\eta^5-C_5H_5)_4(\mu_3-S)_4]$ (6) and $[Co_4(CO)_{12}(\mu_3-Sb)_4]$ (108) are readily seen to have 24 electrons in the orbitals which describe the Co-Co interactions by arguments similar to those above or, considering cyclopentadiene as $C_5H_5^-$, by observing that each cubane cluster contains 4 Co^{III} (d^6) atoms. Similarly, the complexes $[Re_4(CN)_{12}(\mu_3-X)_4]^{4-}$ (where X = S or Se) have 12 electrons in the metal-metal bonding orbitals, corresponding to 4 Re^{IV} (d^3) atoms, and a single metal-metal bond over each Re----Re separation (39), whereas $[Re_4(CO)_{12}(\mu_3-SMe)_4]$ has a 24 electron Re_4 configuration, corresponding to 4 Re^I (d^6) atoms, and no net metal-metal bonding. Other 24 electron M_4 cubane clusters include the following: $[W_4(CO)_{12}(\mu_3-OH)_4]^{4-}$, $[Mo_4(CO)_8(NO)_4(\mu_3-OH)_4]$ (27), $[Fe_4(CO)_{12}(\mu_3-AsMe)_4]$ (50), and $[Os_4(CO)_{12}(\mu_3-O)_4]$ (8); $[Ru_4(\eta^6-C_6H_6)_4(\mu_3-OH)_4]^{4+}$ (17); $[Pt_4R_{12}(\mu_3-X)_4]$ (where R=Me or Et; X=OH, SH, Cl, or I) (117-119); $[M_4(AR_3)_4(\mu_3-X)_4]$ (where M=Cu, Ag, or Au; A=P or As) (123-127), $[Zn_4Me_4(\mu_3-OMe)_4]$ (129) and $[Hg_4Me_4(\mu_3-OSiMe_3)_4]$ (132). The Cu, Ag, Au, Zn and Hg complexes have 8 electrons in the 4 M-L σ-bonding orbitals and 16 in 8 M non-bonding orbitals.

Cubane clusters for which the ground state electronic configuration would be orbitally degenerate in $\underline{T_d}$ symmetry, adopt a lower symmetry in which the degeneracy is removed, as expected from the Jahn-Teller theorem. An orbitally degenerate ground state is expected for cubane clusters which include an M_4 configuration of $11\{(a_1)^2(e)^4(t_2)^5\}, 13\{(a_1)^2(e)^4(t_2)^6(t_1)^1\}$, 14, 16, 17, $19\{(a_1)^2(e)^4(t_2)^6(t_1)^6(t_2)^1\}$, 20, 22 or 23 electrons.

As indicated in Table 9, symmetries lower than $\underline{T}_d$ are adopted by the cubane clusters with M_4 configurations of 11, 16, 19, 20 and 23 electrons. The 20 electron configuration corresponds to a net 2 M-M bonds over the cluster, therefore it is easy to rationalise why $[Fe_4(\eta^5-C_5H_5)_4(\mu_3-S)_4]$ (46) and $[Co_4(\eta^5-C_5H_5)_4(\mu_3-P)_4]$ (16) have a central core with $\underline{D}_{2d}$ symmetry, involving two short (2.650(6) and 2.504(2) Å) and four long (3.363(10) and 3.630(2) Å) metal-metal separations.

Although this approach is very successful in describing the extent of the metal-metal bonding interactions over the range of cubane clusters presently known, there are certain details which do not follow from this simple treatment: (i) A comparison of the dimensions of $[Fe_4(\eta^5-C_5H_5)_4(\mu_3-CO)_4]^+$ (44) with those of the corresponding neutral molecule (41), for example the average Fe---Fe separations are 2.48(1) and 2.52(1), respectively, together with the decrease of $\nu(C-O)$ from 1700 to 1620 cm^{-1} upon reduction, indicate that the electron is added to an orbital which is primarily Fe-Fe and C-O antibonding. This contrasts with the prediction that the electron would be added to the t_2 Fe-Fe bonding orbital. (ii) The Fe_4 core of $[Fe_4(\eta^5-C_5H_5)_4(\mu_3-S)_4]^{2+}$ is expected to have the $\{(a_1)^2(e)^4(t_2)^6(t_1)^6\}$ 18 electron configuration, thus a regular $\underline{T}_d$ structure would be expected for this Fe_4S_4 core, each Fe---Fe separation corresponding to a bond of order 0.5. However, the cubane framework of this complex adopts a $\underline{D}_{2d}$ geometry with four Fe--Fe separations of 2.834(3) Å, each perhaps corresponding to a bond of order 0.75, and two of 3.254(3) Å (47). $[Fe_4\{S_2C_2(CF_3)_2\}_4(\mu_3-S)_4]^{2-}$ has an Fe_4S_4 cubane core of similar dimensions (15) and, with 2 Fe^{IV} (d^4) and 2 Fe^{III} (d^5) centres, this Fe_4S_4 system also has 18 electrons in the metal-metal orbitals (47). (iii) $\left[Co_4(\eta^5-C_5H_5)_4(\mu_3-P)_4\right]$ has 4 P---P separations of 2.568(2) $\overset{o}{A}$ and thus may include some direct bonding over these approaches, which is not allowed for in the simple bonding scheme. The studies of Dahl (4, 16, 44, 47, 110) should be consulted for a more detailed molecular orbital treatment of these clusters.

$\left[M_4(LL)_4(ROH)_4(\mu_3-OMe)\right]$ (where M = Co or Ni, LL = MeCOCHCOMe, and R = Me (112); or M = Ni, LL = o-O.C$_6$H$_4$. CHO, and R = Me or Et (113, 114)) have 28 (Co^{II}) or 32 (Ni^{II}) electrons associated with the M_4 core. Therefore, they involve no net metal-metal bonding and the (4 or 8) electrons, remaining after filling the metal-metal orbitals, are accommodated in M-L anti-bonding orbitals. The latter are expected to be primarily metal in character and derive from the $d_{x^2-y^2}$, and d_{z^2} orbitals of each cobalt atom and constitute the set $(e+t_1+t_2)$ in $\underline{T}_d$ symmetry. A study of the magnetic properties which result from the partial occupancy of these orbitals has indicated that the nickel pentane-2,4-dionato-cluster has an electronic ground state with eight unpaired electrons, presumably corresponding to this ....$(e)^2(t_1)^3(t_2)^3$ configuration.

Confusion arises in the description of the electronic structure of the $\left[Fe_4(SR)_4(\mu_3-S)_4\right]^{n-}$ (n = 1-4) complexes, primarily because of the partial occupancy of the metals' orbitals and the consequent necessity to take account of magnetic couplings between unpaired electrons on the different Fe centres, which may be as, or even more, important than the energy separations resulting from covalent bonding interactions. There is no simple way of incorporating the two types of interaction within the same model and, to facilitate comparisons with other cubane systems, only a molecular orbital approach will be

considered. In the case of $[Fe_4(SR)_4(\mu_3-S)_4]^{2-}$
complexes, after the 12 Fe-S and 4 Fe-SR bonding
orbitals are filled, 22 valence electrons remain to
be assigned. The Fe---Fe separations of ca. 2.75 $\overset{\circ}{A}$
although slightly longer than the Fe-Fe single bond
distance in $[Fe_4(\eta^5-C_5H_5)_4(\mu_3-CO)_4]$ and
$[Fe_4(NO)_4(\mu_3-S)_4]$, clearly indicate a strong, direct
iron-iron net bonding interaction. A simple view is
obtained by accommodating 12 electrons in the
(a_1+e+t_2) metal-metal bonding orbitals and the
remaining 10 in the $(e+t_1+t_2)$ metal 'non-bonding'
orbitals. Thus the remaining discussion apparently
centres upon the occupancy of these latter orbitals,
which derive from the d_{x2-y2} and d_{z2} orbitals on the
individual iron centres and are of the correct
symmetry for π-bonding with the RS^- and S^{2-} groups.

An SCW-Xα-SW calculation has been accomplished
for $[Fe_4(SMe)_4(\mu_3-S)_4]^{2-}$ (100). The principal
conclusion reached was that the electronic configuration
fills up to a $(t_2)^4$ level, which was calculated to be
predominantly Fe_4 (antibonding) in character with
sulphur 3p contributions from the (Me)S and (μ_3-S)
groups of 18 and 7%, respectively. The displacement
from T_d to approximate D_{2d} symmetry observed for the
$[Fe_4(S\overline{R})_4(\mu_3-S)_4]^{2-}$ (where R = Ph (62), CH_2Ph (11),
$CH_2CH_2CO_2^-$ (63), or CH_2CH_2OH (64)) complexes and the
corresponding natural centres (57, 58), was suggested
to be consistent with a Jahn-Teller distortion,
removing the degeneracy of the t_2 level to give an
......$(e)^4(b_2)^0$ configuration. The ground state
therefore has a net spin of zero, in agreement with
the diamagnetic nature of the centres at absolute zero
(76) and the lack of any e.p.r. activity (77). The
paramagnetism observed at higher temperatures was
suggested to occur because S$\neq$0 states become thermally

populated by the excitation of electrons across the very small e-b_2 energy gap. This description is also consistent with the $S=\frac{1}{2}$ electronic ground state of $[Fe_4(SR)_4(\mu_3-S)_4]^{3-}$ complexes and their biological counterparts. However, it is not clear why a centre of the latter type should adopt a structure in which the Fe_4S_4 core is elongated (61), in contrast to the compressed nature of this core in the dianions (62); simple considerations suggest that this opposite distortion would reverse the e, b_2 ordering and the 5 electrons in the t_2 level would give rise to the $.....(b_2)^2(e)^3$ orbitally degenerate ground state for $[Fe_4(SR)_4(\mu_3-S)_4]^{3-}$ complexes. Also, this model suggests that $[Fe_4(SR)_4(\mu_3-S)_4]^-$ type centres should have the $.....(t_2)^3$ electronic ground state, which is at variance with the Mössbauer results obtained for the oxidised $\{Fe_4S_4\}$ (HiPIP$_{ox}$) centre of <u>Chromatium vinosum</u> which indicate (77, 85) that the ground state has a net spin of $\frac{1}{2}$.

Some advantages seem to accrue from the simple bonding scheme outlined earlier, if the configuration of $[Fe_4(SR)_4(_3-S)_4]^{2-}$ complexes fill up to a $.....(t)^6(e)^4(t)^0$ occupancy of the iron 'non-bonding' orbitals. (i) The electronic ground state of the 1-, 2-, and 3- clusters are predicted to be $S=\frac{1}{2}$, 0, and $\frac{1}{2}$ respectively. (ii) As these 'non-bonding' orbitals are expected to be of a very similar energy, the systems are expected to exhibit a significant paramagnetism at normal temperatures, deriving from thermal promotions from the e to the higher t_2 level. (iii) These orbitals are not appreciably involved in the Fe_4S_4 cubane core and thus any distortion from $\underline{T_d}$ symmetry due to orbital degeneracy therein is expected to be small, as is observed. $[Fe_4(SPh)_4(\mu_3-S)_4]^{3-}$ and analogous centres are expected to have an orbitally degenerate $(t)^1$ ground state and thus the adoption (61) of a structure with a symmetry lower than $\underline{T_d}$ is consistent with this approach. $\left[Fe_4(SR)_4(\mu_3\text{-}S)_4\right]^{2-}$ centres are expected to have the .....$(t)^6(e)^4(t)^0$ configuration and this should not, in the first instance undergo a Jahn-Teller distortion. However, at room temperature, the thermal population of states deriving from configurations such as $(t)^6(e)^3(t)^1$ may be responsible for the distortion observed (62) and there are no grounds for expecting the sense of this distortion to be the same as in a 3- ion. (iv) This simple approach emphasises that changes in the occupancy of the frontier orbitals should not lead to much structural reorganisation of the cluster, a prerequisite for the effecient function of these centres in electron transfer processes of biological systems.

Nevertheless, an accurate description of the electronic structures of these $\left[Fe_4(SR)_4(\mu_3\text{-}S)_4\right]^{n-}$ centres is not yet available and, given their importance, an improvement in this situation would be most welcome.

I am pleased to record my gratitude to Drs. K. Wade and G. Christou for their constructive contributions to discussions which helped in the preparation of this review.

334.

5. REFERENCES

1. P. Eaton and T. Cole, J. Amer. Chem. Soc., 86,
       3157 (1964); E. B. Fleischer, J. Amer. Chem.
       Soc., 86, 3889 (1964).
2. J. C. Barborak, L. Watts and R. Pettit, J. Amer.
       Chem. Soc., 88, 1328 (1966).
3. Definition of metal cluster used here.
4. R. S. Gall, C. T.-W. Chu and L. F. Dahl, J. Amer.
       Chem. Soc., 96, 4019 (1974).
5. R. S. Nyholm, M. R. Truter and C. W. Bradford,
       Nature, 228, 648 (1970).
6. G. L. Simon and L. F. Dahl, J. Amer. Chem. Soc.,
       95, 2164 (1973).
7. M. R. Churchill, B. G. DeBoer and S. J. Mendak,
       Inorg. Chem., 14, 2041 (1975).
8. D. Bright, J. Chem. Soc. Chem. Comm., 1169 (1970).
9. M. R. Churchill and F. J. Rotella, Inorg. Chem.,
       16, 3267 (1977).
10. A. S. Foust and L. F. Dahl, J. Amer. Chem. Soc.,
       92, 7337 (1970).
11. B. A. Averill, T. Herskovitz, R. H. Holm and
       J. A. Ibers, J. Amer. Chem. Soc., 95, 3523
       (1973).

12a. J. P. Fackler, Jr. and D. Coucouvanis, J. Amer.
      Chem. Soc., <u>88</u>, 3913 (1966); L. E. McCandlish,
      E. C. Bissell, D. Boucouvanis, J. P. Fackler
      and K. Knox, J. Amer. Chem. Soc., <u>90</u>, 7357
      (1968).

12b. L. D. Lower and L. F. Dahl, J. Amer. Chem. Soc.,
      <u>98</u>, 5046 (1976).

13.   J. Deutscher, S. Fadel and M. L. Ziegler, Angew.
      Chem. Intl. Ed., <u>16</u>, 704 (1977).

14.   L. F. Dahl, G. L. Davis, D. L. Wampler and R. West,
      J. Inorg. Nucl. Chem., <u>24</u>, 357 (1962);
      V. A. Maroni and T. G. Spiro, J. Amer. Chem.
      Soc., <u>89</u>, 45 (1967).

15.   I. Bernal, B. R. Davis, M. L. Good and S. Chandra,
      J. Coord. Chem., <u>2</u>, 61 (1972).

16.   G. L. Simon and L. F. Dahl, J. Amer. Chem. Soc.,
      <u>95</u>, 2175 (1973).

17.   R. O. Gould, C. L. Jones, D. R. Robertson and
      T. A. Stephenson, J. Chem. Soc. Chem. Comm.,
      222 (1977).

18.   E. Weiss and E. A. C. Lucken, J. Organometal. Chem.,
      <u>2</u>, 199 (1964).

19.   E. Weiss, H. Alsdorf, H. Kühr and H.-F. Grützmacher,
      Chem. Ber., <u>101</u>, 3777 (1968); E. Weiss,
      K. Hoffmann and H.-F. Grützmacher, Chem. Ber.,
      <u>103</u>, 1190 (1970).

20.   D. Mootz, A. Zinnius and B. Böttcher, Angew. Chem.
      Int. Ed., <u>8</u>, 378 (1969); N. A. Bell and
      G. E. Coates, J. Chem. Soc. (A), 1069 (1966);
      G. E. Coates and M. Tranah ibid., 236 (1967).

21.   T. R. R. McDonald and W. S. McDonald, Proc. Chem.
      Soc., 366 (1962); Acta Cryst., <u>B28</u>, 1619 (1972);
      A. Tzschach and A. Balszuweit, Z. Chem., <u>8</u>,
      121 (1968).

22.   C. O. Quicksall and T. G. Spiro, Inorg. Chem., <u>9</u>,
      1045 (1970).

336.

23. O. E. Esval, Thesis, Univ. of Carolina, (see Ref. 14), (1962).

24. A. J. Barry, W. H. Daudt, J. J. Domicone and J. W. Gilkey, J. Amer. Chem. Soc., 77, 4248 (1955).

25. H. Schumann and H. Benda, Angew. Chem. Int. Ed., 7, 813 (1968).

26. W. Hieber, K. Englert and K. Rieger, Z. anorg. Chem., 300, 295 (1959); W. Hieber and K. Englert, Z. anorg. Chem., 300, 304 (1959).

27. V. Albano, P. Bellon, G. Ciani and M. Manassero, J. Chem. Soc. Chem. Comm., 1242 (1969); U. Sartorelli, L. Garlaschelli, G. Ciani and G. Bonora, Inorg. Chim. Acta, 5, 191 (1971).

28. See H. Vahrenkamp, Angew. Chem. Int. Ed., 14, 322-citation (1975) E. O. Fischer and K. Ulm unpublished results; K. Ulm, Dissertation Universität München, (1961).

29. See H. Vahrenkamp, Angew. Chem. Int. Ed., 14, 322-citation (1975) L. F. Dahl, Lecture VIth Int. Conf. Organometal. Chem., Amherst, Mass., (USA), Aug. 1973.

30. L. J. DeHayes, H. C. Faulkener, W. H. Daub, Jr. and D. T. Sawyer, Inorg. Chem., 14, 2111 (1975).

31. G. Christou, C. D. Garner and F. E. Mabbs, Inorg. Chim. Acta, 29, L189 (1978); G. Christou, C. D. Garner, F. E. Mabbs and T. J. King, J. Chem. Soc. Chem. Comm., 740 (1978).

32. S. R. Acott, G. Christou, C. D. Garner, F. E. Mabbs, M. G. B. Drew, T. J. King, J. D. Rush and C. E. Johnson, results in press.

33. T. E. Wolff, J. M. Berg, C. Warrick, K. O. Hodgson,
    R. H. Holm and R. B. Frankel, J. Amer. Chem.
    Soc., 100, 4630 (1978).

34. S. P. Cramer, K. O. Hodgson, W. O. Gillum and
    L. E. Mortenson, J. Amer. Chem. Soc., 100,
    3398 (1978); S. P. Cramer, W. O. Gillum,
    K. O. Hodgson, L. E. Mortenson, E. I. Stiefel,
    J. R. Chisnall, W. J. Brill and V. K. Shah,
    J. Amer. Chem. Soc., 100, 3814 (1978).

35. G. Christou, C. D. Garner, T. J. King, C. E. Johnson
    and J. D. Rush, J. Chem. Soc. Chem. Comm.,
    (1979) in press.

36. A. Müller, H. Bögge and E. Königer-Ahlborn,
    J. Chem. Soc. Chem. Comm., 739 (1978).

37. P. S. Braterman, J. Chem. Soc. (A), 2907 (1968).

38. W. Harrison, W. C. Marsh and J. Trotter, J. Chem.
    Soc. Dalton Trans., 1009 (1972).

39. M. Laing, P. M. Kiernan and W. P. Griffith,
    J. Chem. Soc. Chem. Comm., 221 (1977).

40. R. B. King, Inorg. Chem., 5, 2227 (1966).

41. M. A. Neuman, Trinh-Toan and L. F. Dahl, J. Amer.
    Chem. Soc., 94, 3383 (1972).

42. N. J. Nelson, N. E. Kime and D. F. Shriver,
    J. Amer. Chem. Soc., 91, 5173 (1969).

43. J. A. Ferguson and T. J. Meyer, J. Chem. Soc.
    Chem. Comm., 623 (1971).

44. Trinh-Toan, W. P. Fehlhammer and L. F. Dahl,
    J. Amer. Chem. Soc., 94, 3389 (1972).

45. R. A. Schunn, C. J. Fritchie, Jr. and C. T. Prewitt,
    Inorg. Chem., 5, 892 (1966).

46. C. H. Wei, G. R. Wilkes, P. M. Treichel and
    L. F. Dahl, Inorg. Chem., 5, 900 (1966).

47. Trinh-Toan, B. K. Teo, J. A. Ferguson, T. J. Meyer
    and L. F. Dahl, J. Amer. Chem. Soc., 99,
    408 (1977).

48. Trinh-Toan, P. Fehlhammer and L. F. Dahl, J. Amer.
    Chem. Soc., 99, 402 (1977).

338.

49.   A. L. Balch, J. Amer. Chem. Soc., 91, 6962 (1969);
        G. N. Schrauzer, V. P. Mayweg, H. W. Finck
        and W. Heinrich, J. Amer. Chem. Soc., 88,
        4604 (1966).

50.   E. Röttinger and H. Vahrenkamp, Angew. Chem. Int.
        Ed., 17, 273 (1978).

51.   Iron-Sulfur Proteins, W. Lovenberg (ed.), Academic
        Press, New York, Vols. I and II, (1973);
        Vol. III (1977) and references therein.

52.   R. H. Holm, Acc. Chem. Res., 10, 427 (1977) and
        references therein.

53.   D. O. Hall, K. K. Rao and R. Cammack, Sci. Prog.
        Oxf., 62, 285 (1975) and references therein.

54.   B. A. Averill and W. H. Orme-Johnson in 'Metal
        Ions in Biological Systems', H. Sigel (ed.),
        Marcel Dekker, New York, Vol. 7, p. 127
        (1977) and references therein.

55.   R. V. Eck and M. O. Dayhoff, Science, 152, 363
        (1966).

56.   D. O. Hall, Adv. Chem. Ser., 162, 227 (1977) and
        references therein.

57.   C. W. Carter, Jr., J. Kraut, S. T. Freer,
        Ng. H. Xuong, R. A. Alden and R. G. Bartsch,
        J. Biol. Chem., 249, 4212 (1974);
        C. W. Carter, Jr., J. Kraut, S. T. Freer and
        R. A. Alden, J. Biol. Chem., 249, 6339 (1974);
        C. W. Carter, Jr., J. Biol. Chem., 252, 7802
        (1977).

58.   E. T. Adman, L. C. Sieker and L. H. Jensen,
        J. Biol. Chem., 248, 3987 (1973); E. T. Adman,
        K. D. Watenpaugh and L. H. Jensen, Proc. Natl.
        Acad. Sci. USA, 72, 4854 (1975); E. T. Adman,
        L. C. Sieker and L. H. Jensen, J. Biol. Chem.,
        25, 3801 (1976).

59. G. N. Schrauzer, G. W. Kiefer, K. Tano and
     P. A. Doemeny, J. Amer. Chem. Soc., 96, 641
     (1974).

60. G. Christou and C. D. Garner, J. Chem. Soc. Dalton
     Trans., in press.

61. E. J. Laskowski, R. B. Frankel, W. O. Gillum,
     G. C. Papaefthymiou, J. Renaud, J. A. Ibers
     and R. H. Holm, J. Amer. Chem. Soc., 100,
     5322, (1978).

62. L. Que, Jr., M. A. Bobrik, J. A. Ibers and
     R. H. Holm, J. Amer. Chem. Soc., 96, 4168
     (1974).

63. H. L. Carrell, J. P. Glusker, R. Job and
     T. C. Bruice, J. Amer. Chem. Soc., 99, 3683
     (1977).

64. G. Christou, C. D. Garner and M. G. B. Drew,
     submitted for publication.

65. G. R. Dukes and R. H. Holm, J. Amer. Chem. Soc.,
     97, 528 (1975).

66. L. Que, Jr., R. H. Holm and L. E. Mortenson,
     J. Amer. Chem. Soc., 97, 463 (1975).

67. W. O. Gillum, L. E. Mortenson, J.-S. Chen and
     R. H. Holm, J. Amer. Chem. Soc., 99, 584
     (1977); D. M. Kurtz, Jr., G. B. Wong and
     R. H. Holm, J. Amer. Chem. Soc., 100, 6777
     (1978).

68. G. Ghristou, C. D. Garner, M. W. W. Adams,
     R. Cammack, K. K. Rao and D. O. Hall,
     submitted for publication.

69. J.-S. Hong and J. C. Rabinowitz, Biochem. Biophys.
     Res. Comm., 29, 246 (1967).

70. B. V. DePamphilis, B. A. Averill, T. Herskovitz,
     L. Que, Jr. and R. H. Holm, J. Amer. Chem.
     Soc., 96, 4159 (1974); J. Cambray, R. W. Lane,
     A. G. Wedd, R. W. Johnson and R. H. Holm,

340.

Inorg. Chem., 16, 2565 (1977).

71. J. G. Reynolds, E. J. Laskowski and R. H. Holm,
J. Amer. Chem. Soc., 100, 5315 (1978).

72. K. Dus, H. DeKlerk, D. Sletten and R. G. Bartsch,
Biochem. Biophys. Acta, 140, 291 (1967).

73. N. A. Stombaugh, J. E. Sundquist, R. H. Burris
and W. H. Orme-Johnson, Biochemistry, 15,
2633 (1976).

74. W. D. Phillips, C. C. McDonald, N. A. Stombaugh
and W. H. Orme-Johnson, Proc. Natl. Acad.
Sci., 71, 140 (1974).

75. M. Cerdonio, R.-H. Wang, J. Rawlings and
H. B. Gray, J. Amer. Chem. Soc., 96, 6534
(1974).

76. T. Herskovitz, B. A. Averill, R. H. Holm,
J. A. Ibers, W. D. Phillips and
J. F. Weiher, Proc. Natl. Acad. Sci. USA,
69, 2437 (1972).

77. R. Cammack, D. P. E. Dickson and C. E. Johnson,
in 'Iron-Sulfur Proteins' W. Lovenberg (ed.),
Academic Press, New York, Vol. III (1977),
p. 283 and references therein.

78. R. N. Mullinger, R. Cammack, K. K. Rao, D. O. Hall,
D. P. E. Dickson, C. E. Johnson, J. D. Rush
and A. Simopoulus, Biochem. J., 151, 75
(1975).

79. G. Christou, C. D. Garner and R. Cammack, submitted
for publication.

80. R. Matthews, S. Charlton, R. H. Sands and G. Palmer,
J. Biol. Chem., 249, 4326 (1974).

81. R. Cammack, Biochem. Soc. Trans., 3, 482 (1975).

82. W. V. Sweeney, A. J. Brearden and J. C. Rabinowitz,
Biochem. Biophys. Res. Comm., 59, 188 (1974).

83. R. Cammack and G. Christou, personal communication.

84. R. H. Holm and J. A. Ibers, in 'Iron-Sulfur
    Proteins', W. Lovenberg (ed.) Academic Press,
    New York Vol. III, (1977), p. 206, and
    references therein.

85. D. P. E. Dickson, C. E. Johnson, R. Cammack,
    M. C. W. Evans, D. O. Hall and K. K. Rao,
    Biochem. J., 139, 105 (1974).

86. K. K. Rao, M. C. W. Evans, R. Cammack, D. O. Hall,
    C. L. Thompson, P. J. Jackson and C. E. Johnson,
    Biochem. J., 129, 1063 (1972).

87. J. D. Rush, C. E. Johnson, G. Christou and
    C. D. Garner, unpublished results (NB see
    definition in text for reference used).

88. R. H. Holm, W. D. Phillips, B. A. Averill,
    J. J. Mayerle, and T. Herskovitz, J. Amer.
    Chem. Soc., 96, 2109 (1974).

89. P. J. Stephens, A. J. Thompson, T. A. Keiderling,
    J. Rawlings, K. K. Rao and D. O. Hall, Proc.
    Nat. Acad. Sci. USA, 75, 5273 (1978).

90. C. W. Carter, Jr., J. Kraut, S. T. Freer, R. A. Alden,
    L. C. Sieker, E. Adman and L. H. Jensen,
    Proc. Nat. Acad. Sci., USA, 68, 3526 (1972).

91. C. L. Hill, J. Renaud, R. H. Holm and
    L. E. Mortenson, J. Amer. Chem. Soc., 99,
    2549 (1977).

92. R. Maskiewicz and T. C. Bruice, J. Chem. Soc. Chem.
    Comm., 703 (1978).

93. S.-P. W. Tang, T. G. Spiro, C. Antanaitis, T. H. Moss,
    R. H. Holm, T. Herskovitz and L. E. Mortenson,
    Biochem. Biophys. Res. Comm., 62, 1 (1975).

94. R. J. Kassner and W. Yang, J. Amer. Chem. Soc.,
    99, 4351 (1977).

95. E. T. Lode, C. L. Murray, W. V. Sweeney and
    J. C. Rabinowitz, Proc. Nat. Acad. Sci. USA,
    71, 1361 (1974).

342.

96. M. W. W. Adams, S. G. Reeves, D. O. Hall,
     G. Christou, B. Ridge and H. N. Rydon,
     Biochem. Biophys. Res. Comm., 79, 1184
     (1977).

97. G. B. Wong. M. A. Bobrik and R. H. Holm. Inorg.
     Chem., 17, 578 (1978).

98. M. A. Bobrik, K. O. Hodgson and R. H. Holm,
     Inorg. Chem., 16, 1851 (1977).

99. R. W. Johnson and R. H. Holm, J. Amer. Chem.
     Soc., 100, 5338 (1978).

100. C. Y. Yang, K. H. Johnson, R. H. Holm and
     J. G. Norman, Jr., J. Amer. Chem. Soc.,
     97, 6596 (1975).

101. T. C. Bruice, R. Maskiewicz and R. C. Job, Proc.
     Nat. Acad. Sci., USA, 72, 231 (1975).

102. A. Schwartz and E. E. van Tamelen, J. Amer. Chem.
     Soc., 99, 3189 (1977).

103. D. L. Erbes, R. H. Burris and W. H. Orme-Johnson,
     Proc. Nat. Acad. Sci., USA, 72, 4795 (1975).

104. B. A. Averill and W. H. Orme-Johnson, J. Amer.
     Chem. Soc., 100, 5234 (1978).

105. J. R. Andreesen and L. G. Ljungdahl, J. Bacteriol.,
     116, 867 (1973); 120, 6 (1974).

106. G. Christou, B. Ridge and H. N. Rydon, J. Chem.
     Soc. Dalton Trans., 1423 (1978).

107. M. A. Bobrik, E. J. Laskowski, R. W. Johnson,
     W. O. Gillum, J. M. Berg, K. O. Hodgson
     and R. H. Holm, Inorg. Chem., 17, 1402
     (1978).

108a. T. Blackmore, J. D. Cotton, M. J. Bruce and
      F. G. A. Stone, J. Chem. Soc. (A), 2931
      (1968).

108b. A. S. Foust and L. F. Dahl, J. Amer. Chem. Soc.,
      92, 7337 (1970).

109. D. E. C. Corbridge, Top. Phosphorus Chem., 3, 57
     (1966).

110. R. S. Gall, N. G. Connelly and L. F. Dahl,
        J. Amer. Chem. Soc., 96, 4017 (1974).

111. V. A. Uchtman and L. F. Dahl, J. Amer. Chem. Soc.,
        91, 3756 (1969).

112. J. A. Bertrand and D. Caine, J. Amer. Chem. Soc.,
        86, 2298 (1964).

113. J. A. Bertrand, A. P. Ginsberg, R. I. Kalplan,
        C. E. Kirkwood, R. L. Martin and
        R. C. Sherwood, Inorg. Chem., 10, 240 (1971).

114. G. N. Schrauzer and J. Kohnle, Chem. Ber., 97,
        1727, (1964).

115. J. E. Andrew and A. B. Blake, J. Chem. Soc. Chem.
        Comm., 1174 (1967); J. Chem. Soc. (A),
        1456 (1969).

116. K. A. Burkov, L. S. Lileć and L. G. Sillén,
        Acta Chem. Scand., 19, 14 (1965);
        G. B. Kolski, N. K. Kildahl and
        D. W. Margerum, Inorg. Chem., 8, 1211 (1969).

117. R. E. Rundle and J. H. Sturdivant, J. Amer. Chem.
        Soc., 69, 1561 (1947).

118. D. O. Cowan, N. G. Krieghoff and G. Donnay,
        Acta Cryst., B24, 287 (1968);
        H. S. Preston, J. C. Mills and
        C. H. L. Kennard, J. Organometal. Chem., 14,
        447 (1968); T. G. Spiro, D. H. Templeton
        and A. Zalkin, Inorg. Chem., 7, 2165 (1968);
        M. N. Hoechstetter and C. H. Brubacker, Jr.,
        8, 400 (1969).

119. R. Graves, J. M. Homan and G. L. Morgan, Inorg.
        Chem., 9, 1592 (1970).

120. S. E. Binns, R. H. Cragg, R. D. Gillard, B. T. Heaton
        and M. F. Pilbrow, J. Chem. Soc. (A), 1227
        (1969).

121. F. J. Hollander and D. Coucouvanis, J. Amer. Chem.
        Soc., 96, 5648 (1974).

122. D. Coucouvanis, D. Swenson, N. C. Baeziger,
      R. Pedelty and M. L. Caffrey, J. Amer.
      Chem. Soc., $\underline{99}$, 8097 (1977).

123. F. G. Mann, D. Purdie and A. F. Wells, J. Chem.
      Soc., 1503 (1936); 1828 (1937); A. F. Wells,
      Z. Kristallogr., $\underline{94}$, 447 (1936).

124. B.-K. Teo and J. C. Calabrese, Inorg. Chem., $\underline{15}$,
      2467 (1976); 2474 (1976).

125. M. R. Churchill and K. L. Kalra, Inorg. Chem.,
      $\underline{13}$, 1065 (1974).

126. M. R. Churchill and K. L. Kalra, Inorg. Chem.,
      $\underline{13}$, 1899 (1974).

127. M. R. Churchill and B. G. DeBoer and
      D. J. Donovan, Inorg. Chem., $\underline{14}$, 617 (1975).

128. G. E. Coates and D. Ridley, J. Chem. Soc.,
      1870 (1965).

129. H. M. M. Shearer and C. B. Spencer, J. Chem. Soc.
      Chem. Comm., 194 (1966).

130. W. H. Eisenhuth and J. R. van Wazer, J. Amer.
      Chem. Soc., $\underline{90}$, 5397 (1968).

131. M. L. Ziegler and J. Weiss, Angew. Chem. Int.
      Ed., $\underline{9}$, 905 (1970).

132. G. Dittmar and E. Hellner, Angew. Chem. Int.
      Ed., $\underline{8}$, 679 (1969).

Transition Metal Clusters
Edited by Brian F.G. Johnson
© John Wiley & Sons Ltd.

CHAPTER 5

Thermochemical Estimation of Metal-to-Metal Bond Enthalpy Contributions in Clusters

J. A. Connor

Department of Chemistry, The University of Manchester, Manchester, U.K.

PAGE

346.

1. POLYNUCLEAR METAL CARBONYLS

1.1. Introduction

The common observation of triangular faces in the
structures of cluster compounds of transition metals
which have close-packed (c.p.) structures in the bulk
is one important basis for the comparison between the
two systems, although there is no a priori relationship between, say, $Rh_6(CO)_{12}$ and a clean rhodium metal
surface.

In the case of transition metal carbonyls, the
value of the comparison is enhanced because it is now
possible to make a variety of physical measurements on
the chemisorption of carbon monoxide on the bulk metal
and to compare the results with those made on a
molecular binary carbonyl of the same metal. For
example, photoelectron spectroscopic studies have
shown (1) that the spectra associated with CO adsorbed
on a solid (semi-infinite) metal (e.g. on Ru(001) or
Ir(001) can be reproduced satisfactorily by a molecular
carbonyl compound containing three $(Ru_3(CO)_{12})$ or four
$(Ir_4(CO)_{12})$ metal atoms. Mononuclear metal carbonyl
compounds (e.g. $W(CO)_6$) produce less satisfactory
models of the spectra from adsorption of CO on W(110).

Measurements of the heat of adsorption of CO on
evaporated polycrystalline transition metal films were
made at an early stage in the development of the

chemistry of solid surfaces. More recently, it has
been shown that the heat of chemisorption of CO is
dependent upon the particular crystal lattice plane
involved. It is important to bear in mind that the
enthalpy of adsorption (whether initial or integral)
of CO on metal surfaces may refer to dissociative
adsorption (to give C + O on the surface) as is
commonly encountered for the earlier members of a
transition series, or to associative adsorption which
is found for the later members of a transition series.
An example of the former is the adsorption of CO on
W(110) (2), and an example of the latter is the
adsorption of CO on Ir(110) (3). Dissociative
adsorption processes usually give rise to larger
enthalpies of adsorption than do associative adsorption,
as might be expected when it is remembered that the
bond dissociation enthalpy of carbon monoxide,
$D_o(CO) = 1072$ kJ mol^{-1} (4). Some representative data
are shown in Table 1A, B. Particular importance
attaches to measurement of the enthalpy of chemisorption of CO on metal films because, by using them
as a reference, it has been shown empirically (5) that
the heat of adsorption $Q(\underline{M},\underline{x})$ of any molecule $\underline{x}$ on any
metal $\underline{M}$ is given by $C^o(\underline{x})Q^o(\underline{M})$ where $C^o(\underline{x})$ is a
coefficient depending only on the molecule and $Q^o(\underline{M})$
depends only on the metal.

TABLE 1A

Initial (A) and integral (B) enthalpy of
adsorption of CO (kJ mol^{-1}) on polycrystalline
evaporated metal films[a]

	A	B		A	B		A	B
Ti	640	628	Zr	632	619			
			Nb	552	485	Ta	561	536
			Mo	310	251	W	527	335
Mn	326	318						
Fe	192	146						
Co	197	192	Rh	192	184			
Ni	176	167	Pd	180	167	Pt	201	184

a. D. Brennan and F. H. Hayes, Phil. Trans. Roy. Soc.,
    258A, 347 (1965).
    D. Brennan and D. O. Hayward, Phil. Trans. Roy.
    Soc., 258A, 375 (1965).

TABLE 1B

Heat of adsorption (kJ mol^{-1}) of CO on single crystals of close-packed metals

		Crystal Face			
		(100)	(110)	(111)	Reference
3d-series	Ni	126	126	111	a
	Cu	69		50	a,f
	Ru	109			f
4d-series	Rh		130		b
	Pd	153	167	142	c
	Ag			27	a
5d-series	Ir		155	146	d
	Pt		130	146	a,e

a. G. Ertl, Angew. Chem., **88**, 423 (1976); Internat. Edn., **15**, 391 (1976).

b. R. A. Marbrow and R. M. Lambert, Surf. Sci., **67**, 487 (1977).

c. H. Conrad, G. Ertl, J. Koch and E. E. Latta, Surf. Sci., **43**, 462 (1974).

d. C. M. Comrie and W. H. Weinberg, J. Chem. Phys., **64**, 250 (1976); J. L. Taylor, D. E. Ibbotson and W. H. Weinberg, J. Chem. Phys., **69**, 4298 (1978).

e. C. M. Comrie and R. M. Lambert, J.C.S. Faraday 1, **72**, 1659 (1976).

f. I. Toyoshima and G. A. Somorjai, Catal. Rev. Sci. Eng., **19**, 105 (1979).

1.2. Thermochemical Measurements on Polynuclear
      Metal Carbonyls

The heat of chemisorption of CO on an evaporated
metal film can be compared to the metal-CO bond
enthalpy derived from thermochemical measurements on
molecular metal carbonyls $M_m(CO)_n$. The standard
enthalpies of formation of most neutral mononuclear
binary carbonyls (m = 1) have been precisely
determined by combustion calorimetry and several of
the more simple polynuclear (m>1) binary carbonyls
have been studied by microcalorimetry (6). The values
of the standard enthalpy of formation in the gas phase
which have been obtained by these methods carry the
experimental uncertainty in the enthalpy of formation
in the standard state (solid, liquid) and the
uncertainty (experimental or estimated) in the
enthalpy of sublimation or vaporisation. The average
uncertainty is 1 per cent (Table 2).

TABLE 2

Standard enthalpy of formation of polynuclear metal carbonyls $[M_m(CO)_n]$ in the gas phase.
Enthalpy of disruption ΔH_D^- and bond description (6).

	Description	$\Delta H_f^O, g$	ΔH_D
$Mn_2(CO)_{10}$	10T + M	-1597.5 ± 5.3	1068
$Re_2(CO)_{10}$	10T + M	-1559 ± 21	2029
$Fe_2(CO)_9$	6T + 6B + M	-1335 ± 25	1173
$Fe_3(CO)_{12}$	10T + 4B + M	-1753 ± 28	1676
$Ru_3(CO)_{12}$	12T + 3M	-1820 ± 28	2414
$Os_3(CO)_{12}$	12T + 3M	-1644 ± 28	2690
$Co_2(CO)_8$	6T + 4B + M	-1172 ± 10	1160
$Co_4(CO)_{12}$	9T + 6B + 6M	-1749 ± 28	2121
$Rh_4(CO)_{12}$	9T + 6B + 6M	-1749 ± 28	2649
$Rh_6(CO)_{16}$	12T + 8B + 11M	-2299 ± 28	3874
$Ir_4(CO)_{12}$	12T + 6M	-1715 ± 26	3051

All values given in kJ mol^{-1}

1.2.1. The Two Centre Electron Pair Bond
Interpretation of the Disruption Enthalpy

The enthalpy of disruption ΔH_D refers to the process

$$M_{\underline{m}}(CO)_{\underline{n}}(g,298K) \rightarrow \underline{m}M(g,298K) + \underline{n}CO(g,298K)$$

and is calculated from

$$\Delta H_D = \underline{m}\Delta H_f^O[M,g] + \underline{n}\Delta H_f^O[CO,g] - \Delta H_f^O[M_{\underline{m}}(CO)_{\underline{n}},g]$$

where $[M,g]$ implies metal atoms in their ground rather than their valence states. The structures of poly-nuclear metal carbonyls can be described in terms of the number of metal-CO terminal (T) bonds, metal-CO bridging (B) bonds and metal-metal (M) bonds which are each considered as two-centre electron pair bonds. If it is assumed that the crystal molecular structures (7) are retained in the gaseous phase then the structures can be described in terms of the bond enthalpy contributions in Table 2, without reference to the length of the bonds involved.

If the results for the polynuclear iron and cobalt carbonyls are considered together with those (6) for the mononuclear binary carbonyls of these metals, then assuming that bond enthalpy contributions are transferable between molecules, the values of these enthalpy contributions (now designated $\bar{T}$, $\bar{M}$ and $\bar{B}$) can be obtained by solving the following simultaneous equations:

$Fe(CO)_5$	5T		= 585	$Co(CO)_4$	4T		= 544
$Fe_2(CO)_9$	6T+6B+M	=	1173	$Co_2(CO)_8$	6T+4B+M	=	1160
$Fe_3(CO)_{12}$	10T+4B+M	=	1676	$Co_4(CO)_{12}$	9T+6B+6M	=	2121

to give $\bar{T} = 117$, $\bar{B} = 65$ and $\bar{M} = 82$ kJ mol^{-1} for iron and $\bar{T} = 136$, $\bar{B} = 65$ and $\bar{M} = 84$ kJ mol^{-1} for cobalt. These values indicate the empirical approximations $\bar{T} \sim 2\bar{B}$ and $\bar{M} \sim 0.66$ $\bar{T}$, the former is consistent with the observations of bridge-terminal exchange of coordinated CO in certain polynuclear systems (8). Finally, these approximations can be summarised by the relation:

$$2\bar{T} \simeq 3\bar{M} \simeq 4\bar{B} \qquad\qquad \underline{\underline{1}}$$

   In relating these enthalpy contributions of the molecular cluster carbonyls to the properties of the bulk metal, it may be noted that the cohesive energy of the bulk is dominated by the d-electron contribution and is determined by the number of nearest neighbours (coordination number), $\underline{z}$ in the bulk and their distance, $\underline{r}$ from the atom (9). The cohesive energy can be represented by the enthalpy of atomisation (sublimation) of the metal, $\Delta H_f^o(M,g)$. On the basis of the model, each metal-metal bond in the bulk contributes $2\Delta H_f^o(M,g)/\underline{z}$ kJ mol^{-1}.

   In body-centered cubic (b.c.c.) metals (e.g. iron) each metal atom is bonded to eight nearest neighbours at a distance $\underline{r}$ and rather less strongly to six next-nearest neighbours at a distance $2\underline{r}/\sqrt{3}$. In a c.p. structure each atom is bonded to twelve nearest neighbours in a (3:6:3) arrangement. For metals having both b.c.c. and c.p. structures such as iron, the ratio $\underline{r}$(c.p.)/$\underline{r}$(b.c.c.) is fairly constant and equal to 1.017. The two lattice types differ so little in energy that it seems justified to consider that their enthalpies of atomisation can be taken as being identical. For b.c.c. structures the value of $2/\underline{z}$ is 5.55 for the eight nearest neighbours (rather than 4), whereas for c.p. structures $2/\underline{z}$ is 6 for the twelve nearest neighbours. The metal-metal bond enthalpy, $\bar{M}$ for the metals already mentioned in Table 2, are shown in Table 3 and compared with values obtained using the approximations $\underline{1}$. These empirical approximations can be extended to:

$$2\bar{T} \simeq 3\bar{M} \simeq 4\bar{B} \simeq 6\Delta H_f^o(M,g)/\underline{z} \qquad \underline{2}$$

While it is not suggested that there is any necessary relation between $\bar{M}$, $\bar{T}$ and $\Delta H_f^o(M,g)$ it is clear that these relationships, $\underline{2}$, provide a useful index of the enthalpy contributions of bonds in polynuclear systems.

TABLE 3

Standard enthalpy of atomisation $\Delta H_f^o[M,g]$ kJ mol^{-1} (4), coordination number $\underline{z}$ of M in the bulk metal under ambient conditions, and $M(i) = 2\Delta H_f^o[M,g]/z$ kJ mol^{-1} for metals which form molecular polynuclear carbonyls, $[M_m(CO)_n]$.

M	$\Delta H_f^o[M,g]$	z^a	$\bar{M}(i)$	$\bar{M}(ii)^b$	$\bar{T}(ii)^b$
Mn	284.5	8(12)	51(47)	67	99
Fe	471.1	8(12)	75(70)	80	117
Co	428.4	12h	71	88	136
Tc	695.0	12h	116	–	–
Ru	651.0	12h	109	115	172
Rh	557.3	12c	93	110	166
Re	775.7	12h	129	127	182
Os	789.9	12h	132	128	192
Ir	665.2	12c	111	127	191

a. h-hexagonal close packed c-cubic close-packed.
b. derived from ΔH_D(Table 2) using the empirical approximations $\underline{2}$, for metal carbonyls shown in Table 2.

1.2.2. Bond Enthalpy-Bond Length Interpretations of the Disruption Enthalpy

Two alternative approaches to the interpretation of ΔH_D values for cluster carbonyls have been proposed by Wade (10,11). These do not rely on the two-centre electron pair bond concept to describe the bonds in polynuclear metal carbonyls and thus enable systems such as $Rh_6(CO)_{16}$ which contains triply (face) bridging carbonyl ligands to be considered more realistically.

1.2.2.1. Enthalpy Independent of Length

The average metal-metal distances in $Rh_4(CO)_{12}$ (2.73 Å) and $Rh_6(CO)_{16}$ (2.76 Å) differ only slightly, so that it is assumed that the metal-metal polyhedron edge bonds in the two clusters are of identical energy $\bar{M}$, whatever their formal bond order. The enthalpy of disruption (Table 2) for $Rh_4(CO)_{12}$ can be described by 6M + 12T, whereas the corresponding quantity for $Rh_6(CO)_{16}$ is described by 12M + 16T without regard to the number of electron pair bonds (i.e. 12M rather than 11M (table 2)). Solution of these equations gives $\bar{M}$ = 86 and $\bar{T}$ = 178 kJ mol^{-1}.

1.2.2.2. Enthalpy Related to Length

The differences in the average metal-metal distance between the bulk metal and various polynuclear metal carbonyls are considered in terms of an empirical logarithmic relationship between bond enthalpy and bond length of the type

$$\bar{M} = A\left[\underline{r}(M-M)\right]^{-\underline{k}}$$

where A and $\underline{k}$ are constants characteristic of the class of neutral polynuclear metal carbonyls. It is possible to derive the following relationships from the earlier description (1.2.1) of the comparison of close-packed and body-centered lattices:

$$\Delta H_D = 4/\underline{r}^{\underline{k}} + 3/(1.155\underline{r})^{\underline{k}} = 6 \ (1.017\underline{r})^{\underline{k}}$$

where $\underline{r}$ is the interatomic distance in a c.p. metal. These equations can be solved to give $\underline{k}$ = 4.5962. Using the known interatomic distances and enthalpies $\bar{M}$(Table 3) in the metals themselves, it is possible to calculate values of $\bar{M}$ (and consequently of $\bar{T}$) in the metal carbonyls. Figure 1 shows the relation between $\bar{M}$ and $\underline{r}$ evaluated for osmium. Table 4 shows values of $\bar{M}$ and $\bar{T}$ for the polynuclear metal carbonyls.

TABLE 4

Enthalpy of disruption $\Delta H_D(kJ \ mol^{-1})$, metal-metal
bond length $\underline{r}(M-M)\overset{o}{A}$, and the bond enthalpy
contributions $\bar{M}$ and $\bar{T}$ $(kJ \ mol^{-1})$ for metals and
polynuclear metal carbonyls.

	$\Delta H_D kJmol^{-1}$	$\underline{r}(M-M)\overset{o}{A}$	$\bar{M}kJmol^{-1}(a)$	$\bar{T} \ kJmol^{-1}$
Mn	285	2.74	47	
$Mn_2(CO)_{10}$	1068	2.92	35	103
Fe	417	2.48	75	
$Fe_2(CO)_9$	1173	2.52	70	123
		2.56	65	
$Fe_3(CO)_{12}$	1676	2.68	52	126
Co	428	2.51	71	
$Co_2(CO)_8$	1160	2.52	70	136
$Co_4(CO)_{12}$	2121	2.49	74	140
Ru	651	2.65	109	
$Ru_3(CO)_{12}$	2414	2.85	78	182
Rh	557	2.69	93	
$Rh_4(CO)_{12}$	2648	2.73	86	178
$Rh_6(CO)_{16}$	3874	2.78	80	182
Re	776	2.74	129	
$Re_2(CO)_{10}$	2029	3.04	80	195
Os	790	2.68	132	
$Os_3(CO)_{12}$	2690	2.88	94	201
Ir	665	2.71	111	
$Ir_4(CO)_{12}$	3051	2.68	117	196

(a). $\bar{M} = A\underline{r}(M-M)^{-4.6}$

    Two necessary consequences of the model employed
are firstly that the values of $\bar{M}$ in Table 4 are
generally less than those in Table 3; <u>mutatis mutandis</u>

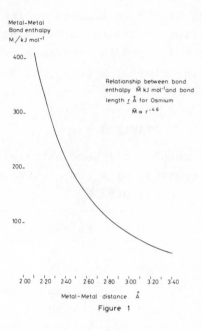

Metal-Metal
Bond enthalpy
M / kJ mol⁻¹

400—

Relationship between bond
enthalpy $\bar{M}$ kJ mol⁻¹ and bond
length $\underline{r}$ Å for Osmium
$\bar{M} \propto r^{-4.6}$

300—

200—

100—

2·00 2·20 2·40 2·60 2·80 3·00 3·20 3·40

Metal-Metal distance Å

Figure 1

the values of $\bar{T}$ are generally greater (Table 4) but
the differences are much less pronounced because of
the numbers of bonds involved, and secondly that the
value of $\bar{T}$ increases with increasing nuclearity of
the cluster $M_m(CO)_n$ (e.g. $\bar{T}[Fe_2(CO)_9] < [Fe_3(CO)_{12}]$) as
the number of CO groups per metal atom $\underline{n}/\underline{m}$ decreases.

1.3. Conclusions

The following are the principal conclusions which
can be drawn from the bond enthalpy contributions in
polynuclear metal carbonyls, <u>irrespective</u> of the
method, (1.2.1) or (1.2.2.2), used to interpret ΔH_D,
the enthalpy of disruption.

1. The metal-CO bond enthalpy, $\bar{T}$ is always greater
than the metal-metal bond enthalpy contribution, $\bar{M}$ in

the systems which have been studied so far.

2. The contribution from metal-metal bonding to the ΔH_D expressed as $P = 100\Sigma\bar{M}/\Delta H_D$ increases with the nuclearity $\underline{m}$ of the cluster $M_m(CO)_n$, as shown in Table 5.

TABLE 5

Contribution of metal-metal bond enthalpy to the enthalpy of disruption of $M_m(CO)_n$ expressed as
$$P = 100\Sigma\bar{M}/\overline{\Delta H}_D$$

	$\underline{n}/\underline{m}$	P(Table 3)	P(Table 4)
M_2, binuclear	4,4.5,5	6	5
M_3, trinuclear	4	15	10
M_4, tetranuclear	3	25	21
M_6, hexanuclear	2.67	32	25

3. There is no purely thermodynamic reason why neutral, binary polynuclear metal carbonyls of elements other than those shown in Tables 3 and 4 should be unstable with respect to decomposition to the metal and carbon monoxide at ambient temperature, unless reactions to form the metal oxide and metal carbide are predominant. The fact that compounds such as $\left[(\eta^5-C_5H_5)_4V_4(CO)_4\right]$ (12) and $\left[Pt_3(CO)_3(\mu_2-CO)_3\right]_n^{2-}$ (n = 2-5) (13) are known, is to be seen as supporting this argument.

These conclusions provide some support for the proposal that, for small clusters at least, the structure of polynuclear metal carbonyls should be considered in terms of fitting a cluster of metal atoms into the vacancies left in a regular polyhedron of carbon monoxide ligands (14), because the energy of the metal-CO interactions, which may be represented by (100-P) (Table 5), dominates any influence the metal-metal interactions may exert.

The absence of thermochemical information on neutral binary clusters of higher nuclearity such as $Os_7(CO)_{21}$ and $Os_8(CO)_{23}$ (15) precludes any statement about the value of P in these systems; however, if the ratio $\underline{n}/\underline{m}$ is a guide, it suggests that P should perhaps be ca. 30. Clusters of even higher nuclearity are usually anionic (e.g. $Rh_{15}(CO)_{27}^{2-}$ (16)): as it is not possible at present to either determine the role of the lattice enthalpy in such systems or to measure the enthalpy of formation of the anion, the precise variation of P with the ratio $\underline{n}/\underline{m}$ is a matter of speculation. If a smooth variation is presumed, then Figure 2 shows the resulting curve. From Figure 2 it appears that the enthalpy contribution of the metal-metal bonds in a neutral metal carbonyl cluster will be dominant (P>50) when $\underline{n}/\underline{m} < \, ^3/_2$, such as in $[M_2(CO)_3]_{\underline{y}}$-type compounds where $\underline{y}$ represents the degree of polymerisation.

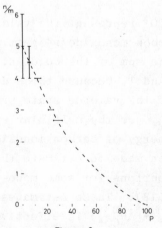

Relationship between ligand/metal ratio n/m for metal carbonyls $[M_m(CO)_n]$ and the contribution from metal-metal bonding to the total enthalpy of disruption, $P = 100 \, \Sigma \bar{M} / \Delta H_D$

Figure 2

If the premises of section (1.1) are reconsidered in the light of the results set out in section (1.2) then in the case of rhodium, for example, there is evidence for a similarity between the surface adsorption of CO and $Rh_6(CO)_{16}$ (Table 6), as far as enthalpies are concerned.

TABLE 6

Enthalpy of chemisorption of CO on rhodium metal compared with $\bar{T}(Rh-CO)$ in $Rh_6(CO)_{16}$ (kJ mol^{-1}).

Rh metal	Adsorption on (110) face	130
	Initial enthalpy (polycrystalline)	192
	Integral enthalpy (polycrystalline)	184
$Rh_6(CO)_{16}$	Electron pair bond interpretation	166
	Bond length–bond enthalpy interpretation	182

The close similarity between the CO of $Rh_6(CO)_{16}$ and CO adsorbed on a Pd(111) single crystal surface has been established by photoelectron spectroscopy (1,17).

Finally it is important to note that the enthalpy change, ΔH^*, in the reaction

$$\underline{M}_{\underline{m}}(CO)_{\underline{n}}\left[g, 298K\right] \rightarrow \underline{m}M^*\left[g, 298K\right] + \underline{n}CO^*\left[g, 298K\right]$$

in which M* and CO* represent the valence states of the metal and carbon monoxide respectively, is not represented by the sum of the bond enthalpy contributions $\bar{M}$ and $\bar{T}$, because these do not take account of either the valence state promotion energy of the metal, ΔH_{M^*}, or the molecular electron reorganisation energy of carbon monoxide, $\Delta H_{CO^*}^M$. Attempts have been made to estimate the values of these two contributions for some mono- and bi-nuclear metal carbonyls (18). These estimates suggest that $\Delta H^*(=\bar{T} + \bar{M} + \Delta H_{M^*} + \Delta H_{CO^*}^M)$ is effectively constant

through a $\underline{d}$-transition series with a value of
approximately 365 (3$\underline{d}$-metals), 415 (4$\underline{d}$-metals) and
460 kJ mol^{-1} (5$\underline{d}$-metals).

2. BINUCLEAR METAL-TO-METAL BONDED SYSTEMS

The least component of any cluster system is
provided by a bond between two metal atoms, although
many such bonds will be involved in a cluster of high
nuclearity.

2.1. Dimetal Decacarbonyls, $M_2(CO)_{10}$ of Manganese, Technetium and Rhenium

The problems which generally arise in this area
are apparent in the simplest binuclear binary
carbonyls, $M_2(CO)_{10}$. In these molecules there are no
bridging carbonyl ligands and the metal-metal bond
between the two $M(CO)_5$ groups is unusually long, being
approximately 10 per cent longer than the corresponding
interatomic distance in the bulk metal. In addition to
the calorimetric measurements (6) mentioned earlier
(section 1.2), the interpretation of which has been
considered (sections 1.2.1 and 1.2.2.2), various
attempts have been made to determine the metal-metal
bond enthalpy in $M_2(CO)_{10}$ molecules by mass spectroscopy (19) and by kinetic measurements on the
reactions of $M_2(CO)_{10}$ in solution (20).
In general, if the excitation energy and the

excess kinetic energy of the fragment ions produced
by electron impact are neglected, then the appearance
potential measured by mass spectrometry can only give
an upper limit to the value of the bond dissociation
energy deduced from such measurements. Where proper
comparison can be made, it would appear that bond
dissociation enthalpies estimated from appearance
potential measurements exceed those derived from
calorimetric measurements by 10-20 per cent. Where
kinetic measurements in solution are concerned, the
identification of the enthalpy of activation for a
bond breaking process with the dissociation enthalpy
of that bond presumes, among other things, that the
solvent plays no part. If there is an exothermic
interaction between the solvent (albeit non-polar)
and the coordinatively unsaturated molecule $M(CO)_5$
which is formed as the M-M bond in $M_2(CO)_{10}$ is broken,
then the activation enthalpy, $\Delta H^{\ddagger}$ will represent an
upper limit value of the metal-metal bond enthalpy.
For the substitution reaction between $M_2(CO)_{10}$ and a
ligand, L in a solvent, S which is represented by the
sequence

$$M_2(CO)_{10} \xrightarrow{} 2M(CO)_5 \xrightarrow{+S} 2\left[SM(CO)_5\right] \xrightarrow[CO]{+L} 2\left[LM(CO)_4 S\right] \xrightarrow{-S} \left[M(CO)_4 L\right]_2$$
$$\quad A \qquad\qquad B \qquad\qquad C \qquad\qquad D \qquad\qquad E$$

the energy profile could be similar to that shown in
Figure 3. The results of the various methods of
determining D(M-M) in these molecules are shown in
Table 6, where comparison is also made with the bulk
metal.

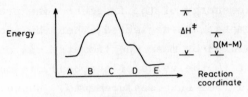

Figure 3

Energy profile for the reaction between
$M_2(CO)_{10}$ and a ligand L in a solvent.

TABLE 6

Estimates of the M-M bond enthalpy contribution
$(kJ \ mol^{-1})$ in $M_2(CO)_{10}$ molecules

	r(M-M)Å	TC1[a]	TC2[b]	EI[c]	K[d]	M̄(metal)[a]
$Mn_2(CO)_{10}$	2.923	67	35	105[e]	154	51(47)
$Tc_2(CO)_{10}$	3.036		72	177	160	116
$Re_2(CO)_{10}$	3.040	127	80	187	166	129
$MnRe(CO)_{10}$				210	163	

a - see Table 3: two centre electron pair bond inter-
    pretation.

b - see Table 4: bond length-bond enthalpy inter-
    pretation.

c - electron impact determination (19).

d - activation enthalpy from reaction kinetic measure-
    ments in solution (20).

e - an other determination of D(Mn-Mn) by electron
    impact gives M = 88 kJ mol^{-1} (D. R. Bidinosti and
    N. S. McIntyre, Can. J. Chem., 48, 593 (1970).

The scatter in the results, particularly for manganese,
is alarming. The lowest value of D(Mn-Mn) (35 kJ mol^{-1})
would suggest that the metal-to-metal bond in $Mn_2(CO)_{10}$
is spontaneously broken at ambient temperature, but

this does not happen. Since the estimate (TC2) depends solely upon the metal-metal bond length (which is unusually long, see above) it seems that the method is only able to provide a lower limit value of the metal-metal bond enthalpy. This difference may be explained, in part at least, by the change in coordination number of the metal atom between manganese metal (45) and $Mn_2(CO)_{10}$. The importance of this contribution to the cohesive energy is described in Chapter 9 (9). On the other hand, as already pointed out, the other methods used (EI,K) are expected to give upper limit values for D(Mn-Mn). The same comments can be made about the values of D(Re-Re). This leaves the values of D(M-M) derived from the electron pair bond interpretation (TC1). These can be criticised because on the one hand (rhenium) there appears to be a negligible decrease in the enthalpy contribution in $Re_2(CO)_{10}$ compared to the metal, and on the other hand (manganese) the enthalpy contribution of the metal-metal bond in $Mn_2(CO)_{10}$ is greater than in the bulk metal, notwithstanding the fact that M-M bond length in both $Mn_2(CO)_{10}$ and $Re_2(CO)_{10}$ is longer than in the metal. However, since three methods (TC1, EI and K) out of the four agree in making D(M-M) in $M_2(CO)_{10}$ the same or greater than in the bulk, it may be that in these cases at least the assumption of an inverse relationship between bond length and bond enthalpy is not justified. If so, the TC1 values of the metal-metal bond enthalpy contributions in $M_2(CO)_{10}$ molecules are to be preferred.

2.2. Diatomic Transition Metal Molecules

The species present in the vapours of pure metals and metal alloys have been studied by high temperature mass spectrometry (Knudsen cell effusion) (21) and, when condensed at low temperatures in a noble gas matrix, by electronic and Mössbauer spectroscopy (22). The description of the bonding in diatomic transition metal molecules, M_2, is a problem of great interest at present (23). Clearly, a proper understanding of M_2 is an important component of the much greater problem which is presented by metal clusters of higher nuclearity and the transition from molecular to bulk metal behaviour. The thermodynamic properties obtained for the solid and the gaseous metal are useful in predicting and understanding the process of crystal growth from the vapour phase in a temperature gradient. The derivation of a precise value of the dissociation energy from purely spectroscopic measurements on transition metal diatomic molecules has been possible in very few, if any, cases up to the present. One reason for this is that the change in internuclear distance is small in most of the transitions studied, another reason follows from the large number of low lying atomic states. Consequently, nearly all the dissociation energies known at present have been determined by mass spectrometric studies of high temperature equilibria. The dissociation energies of some homonuclear and heteronuclear diatomic d-transition metal diatomic molecules are shown in Table 7.

TABLE 7

Dissociation energies, D_o(M-M)(kJ mol^{-1}) of diatomic
transition metal molecules

a. Homonuclear[a]

Sc_2	159±21	Y_2	156±21		
Ti_2	134±21				
V_2	241±13	Nb_2	503±10[g]		
Cr_2	151±29	Mo_2	404±20[b]		
Mn_2	42±29				
Fe_2	121±21				
Co_2	163±25	Rh_2	276±21[c]		
Ni_2	258[d]	Pd_2	105±21		
Cu_2	188.2±6.3	Ag_2	159.5±6.3	Au_2	221.8±6.3

b. Heteroatomic[a]

CrAu	211±15	CoAu	218±21	MnAg	112±21
MnAu	182±13	NiAu	248±21	MoNb	448±25[e]
FeAu	207±25	CuAu	232±9	VAu	238±7[f]

a – values taken from (21).

b – S. K. Gupta, R. M. Atkins and K. A. Gingerich,
Inorg. Chem., 17, 3211 (1978).

c – V. Piacenti, G. Balducci and G. Bardi, J. Less-
Common Metals, 37, 123 (1974).

d – E. Rutner and G. L. Haury, J. Chem. Eng. Data, 19,
19 (1974).

e – S. K. Gupta and K. A. Gingerich, J. Chem. Phys.,
69, 4318 (1978).

f – S. K. Gupta, M. Pelino and K. A. Gingerich, J. Chem.
Phys., 70, 2044 (1979).

g – S. K. Gupta and K. A. Gingerich, J. Chem. Phys.,
1979, in press.

When the values of $D_o(M_2)$ for the 3d-transition metals
are corrected for the promotion energy E_{vs}($3d^n4s^2 \rightarrow$
$3d^{n+1}4s$) from the electronic ground state of M to the
valence state, the dissociation energies D_{vs}($=D_o+2E_{vs}$)
are all in the range (240±40) kJ mol^{-1}.

From the table, the observation that
$D(V_2)>D(Cr_2)$ and that $D(Mo_2)>D(Cr_2)$ would suggest that
$D(Nb_2)>D(MoNb)>D(Mo_2)$. Furthermore, $D(MoNb)$ should be
roughly midway between $D(Mo_2)$ and $D(Nb_2)$, so that
$D(Nb_2)$ is expected to be approximately 502 kJ mol^{-1}
(24). The dissociation energies of certain diatomic
molecules can be calculated (25) with a knowledge of
atomic ground states, promotion energies and the
values of the bonding enthalpy per s-, p or d-electron
(26). Dissociation energies of diatomic heteronuclear
4d-transition metal molecules calculated in this way
(27) are shown in Table 8.

TABLE 8

Calculated dissociation energies (kJ mol^{-1}) of
heteronuclear diatomic transition metal molecules (27)

ZrRu	602	NbRu	661
ZrRh	498	NbRh	561
ZrPd	318	NbPd	356

The high values observed for the dissociation
enthalpy of many diatomic transition metal molecules when
compared to the calculated single bond enthalpy in the
bulk metal (compare $D(Mo_2)$404 kJ mol^{-1} (Table 7) with
$\bar{M}(Mo)$118 kJ mol^{-1} calculated from $\Delta H_f^o(Mo,g)$656 kJ mol^{-1}
(4)) lead to the suggestion that there is multiple
bonding in these diatomic molecules. Some recent
molecular orbital calculations using Extended Huckel
and SCF Xα-SW methods have indicated a bond order of
five for Nb_2 (28) and six for Cr_2, CrMo and Mo_2 (29,30).
Other methods (GVB-CI (31), Density functional theory
(23)) applied to 3d-diatomic molecules indicate that
the two metal atoms are joined by a single bond which
is formed by overlap of the 4s atomic orbitals on

each atom. The differences between the two approaches are clearly demonstrated by the results obtained from the calculations of dichromium, shown in Table 9.

TABLE 9

Valence molecular orbital electron configuration
of the ground state of dichromium, Cr_2

Density functional formalism (23)

$$1\sigma_g^2 \; 2\sigma_g^1 \; \pi_u^2 \; \delta_g^2 \; \delta_u^{*2} \; \pi_g^{*2} \; 1\sigma_u^{*1} \qquad r_e = 2.84 \text{ Å}$$
$$\text{Bond order} = 1$$

SCF-Xα-SW method (30)

$$1\sigma_g^2 \; \pi_u^4 \; 2\sigma_g^2 \; \delta_g^4 \qquad r_e = 1.90 \text{ Å}$$
$$\text{Bond order} = 6$$

These differences may be resolved by taking account of the fact that the spin-flip energies of elements towards the middle of a transition series are large (23) and r_e is also large. However, in binuclear and cluster compounds all spins are usually saturated by bond formation with the result that the attractive force is of d-character (9). The Cr-Cr distance in dichromium can be calculated from the equation (section (1.2.2)

$$E(M-M) = A\left[r(M-M)\right]^{-4.6} \qquad \underline{3}$$

using r(Cr-Cr = 2.498 Å and $\bar{M}$ = 72 kJ mol^{-1} in the bulk metal and $D(Cr_2)$ from Table 7 to give r(Cr-Cr) = 2.12 Å in dichromium. The same treatment applied to Mo_2 gives a calculated separation of 2.08 Å in fair agreement with the value (2.12 Å) calculated for the potential energy minimum of the molecule (28,29).

The electronic spectra of a number of matrix isolated transition metal diatomic molecules have been recorded and assigned with the aid of molecular orbital calculations. The lowest energy absorption is usually assigned to the electronic transition $1\sigma_g \leftarrow 2\sigma_u$ to the most tightly bound valence molecular orbital. The results are shown in Table 10.

Similar low energy absorptions assigned to excitation of electrons involved in metal-metal bonding are observed in the electronic spectra of polynuclear metal carbonyls, some examples of which are shown in Table 11.

The position of the σ-σ* transition has been correlated with M-M bond strength and used to imply little ionic bonding in heterodinuclear complexes (32).

TABLE 10

Lowest energy electronic transition (kJ mol^{-1}) for matrix isolated diatomic transition metal molecules

Sc_2	181[a]			
Ti_2	192[a]			
V_2	204[b]		Nb_2	181[g]
Cr_2	260[c]	CrMo 246[f]	Mo_2	232[g]
Mn_2	173[d]			
Fe_2	220[d]			
			Rh_2	348[h]
Ni_2	261[e]			
			Ag_2	299[i]

$\left[1J\ mol^{-1} = 83.5911\ cm^{-1} \right]$

a – R. Busby, W. Klotzbücher and G. A. Ozin, J.Amer. Chem. Soc., 98, 4013 (1976).

b – A. Ford, H. Huber, W. Klotzbücher, E. P. Kündig, M. Moscovitz and G. A. Ozin, J.Chem. Phys., 66, 524 (1977).

c – G. A. Ozin and W. Klotzbücher, J.Mol. Catal., 3, 195 (1978).

d – T. C. DeVore, A. Ewing, H. F. Franzen and V. Calder, Chem. Phys. Letters, 35, 78 (1975).

e – J. Hulse and M. Moscovitz, J. Chem. Phys., 66, 3988, (1977).

f – W. Klotzbücher, G. A. Ozin, J. G. Norman and H. J. Kolari, Inorg. Chem., 16, 2871 (1977).

g – W. Klotzbücher and G. A. Ozin, Inorg. Chem., 16, 984 (1977).

h – A. J. L. Hanlan and G. A. Ozin, Inorg. Chem., 16, 2848 (1977).

i – G. A. Ozin and H. Huber, Inorg. Chem., 17, 155 (1978)

TABLE 11

Low energy electronic transition ($\sigma \rightarrow \sigma^*$)(kJ mol^{-1}) for some metal-metal bonded compounds

a. Homonuclear

	$h\nu(\sigma \rightarrow \sigma^*)$	Ref
$[Cr_2(CO)_{10}]^{2-}$	325	a
$[W_2(CO)_{10}]^{2-}$	341	a
$Mn_2(CO)_{10}$	350	a
$Tc_2(CO)_{10}$	379	a
$Re_2(CO)_{10}$	386	a
$Fe_3(CO)_{12}$	274	b
$Ru_3(CO)_{12}$	306	b
$Os_3(CO)_{12}$	363	b
$Co_2(CO)_8$	342	[32]
$Mo_2(CO)_6(C_5H_5)_2$	308	c
$W_2(CO)_6(C_5H_5)_2$	330	c
$Co_2(CO)_6(P(OMe)_3)_2$	330	d
$Co_2(CO)_6(PCy_3)_2$	316	d

b. Heteronuclear

	$h\nu(\sigma \rightarrow \sigma^*)$	Ref
$[MnCr(CO)_{10}]^-$	350	a
$MnRe(CO)_{10}$	371	a
$[MnFe_2(CO)_{12}]^-$	342	b
$Mn_2Fe(CO)_{14}$	278	b
$Re_2Fe(CO)_{14}$	315	b
$MnReFe(CO)_{14}$	297	b
$FeRu_2(CO)_{12}$	306	b
$C_5H_5(CO)_3MoMn(CO)_5$	321	e
$C_5H_5(CO)_3MoRe(CO)_5$	328	e
$C_5H_5(CO)_3MoFe(CO)_2C_5H_5$	302	f
$C_5H_5(CO)_3MoCo(CO)_4$	338	f
$C_5H_5(CO)_3WRe(CO)_5$	362	e
$C_5H_5(CO)_3WCo(CO)_4$	350	f

$[1 \text{J mol}^{-1} = 83.5911 \text{ cm}^{-1}]$

For references please see next page

TABLE 11 Continued.

a - R. A. Levenson and H. B. Gray, J.Amer. Chem. Soc.,
   97, 6042 (1975).

b - D. R. Tyler, R. A. Levenson and H. B. Gray, J.Amer.
   Chem. Soc., **100**, 7888 (1978).

c - M. S. Wrighton and D. S. Ginley, J.Amer. Chem. Soc.,
   97, 4276 (1975).

d - A. J. Poë and R. A. Jackson, Inorg. Chem., **17**, 2330,
   (1978).

e - D. S. Ginley and M. S. Wrighton, J.Amer. Chem. Soc.,
   97, 4908 (1975).

f - H. B. Abrahamson and M. S. Wrighton, Inorg. Chem.,
   17, 1003, (1978).

2.3.   Multiple Metal-to-Metal Bonds

The preparation and characterisation of compounds
which contain two metal atoms linked by a multiple
bond have been matters of intense interest in the
recent past (33,34), but thermochemical data relating
to the enthalpies of these multiple bonds are sparse.
Species containing $M\overset{n}{=}M$ bonds, where $\underline{n}$ is the formal
bond order and is ≥ 3 in the examples to be considered
here, all can be represented by the general formula
$[X_x M\overset{n}{=}MX_x]^{\pm y}$. It is important that the ligands X should
not bridge the M-M bond if possible. If only neutral
molecules are considered, ($y = 0$), the problem of
determining $D(M\overset{n}{=}M)$, the enthalpy contribution of the
multiple metal-metal bond, is reduced to those of
measuring the enthalpy of formation, $\Delta H_f^O[M_2 X_{2x}, g]$ and
estimating $\bar{D}(M-X)$ for the mononuclear compound from
$\Delta H_f^O[MX_x, g]$.

2.3.1. Hexakis (Dimethylamido) Dimetal Compounds of Molybdenum and Tungsten

Based on the requirements just outlined, the most suitable series of compounds known at present contains the mononuclear compounds $Mo(NMe_2)_4$ and $W(NMe_2)_6$, and the binuclear compounds $M_2(NMe_2)_6$ (M = Mo,W) in which there is a triple bond between the metal atoms as indicated in Figure 3.

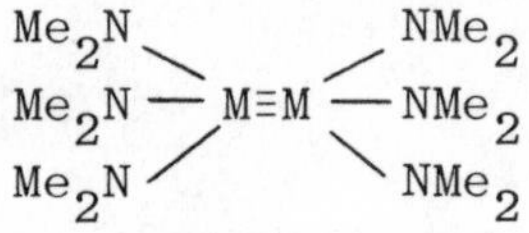

$$
\begin{array}{ccc}
Me_2N & & NMe_2 \\
Me_2N - & M\equiv M & - NMe_2 \\
Me_2N & & NMe_2
\end{array}
$$

The standard enthalpy of formation of each of these compounds in the gas phase has been measured by calorimetry (35) with the results shown in Table 12.

TABLE 12

Standard enthalpy of formation, $\Delta H_f^O(g)$, and derived enthalpy of disruption ΔH_D for $M(NMe_2)_x$ and $M_2(NMe_2)_6$ compounds (kJ mol^{-1})

	$\Delta H_f^O(g)$	ΔH_D
$Mo(NMe_2)_4$	131.4 ± 8	1021.6 ± 19
$Mo_2(NMe_2)_6$	128.2 ± 13	1928.6 ± 28
$W(NMe_2)_6$	268.0 ± 14	1332.5 ± 29
$W_2(NMe_2)_6$	132.5 ± 11	2327.9 ± 29

The structures of all these molecules are known from X-ray crystallographic studies (34,36). The values of the mean bond dissociation enthalpy

$\bar{D}(\text{Mo-NMe}_2) = 255 \text{ kJ mol}^{-1}$ in $\text{Mo(NMe}_2)_4$, and
$\bar{D}(\text{W-NMe}_2) = 222 \text{ kJ mol}^{-1}$ in $\text{W(NMe}_2)_6$ are easily
calculated from the values of ΔH_D for these molecules.
It is well established for other transition metal
compounds such as the molecular halides, that the
value of $\bar{D}(\text{M-X})$ varies inversely with the formal
oxidation number of M. It is possible to determine
the dependence of $\bar{D}(\text{M-NMe}_2)$ values on oxidation number
from the known correlations, using the measured values
of $D(\text{M-NMe}_2)$ (4,37). For the binuclear compounds,
$M_2(\text{NMe}_2)_6$

$$\Delta H_D = D(M\overset{3}{-}M) + 6D(\text{M-NMe}_2)$$

so that the value of $D(M\overset{3}{-}M)$ will be very sensitive to
the particular value of $D(\text{M-Me}_2)$ that is chosen, as
Table 13 shows.

TABLE 13

Metal-ligand and the corresponding metal-metal
bond enthalpy contribution (kJ mol^{-1}) for various
formal oxidation numbers of the metal atom in
$M_2(\text{NMe}_2)_6$ compounds

Formal oxidation number	3	4	5	6
$\bar{D}(\text{Mo-NMe}_2)$	288	255[a]	223	190
$D(\text{Mo}\overset{3}{-}\text{Mo})$	200	396	592	788
$\bar{D}(\text{W-NMe}_2)$	331	295	258	222[a]
$D(\text{W}\overset{3}{-}\text{W})$	340	558	775	995

a. Experimental value

Clearly there is no a priori reason to prefer any one
of the values of $D(M\overset{3}{-}M)$ in Table 13, but the following
arguments lead to the conclusion that bond enthalpy
transfer from $M(\text{NMe}_2)_4$ is probably to be favoured.

The metal atoms in $M_2(NMe_2)_6$ have formal oxidation numbers of 3, which would lead to the lowest values of $D(M \overset{3}{-} M)$ and, in this case, $D(M \overset{3}{-} M) \leq \bar{D}(M-NMe_2)$. Each metal atom in $M_2(NMe_2)_6$ has a valence of 6 (Figure 3), which would imply that the highest values of $D(M \overset{3}{-} M)$ are correct. However, such values are unacceptably high, being comparable only with the dissociation energies (4) of $CO(1075)$, $N_2(946)$ and $C_2H_2(955$ kJ $mol^{-1})$. Each metal atom in the $M_2(NMe_2)_6$ molecules is 4-coordinate, so that the environment of the NMe_2 groups is sterically unhindered as in $Mo(NMe_2)_4$. This is in contrast to the steric strain which is evident in the structure of $W(NMe_2)_6$ (36). There appears to be no basis for considering an oxidation number of five, except as the average value, that is $D(Mo \overset{3}{-} Mo) = (592 \pm 196)$ kJ mol^{-1} for example (35).

2.3.2. μ^2-Acetato Compounds of Chromium and Molybdenum

These compounds, which contain a quadruple bond between the metal atoms which are five-coordinate and have the formal oxidation number 2, also have bidentate acetate ligands which bridge the metal-to-metal bond (33). Calorimetric measurements in solution (38) have led to the standard enthalpies of formation of the crystalline solids which, with estimated values for the enthalpies of sublimation, give the enthalpies of formation in the gaseous phase shown in Table 14.

TABLE 14

Standard enthalpy of formation in
the gas phase of certain chromium
and molybdenum compounds (kJ mol^{-1})

	$\Delta H_f^O(g)$
$Cr_2(OAc)_4$	-2153
$CrMo(OAc)_4$	-1969
$Mo_2(OAc)_4$	-1826
$Mo_2(OAc)_2(acac)_2$	-1660
$Mo(acac)_3$	-1199

Assuming transferability of the bond enthalpy
contributions $D(Mo\overset{4}{-}Mo)$, $\bar{D}(Mo-O[Ac])$ and $\bar{D}(Mo-O[acac])$
leads to an expression for $D(Mo\overset{4}{-}Mo)$

$$= 2\Delta H_f^O(Mo,g)/3 + \Delta H_f^O[Mo_2(OAc)_4,g] - 2\Delta H_f^O[Mo_2(OAc)_2$$
$$(acac)_2,g] + 4\Delta H_f^O[Mo(acac)_3,g]$$

The values of $D(M-M)$ in the other compounds can be
derived from similar equations and the results are
shown in Table 15.

TABLE 15

Enthalpy contributions, $D(M-M)$(kJ mol^{-1})
in metal acetates $M_2(OAc)_4$ (38)

	$D(M-M)$
Cr-Cr	210
Cr-Mo	253
Mo-Mo	337

The influence exerted by the bridging acetate ligands
in strengthening (or weakening) the metal-to-metal
bond is not known. The fact that $D(Cr-Mo)$ is less

than the mean value calculated as D(Cr-Mo) = $0.5\left[D(Cr-Cr) + D(Mo-Mo)\right] = 273$ kJ mol^{-1}, may suggest that the nature of the metal-metal bond is not identical in the two homonuclear systems.

2.3.3. Octahalodimetallate Anions of Molybdenum and Rhenium

Studies of the photolytic cleavage of $\left[Re_2Cl_8\right]^{2-}$ ion in acetonitrile solution led to the suggestion that D(Re$\overset{4}{-}$Re) is ca. 300 kJ mol^{-1} (39). Applying the Birge-Sponer extrapolation to the frequencies of the vibrational progression in ν(M-M) observed in the resonance Raman spectra of the $\left[M_2X_8\right]^{Z-}$ (X = Cl, Br; M = Mo, Z = 4; M = Re, Z = 2) ions leads to estimates of D(M$\overset{4}{-}$M) ~500 kJ mol^{-1} in these systems (40).

2.3.4. Estimates of Metal-to-Metal Bond Enthalpies in Other Systems

The results of determinations by electron impact, reaction kinetic and solution calorimetric methods are presented in Table 16. For the reasons which have been outlined earlier (section 2.1) it is to be expected that the values of D(M-M) deduced from appearance potential measurements and from reaction kinetics in solution are upper limit values. In general there is a correlation between the value of D(M-M) determined from reaction kinetics in solution and the energy of the ($\sigma\rightarrow\sigma*$) transition (Table 11) in the electronic spectra of these compounds in solution, especially where closely related series of compounds are considered (41). However, the precise relationship

81.

TABLE 16

Selected values of metal-metal bond
enthalpies D(M-M) kJ mol^{-1} in certain binuclear
metal carbonyl complexes determined by electron
impact (EI), reaction kinetic (K) and
solution calorimetric (C) methods

	Method	D(M-M)	Ref
$[C_5H_5Mo(CO)_3]_2$	C	160	37
$Mn_2(CO)_8(PPh_3)_2$	K	120	20
$MnRe(CO)_8(PPh_3)_2$	K	155	20
$Tc_2(CO)_8(PPh_3)_2$	K	138	20
$Re_2(CO)_8(PPh_3)_2$	K	162	20
$[C_5H_5W(CO)_3]_2$	EI	234	6
$Mn_2(CO)_9(PPh_3)$	K	151	20
$Mn_2(CO)_8(P(OMe)_3)_2$	K	153	20
$Mn_2(CO)_8(PEt_3)_2$	K	136	20
$Mn_2(CO)_8(PCy_3)_2$	K	99	20
$[C_5H_5Fe(CO)_2]_2$	K	96	6

between these quantities and the gas phase
dissociation enthalpy remains to be determined
because the nature of the interaction between the
solvent and the metal-to-metal bond is not well
understood.

2.4. Summary of Results

The preceding survey of bond enthalpy
contributions in binuclear transition metal compounds
has emphasized the problem of the interpretation of
results. This may be seen particularly clearly in
the various di-molybdenum compounds which have been
mentioned (Table 17).

TABLE 17

Enthalpy contribution of (Mo-Mo) bonds
D(Mo-Mo) (kJ mol^{-1}), bond lengths r(Mo-Mo)$\overset{\circ}{A}$
and values calculated from $D = Ar^{-4.6}$

	D(Mo-Mo)	r(Mo-Mo)	D(calc'd)
Mo metal	118	2.73	118
$[C_5H_5Mo(CO)_3]_2$	160	3.23	54
$Mo_2(NMe_2)_6$	>396	2.21	312
$[Mo_2Cl_8]^{4-}$	500	2.13	370
$Mo_2(OAc)_4$	337	2.09	403
Mo_2	404	(2.12)	378

Comparison of the values of D(Mo-Mo) deduced from
experiment with those calculated from the logarithmic
relation, eqn. 3, which does not involve any concept
of bond order or formal oxidation number, shows major
discrepancies which may indicate that the logarithmic
relation is probably too simple to be used, either in
systems involving bonds between metal atoms of bond
order greater than one, or for formal single bonds
where the bond length is significantly greater than

that in the bulk metal, but where neither of these restrictions are present (e.g., many polynuclear metal carbonyls) the logarithmic relation may be useful.

Comparison with the bond dissociation enthalpies (4) of carbon-carbon bonds in ethane (369) and acetylene (955 kJ mol^{-1}) shows that the ratio of values of D(Mo-Mo) in $[C_5H_5Mo(CO)_3]_2$ and in $Mo_2(NMe_2)_6$ is quite similar (2.59 for carbon, >2.48 for molybdenum).

3. GENERAL SURVEY OF CONCLUSIONS

Apart from the acute lack of thermochemical data on cluster compounds of transition metals, the conclusions which can be drawn from the available information are subject to interpretation and are not unambiguous for the most part.

The structures of polynuclear transition metal carbonyls $[M_m(CO)_n]$ appear to be determined by the ligand sheath at low nuclearity $\underline{m}$, when $\underline{n}/\underline{m}>2$. The orientation of the metal polyhedron within the ligand polyhedron is determined by the latter at low nuclearity $\underline{m}$. It is further suggested that at higher nuclearity and when $\underline{n}/\underline{m}<1.5$, the structure will be dictated by the cluster of metal atoms. Only one example of such a system, the anion $[Pt_{19}(CO)_{12}(\mu\text{-}CO)_{10}]^{4-}$ is known (42) at present. Because the structure of a low nuclearity metal carbonyl cluster is determined by the ligands as far as themodynamic considerations are concerned, then it is to be expected that changes in the ligand environment will induce structural changes in the metal cluster, implying that these clusters will exhibit dynamic structural behaviour (8). As the nuclearity increases and $\underline{n}/\underline{m}$ decreases, it seems reasonable to propose that the dynamic behaviour of the metal cluster with respect to its environment will

decrease in importance. However, as the anion
$[Pt_9(CO)_{18}]^{2-}$ demonstrates (43), rearrangement of the
metal polytope is a process which cannot be overlooked.
There is no measure of the themodynamic quantities
involved in such fluxional processes at present,
although studies on matrix isolated species such as
Ag_n (n = 3, 4, 5 etc.) are just beginning to be made
(44).

It is clear that a more precise description of
the bonds between metal atoms is required, whether in
simple diatomic molecules or in binuclear compounds
such as $Mn_2(CO)_{10}$ or $Mo_2(NMe_2)_6$, before certain
conflicting interpretations of the bond enthalpy
contributions in these systems can be resolved, and
the thermodynamics of metal clusters thereby understood.

386.

4. REFERENCES

1. E. W. Plummer, W. R. Salaneck and J. S. Miller,
   Phys. Rev. B., 18, 1673 (1978).
2. C. Leung, M. Vass and R. Gomer, Surf. Sci., 66,
   67 (1977).
3. J. L. Taylor, D. E. Ibbotson and W. H. Weinberg,
   J. Chem. Phys., 69, 4298 (1978).
4. L. V. Gurvich, G. V. Karachevtziev, V. N. Kondratiev,
   Yu. A. Lebedev, V. A. Medvedev, V. K. Potapov
   and Yu. S. Hodiev, Dissociation Energies of
   Chemical Bonds. Ionisation Potentials and
   Electron Affinities, Moscow, Nauka (1974).
5. J. J. Burton, Surf. Sci., 66, 647 (1977).
6. J. A. Connor, Topics Curr. Chem., 71, 71 (1977).
7. P. R. Raithby, this volume, Chapter 2..
8. B. F. G. Johnson and R. E. Benfield, this
   volume, Chapter 6.
9. R. G. Woolley, this volume, Chapter 9.
10. K. Wade, Inorg. Nucl. Chem. Letters, 14, 71 (1978).
11. C. E. Housecroft, K. Wade and B. C. Smith, J. Chem.
    Soc. Chem. Comm., 765 (1978).
12. W. A. Herrmann, J. Plank and B. Reiter, J. Organo-
    metal. Chem., 164, C25 (1979).
13. J. Calabrese, L. F. Dahl, A. Cavalieri, P. Chini,
    G. Longoni and S. Martinengo, J. Amer. Chem.
    Soc., 96, 2616 (1974).

14. B. F. G. Johnson, J. Chem. Soc. Chem. Comm., 211
    (1976); B. F. G. Johnson and R. E. Benfield,
    J. Chem. Soc. Dalton, 1554 (1978).

15. C. Eady, B. F. G. Johnson, J. Lewis, R. Mason,
    P. B. Hitchcock and K. M. Thomas, J. Chem.
    Soc. Chem. Comm., 385 (1977).

16. S. Martinengo, G. Ciani, A. Sironi and P. Chini,
    J. Amer. Chem. Soc., 100, 7096 (1978).

17. H. Conrad, G. Ertl, H. Knözinger, J. Kuppers and
    E. E. Latta, Chem. Phys. Letters, 42, 115
    (1976).

18. G. Battiston, G. Sbrignadello, G. Bor and
    J. A. Connor, J. Organometal. Chem., 131,
    145 (1977).

19. G. A. Junk and H. J. Svec, J. Chem. Soc. (A),
    2102, (1970).

20. R. A. Jackson and A. Poë, Inorg. Chem., 17, 997
    (1978).

21. K. A. Gingerich, J. Cryst. Growth, 9, 31 (1971).

22. M. Moscovitz and G. A. Ozin, in Cryochemistry,
    Wiley, New York 261 (1976).

23. J. Harris and R. O. Jones, J. Chem. Phys., 70,
    830   (1979); see also P. R. Scott and
    W. G. Richards, Molecular Spectroscopy, 3,
    70 (1976); C. J. Cheetham and R. F. Barrow,
    Adv. High Temp. Chem., 1, 7 (1967).

24. S. K. Gupta and K. A. Gingerich, J. Chem. Phys.,
    69, 4318 (1978).

25. K. A. Gingerich, Chem. Phys. Letters, 23, 270
    (1973).

26. L. Brewer in Phase Stabilities of Metals and
    Alloys (P. Rudman, J. Stringer and R. I. Jaffee,
    editors) McGraw Hill, New York, 39 (1967).

27. K. A. Gingerich, J. Chem. Soc. Faraday 1, 70, 471
    (1974).

28. W. Klotzbücher and G. A. Ozin, Inorg. Chem., <u>16</u>,
      984 (1977).

29. J. G. Norman, H. J. Kolari, H. B. Gray and
      W. C. Trogler, Inorg. Chem., <u>16</u>, 987 (1977).

30. W. Klotzbücher G. A. Ozin, J. G. Norman and
      H. J. Kolari, Inorg. Chem., <u>16</u>, 2871 (1977).

31. T. H. Upton and W. A. Goddard, J. Amer. Chem.
      Soc., <u>100</u>, 5659 (1978).

32. H. B. Abrahamson, C. C. Frazier, D. S. Ginley,
      H. B. Gray, J. Lilienthal, D. R. Tyler and
      M. S. Wrighton, Inorg. Chem., <u>16</u>, 1554 (1977).

33. F. A. Cotton, Acc. Chem. Res., <u>11</u>, 232 (1978);
      W. C. Trogler and H. B. Gray, Acc. Chem. Res.,
      <u>11</u>, 232 (1978).

34. M. H. Chisholm and F. A. Cotton, Acc. Chem. Res.,
      <u>11</u>, 356 (1978).

35. J. A. Connor, G. Pilcher, H. A. Skinner,
      M. H. Chisholm and F. A. Cotton, J. Amer.
      Chem. Soc., <u>100</u>, 7738 (1978).

36. D. C. Bradley and M. H. Chisholm, Acc. Chem. Res.,
      <u>9</u>, 273 (1976); M. H. Chisholm, F. A. Cotton
      and M. W. Extine, Inorg. Chem., <u>17</u>, 1329
      (1978).

37. F. A. Adedeji, K. J. Cavell, S. Cavell,
      J. A. Connor, G. Pilcher, H. A. Skinner and
      M. T. Zafarani-Moattar, J. Chem. Soc.
      Faraday 1, <u>75</u>, 603 (1979).

38. K. J. Cavell, C. D. Garner, G. Pilcher and
      S. Parkes, J. Chem. Soc. Dalton, 1714
      (1979).

39. G. L. Geoffroy, H. B. Gray and G. S. Hammond,
      J. Amer. Chem. Soc., <u>96</u>, 5565 (1974).

40. R. J. H. Clark and N. R. D'Urso, J. Amer. Chem.
      Soc., <u>100</u>, 3088 (1978).

41. A. J. Poë and R. A. Jackson, Inorg. Chem., <u>17</u>,
        2330 (1978).

42. D. M. Washecheck, E. J. Wncherer, L. F. Dahl,
        A. Ceriotti, G. Longoni, M. Manassero,
        M. Sansoni and P. Chini, J. Amer. Chem. Soc.,
        <u>101</u>, 6110 (1979).

43. C. Brown, B. T. Heaton, P. Chini, A. Fumagalli
        and G. Longoni, J. Chem. Soc. Chem. Comm.,
        309 (1977).

44. G. A. Ozin and H. Huber, Inorg. Chem., <u>17</u>, 155
        (1978).

45. J. A. Oberteufer and J. A. Ibers, Acta Cryst.,
        <u>26B</u>, 1499 (1970).

Transition Metal Clusters
Edited by Brian F.G. Johnson
© 1980 John Wiley & Sons Ltd.

CHAPTER 6

Some Reactions of Metal Clusters

A.J. Deeming
*Department of Chemistry, University College London,
London, U.K.*

1. INTRODUCTION

Essentially all reactions of mononuclear complexes are available to clusters in principle but it is relatively rare for a reaction to occur at a single metal centre within a cluster. One cannot usually ignore the sensitive way electronic and steric effects are relayed from one part of a cluster to another, so that the cluster as a whole must be considered even for a reaction at a single metal centre as in the substitution of CO by tertiary phosphines at successive metal atoms in $Ir_4(CO)_{12}$ or $Ru_3(CO)_{12}$ (1,2). Most reactions of clusters explicitly involve the combined influence of the metal atoms and notably ligands adopt bridging rather than terminal sites. Terminal hydrides in clusters are sufficiently rare to justify reference to this in the title of a paper, and some ligands are only found in clusters because they need stabilisation by bonding to several metal atoms. One should therefore expect reactivity uniquely characteristic of clusters. Results of recent years have amply justified this expectation but with a few exceptions little of real practical value has been achieved by way of useful new reactions. But there are lots of new reactions and plenty of promise. A difficulty is the inability to predict the chemistry of clusters and often to adequately rationalise the chemistry once discovered. For example, we do not know how to control conditions that are required for alkynes in clusters to undergo oxidative addition to give alkynyl ligands, various hydrogen transfer reactions, triple-bond fission, oligomerisation and so on. Generally the reaction possibilities are too extensive for a simple analysis. Another difficulty is that for a cluster to react readily it must be labile, but unfortunately lability within the coordination shell and within the metal skeleton usually go together. Where substitutions are facile as with first row metal systems then the cluster is also readily broken down or reconstructed so that for cluster catalysis, for example, it is very difficult to establish the nuclearity of the catalyst or whether a cluster is involved at all. This could be unambiguously demonstrated by catalytic asymmetric induction using a cluster, the chirality of which resulted purely from the symmetry of the cluster framework as for a tetrahedral cluster made up of four different atoms. One such cluster is $CoFeCrS(CO)_8(\eta^5-C_5H_5)$, the first four atoms in the formula forming a tetrahedron, but this complex is still to be resolved and the point to be tested (3). With more inert clusters involving third row transition metals correspondingly higher reaction temperatures are usually required so that mechanistic studies are no easier. Our knowledge of most reactions depends solely on the nature of stable isolable products and often those formed only in low yields. Most of the chemistry of ligands in clusters is confined so far to carbonyl clusters of Group VIII metals and then predominantly those of ruthenium and osmium. Most clusters of the early transition metals have electronegative ligands (halides and oxyanions) and since nothing is known of the chemistry of these clusters with other ligands this is an area worth examination. The reactions of Cu, Ag and Au clusters are also limited and few reactions are known in which the cluster remains as a unit.

In an account of this size a comprehensive treatment is neither possible nor desirable so only the main types of reaction are included with only a few examples of each given. Naturally this selectivity is strongly controlled by the author's interests and perhaps blinkered view of what is important.

Reactions of clusters may be considered in three categories:

1.1. Reactions Involving Major Modifications of the Metal Framework

There are very many reactions of this sort and unfortunately these nearly always prove to be difficult to analyse and predict. Mononuclear metal carbonyls undergo simple substitution reactions by loss and gain of ligands although in many cases "simple" is a better description of the products than the mechanism. Clusters on the other hand commonly undergo changes in structure or nuclearity as ligands coordinate or dissociate so that products may not be simply related to the starting materials at all. Thermal loss of CO from $Os_3(CO)_{12}$ gives a range of high nuclearity clusters such as $Os_5(CO)_{16}$ and $Os_6(CO)_{18}$ (4). But reactions other than loss or gain of ligands lead to changes in nuclearity such as the base treatment of $Rh_4(CO)_{12}$ to give high nuclearity anions such as $[Rh_6(CO)_{15}]^{2-}$, $[Rh_{12}(CO)_x]^{2-}$ (x = 30 or 34), and

$[Rh_7(CO)_{16}]^{3-}$ (5). The initial nucleophilic addition of OH^- or OMe^- at coordinated CO is followed by various cluster transformations and related but different changes of nuclearity occur when $Ir_4(CC)_{12}$ is treated similarly (6). The treatment of $Ru_3(CO)_{12}$ with H_2O leading to $H_4Ru_4(CO)_{12}$ and $H_2Ru_4(CO)_{13}$ may be related (7). Even where the nuclearity does not change in a reaction, gross changes in the cluster framework may have occurred as in the reaction of $Os_6(CO)_{18}$ with ethylene to give $Os_6(CO)_{16}(CCH_3)_2$ and $Os_6(CO)_{16}(CH_3C=CCH_3)$ (8) or in its reduction to $[Os_6(CO)_{18}]^{2-}$ (9). Each of the frameworks of these Os_6 clusters are of different geometry.

1.2. Reactions Leading to Modifications in Ligand to Cluster Bonding without Cluster Reforming

There are a few particularly inert or stable cluster skeletons such as triangular Os_3 or tetrahedral Co_3C which can remain intact through various reactions. There are analogies with benzene chemistry in that the structures of intermediates, in particular the electronic structures, may differ considerably from those of starting materials or products but the overall reaction is substitution rather than addition, or ring opening. Triosmium clusters have a very notable chemistry in which the Os_3 triangle remains while very extensive ligand reactions are occurring. One of the earliest examples of this is the reaction of $Os_3(CO)_{12}$ with PPh_3 to give $Os_3(CO)_{12-x}(PPh_3)_x$ (x = 1, 2 or 3) as well as $HOs_3(CO)_9(PPh_3)(Ph_2PC_6H_4)$, $Os_3(CO)_8(PPh_2)(Ph)(PhPC_6H_4)$, $Os_3(CO)_7(PPh_2)_2(C_6H_4)$, $HOs_3(CO)_8(Ph_2PC_6H_4)(PPh_3)$, $HOs_3(CO)_7(PPh_2)(C_6H_4)(PPh_3)$ and $HOs_3(CO)_7(PPh_2)(Ph_2PC_6H_4C_6H_3)$ (10). All these

complexes contain a triangle of osmium atoms but a
remarkable set of new ligands which are variously
attached to the cluster is generated from the original
PPh_3 ligand. This early work suggested that the
ligand chemistry of Os_3 clusters would prove to be a
particularly exciting and important area and the seven
years following these early reports has proved this to
be so. All organic molecules except perhaps saturated
hydrocarbons are expected to give triosmium derivatives.
Indeed many of the important features of the ligand
chemistry of clusters are best illustrated by examples
from triosmium chemistry and accordingly we will give
many such examples in the rest of this chapter.
Higher clusters of osmium are also demonstrating a
rich ligand chemistry but cluster reformation is an
added but interesting complication in these cases.

1.3. Reactions in which the Cluster Nature of the Substrate is Incidental

It is not easy to cite examples of these.
Sometimes the reaction centre is well removed from
the cluster itself as with reactions at the group R
in the complexes $Co_3(CR)(CO)_9$, such as the Friedel-
Crafts acylation of $Co_3(CPh)(CO)_9$ go give exclusively
the 4-acyl substituted phenyl (11). However, the
$Co_3(CO)_9C$ substituent at the benzene ring has the
effect of stabilising the intermediate carbonium ion
and this is undoubtedly due to the particular
electronic structure of the Co_3C cluster which is
appropriately modified as the carbonium centre is
generated leading to positive charge delocalisation
towards the metal atoms. The above chemistry is
related to the special stability found for compounds

such as $[Co_3(CO)_9(C-CH_2)]^+$ which might be regarded as a stabilised carbonium ion (12). The chemistry of these organic ligands is fundamentally related to the structure and bonding of the cluster.

2. SIMPLE ADDITIONS TO CLUSTERS

Not unexpectedly, clusters react by unimolecular
or bimolecular pathways. The addition of a substrate
to a cluster is on the whole easy to study especially
in cases, as described in this section, where the
adduct may be isolated or at least examined
spectroscopically. Many reactions are clearly multi-
step but with the first being in many cases a simple
addition of either an electrophile or a nucleophile.
In the almost complete absence of any mechanistic
approach to cluster chemistry, a study of their simple
addition reactions at least provides some basis upon
which to develop our knowledge of bimolecular pathways.

2.1. Electrophiles

2.1.1. Protonation at Metal Atoms

It is expected that electrophilic addition at
clusters in the absence of any other process should
usually occur with no change in the skeletal geometry
since there is no change in the number of electrons
associated with the cluster. This applies only strictly

to small clusters since high nuclearity clusters can adopt different geometries compatible with a given number of electrons, for example the series $[Os_6(CO)_{18}]^{2-}$, $[HOs_6(CO)_{18}]^-$ and $H_2Os_6(CO)_{18}$ and the ruthenium analogues have different geometries depending upon the charge and the metal (see Chapter 3). The metal atoms are regions of high electron density in that they are the normal sites for protonation and reversible protonation and deprotonation is, of course, a common feature of anions and hydrido-anions of metal carbonyls. In essentially all cases the hydrido-ligands occupy bridging positions in the cluster.

$$[Re_3(CO)_{12}]^{3-} \underset{-H^+}{\overset{+H^+}{\rightleftharpoons}} [HRe_3(CO)_{12}]^{2-} \underset{-H^+}{\overset{+H^+}{\rightleftharpoons}} [H_2Re_3(CO)_{12}]^-$$

$$\underset{-H^+}{\overset{+H^+}{\rightleftharpoons}} H_3Re_3(CO)_{12} \tag{13}$$

$$Os_3(CO)_{12} \underset{-H^+}{\overset{+H^+}{\rightleftharpoons}} [HOs_3(CO)_{12}]^+ \tag{14,15}$$

$$Os_3(CO)_9L_3 \underset{-H^+}{\overset{+H^+}{\rightleftharpoons}} [HOs_3(CO)_9L_3]^+ \underset{-H^+}{\overset{+H^+}{\rightleftharpoons}} [H_2Os_3(CO)_9L_3]^{2+} \tag{15}$$

$(L= PEt_3)$

The above examples provide isoelectronic series of compounds but, as expected of an isoelectronic pair, the one with the higher negative charge is more readily protonated. High positive charges are only encountered with the more basic ligands, e.g. with $Os_3(CO)_9(PEt_3)_3$ rather than $Os_3(CO)_{12}$, or in strongly acidic media such as H_2SO_4 or better still HSO_3F/SO_2.

A notable feature of these protonations is the unusually high kinetic isotope effects, for example, for protonation of the anions $[MCo_3(CO)_{12}]^-$ $k_H/k_D = 16.8 \pm 1.0$ (M = Fe); 15.4 ± 1.0 (M = Ru) and

16.2 ± 1.0 (M = Os) and for $Os_3(CO)_{12}$ $k_H/k_D = 11 ± 2$
(14). The importance of tunnelling versus zero point
energy effects $[\nu(OH) >> \nu(MHM)$ for transfer of a
proton from an oxyacid to a metal bridging site in a
cluster] in giving this high isotope effect remains
to be assessed. This might become clearer when it is
known whether proton transfer from the medium to the
metal atoms through the carbonyl shell is direct or
whether there are intermediates with protonated CO
groups. The rates of protonation and deprotonation
of clusters are very much less in general than the
rates of hydride site exchange within the cluster and
there is no reported case of coalescence of cluster
hydride and solvent [1]H n.m.r. signals. Hydride site

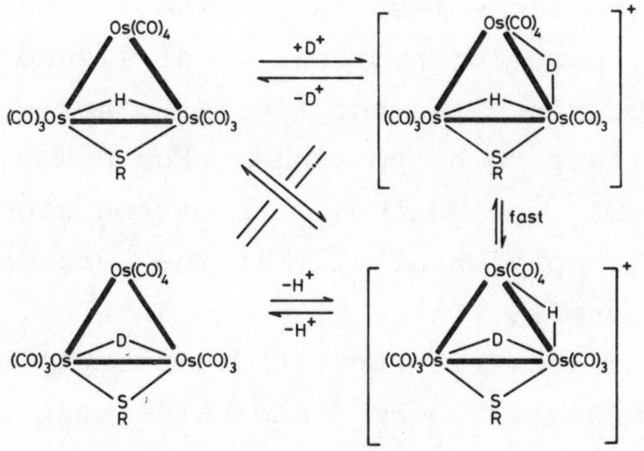

Figure 1

exchange is slow on an n.m.r. time scale in the example
in Figure 1 (the two isomers give separate hydride
signals although equilibrium between them is rapidly
reached nonetheless), but there is no exchange between
the metal hydride and the solvent (D_2SO_4) to give
$DOs_3(SR)(CO)_{10}$ until the solvent is diluted with
D_2O (16). It is not generally possible to selectively

deuterate particular hydride sites in a cluster by exchange with deuterated acid.

The metal atoms are also preferentially protonated in clusters containing alkene or alkyne derived ligands (17,18) and indeed most other ligands. This is to be expected since the nucleophilic character of unsaturated hydrocarbons is often drastically reduced on coordination to transition metals in low oxidation states and this is so even in mononuclear systems, e.g. (diene)$Fe(CO)_3$, where the metal centre is initially protonated in preference to the alkene.

2.1.2. Addition of Other Nucleophiles at CO

In spite of the above results one should ask whether there are any cases of protonation at ligand sites, particular CO, since the addition of other electrophiles shows this to be possible. For example, $AlBr_3$ adds to $Fe_3(CO)_{12}$ or $Ru_3(CO)_{12}$ at oxygen atoms of CO groups and then preferentially at the more basic bridging carbonyls. Thus, although $Ru_3(CO)_{12}$ has no bridging CO, a low frequency $v(CO)$ absorption in $Ru_3(CO)_{12} \cdot AlBr_3$ at 1535 cm^{-1} indicates that a bridging carbonyl structure as in $Fe_3(CO)_{12} \cdot AlBr_3$ has been generated (19). Many complexes are derived formally or in reality by addition at the μ^3-CO of $[Co_3(CO)_{10}]^-$ and for the example shown in Figure 2 the products have all been structurally characterised (20,21). Of the three metals, Ti, Zr and Hf, titanium is expected to form the weaker M-O bond {for example, $\bar{E}$(M-O) for $M(OPr^i)_4 = 481(Ti)$, $552(Zr)$ and $573(Hf)$ kJ mol^{-1} (22)} so that it can be seen that the stronger M-O bond is associated with the longer and apparently weaker C-O

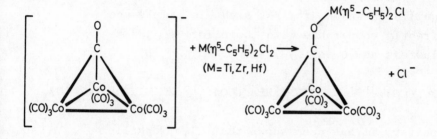

μ^3-C-O distance   1·22 Å (M=Ti)
                     1·276(6) (M=Zr)
                     1·32(3)   (M=Hf)

Figure 2

bond. Thus electrophilic addition has a CO bond
weakening effect that can be very pronounced for
strong Lewis acids and this is a result significant to
CO reduction in clusters (see Section 5.2).

2.1.3. Addition of Protons at CO

One series of protonation/deprotonation
equilibria which at first sight is simply and
directly comparable with those for Os_3 and Re_3
clusters above is:

$$\left[Fe_3(CO)_{11}\right]^{2-} \xrightleftharpoons[-H^+]{+H^+} \left[HFe_3(CO)_{11}\right]^- \xrightleftharpoons[-H^+]{+H^+} H_2Fe_3(CO)_{11}$$

However, it has been shown that electrophilic
methylation of $\left[HFe_3(CO)_{11}\right]^-$, or of $\left[Fe_3(CO)_{11}\right]^{2-}$
followed by protonation, is at the oxygen atom of the
μ^2-CO to give $HFe_3(C-OMe)(CO)_{10}$ which is structurally
closely comparable with the compounds from which it
is derived (23). This immediately raises doubts
about the structure of $H_2Fe_3(CO)_{11}$ and it was
subsequently shown that this should be reformulated
as $HFe_3(C-OH)(CO)_{10}$, the first and only example so
far of protonation at oxygen of a CO group (24). The
two hydrogen atoms in $HFe_3(C-OH)(CO)_{10}$ give [1]H n.m.r.
signals at -18.4 p.p.m. (Fe-bound) and +15.0 p.p.m.
(O-bound) and low field signals are hence found for H
attached either at C or O atoms, that is for formyl
and hydroxymethylidene ligands respectively. Although
the structure was not confirmed by diffraction studies,
the [13]C n.m.r. spectra shown in Figure 3 of the COH
and COMe complexes are so similar as to make structural
assignment unambiguous, the structure of the COMe

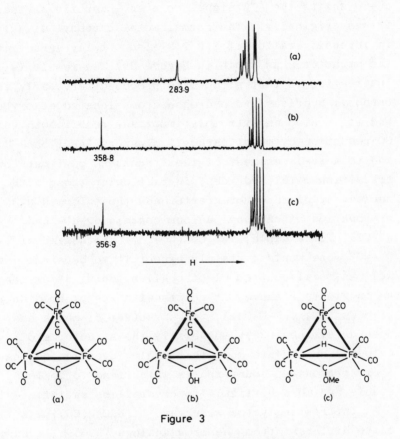

Figure 3

^{13}C n.m.r. spectra at low temperatures for (a),(b) and (c)
recorded at low temperatures in CD_2Cl_2. Shifts (downfied
from TMS) are given for the μ^2-CO, μ^2-COH and μ^2-COMe ligands.
[Published with permission from J. Amer. Chem. Soc., 1978, 100, 5239]

complex having been fully established by X-ray
diffraction. As expected, electrophilic addition at
μ^2-CO shifts the ^{13}C signal to even lower fields than
it was originally. The normal metal carbonyl signals
in intensity ratio 1:1:2:2:2:2 also totally agree with
the structures as shown in Figure 3. The result is
interesting in that structures proposed very early for
carbonyl hydrides had hydrogen atoms located at oxygen
and it is also an extremely important result both to
CO reduction by metal carbonyl clusters (section 5.2)
and to a re-assessment of the structures of first row
transition metal hydrides, in particular those with
nuclearity of three or greater and those formed by
protonation of carbonyl anions containing μ^2- and
μ^3-CO. For example, is $[Co_3(CO)_{10}]^-$ protonated at
μ^3-CO? One would certainly expect it to be since, as
well as metal-centred electrophiles adding at oxygen
as in Figure 2 above, it reacts with acetyl bromide to
give $Co_3(CO)_9(\mu^3$-C-OAc), the structure of which has
been determined (25). Generally the structures of
carbonyl anions are more thoroughly investigated than
their protonated counterparts which tend to be less
stable and more difficult to crystallise satisfactorily.

Interestingly the structure of $H_2Os_3(CO)_{11}$ is
quite different from the iron analogue (26). It seems
likely that their different structures are thermodynamically selected since the osmium complex adopts
the same structure with one terminal and one μ^2-hydride
whether formed by protonation of $[HOs_3(CO)_{11}]^-$ (27) or
by CO addition to $H_2Os_3(CO)_{10}$ (28,29). Three not
necessarily independent factors seem to account for
the differences: the greater strength of Os-H than
Fe-H bonds, the reduced facility of osmium clusters to
accommodate bridging CO (protonated or otherwise) and
the greater basicity of osmium compared with iron in

its low oxidation state compounds.

2.1.4. Other Electrophilic Additions

Protonation and methylation of $[HFe_3(CO)_{11}]^-$
occur at the same site but, whereas the first and
second protonations of $[HOs_3(S)(CO)_9]^-$ occur at the
metal, electrophilic methylation using Me_3O^+ occurs
at sulphur to give $HOs_3(\mu^3\text{-SMe})(CO)_9$ (30) which is
presumed to have a structure related to that
established for $HFe_3(\mu^3\text{-SPr}^i)(CO)_9$ (31) (Figure 4).
There is a related addition of $Cr(CO)_5$, derived from
$Cr(CO)_6$, at a very similar μ^3-S ligand in $Co_2Fe(S)(CO)_9$

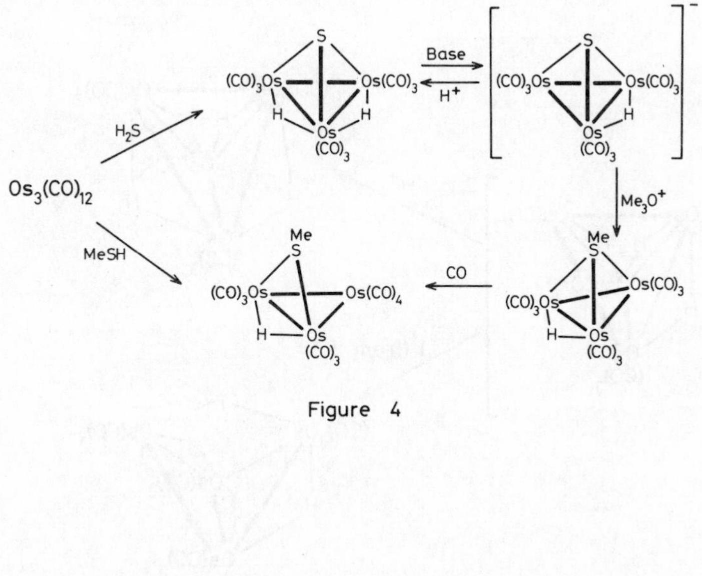

Figure 4

408

to give $Co_2Fe\{\mu^3\text{-}SCr(CO)_5\}(CO)_9$ with a sulphide bound to four metal atoms (32).

There are examples of electrophilic additions, other than of protons at metal atoms in clusters. Both H^+ and I^+ (from I_2) add at the osmium atoms of $[H_3Os_4(CO)_{12}]^-$ and the structural difference between the related products shown in Figure 5 is due to the iodine atom being a 3e-donating bridging ligand so that a butterfly metal geometry is obtained for $H_3IOs_4(CO)_{12}$ (33). Electrophiles larger than H^+ seem unable to penetrate the CO shell to attack the metal atoms directly but it is not known to what extent Os-Os bond cleavage precedes Os-I bond formation, in this particular case.

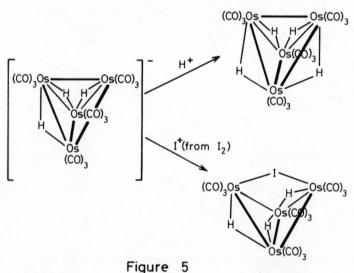

Figure 5

Electrophilic addition X^+(X = H or I) to $[H_3Os_4(CO)_{12}]^-$ to give $H_3XOs_4(CO)_{12}$

2.2. Nucleophiles

2.2.1. Addition at Metal Atoms

Simple addition of nucleophiles at metal atoms in
clusters increases the number of electrons associated
with the cluster and should modify the geometry or
nuclearity of the cluster and this often leads to a
cascade of new cluster products not always easily
related to the starting materials. Nucleophilic
addition seems to occur as the second order component
of substitution reactions of clusters but the nature
of intermediates formed is obscure. There are,
however, a few cases where simple addition leads no
further and two examples are given in Figure 6. Such
additions are expected to be rare for small clusters,
which makes $H_2Os_3(CO)_{10}$ particularly interesting since
it can add a variety of 2e-donors (28,29). This is
possible because the double Os=Os bond of $H_2Os_3(CO)_{10}$
(34) may be reduced on addition, increasing in length
from 2.681 to 2.989 (L = CO (26)) or 3.019 Å (L = PPh_3
(35)) (hydride-bridged Os-Os bonds) in the adducts.
This ability to form simple adducts enables $H_2Os_3(CO)_{10}$
to react at room temperature or below with many important
molecules such as unsaturated hydrocarbons and although
subsequent reactions follow in many cases, simple
nucleophilic addition seems to be the first step in all
cases. Thus $H_2Os_3(CO)_{10}$ can catalyse the isomerisation
of alkenes at room temperature without having to create
coordination sites by dissociation (29), it can undergo
associative CO-substitution (29) and is a very good
starting material for a good many triosmium clusters.

410.

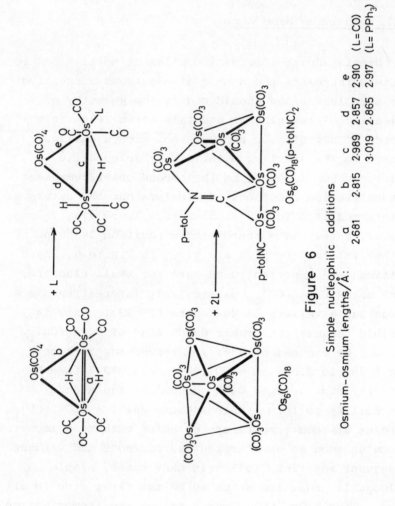

Figure 6

Simple nucleophilic additions

Osmium–osmium lengths/Å:

	a	b	c	d	e	
	2.681	2.815	2.989	2.857	2.910	(L=CO)
			3.019	2.865	2.917	(L= PPh₃)

The nucleophile added to $H_2Os_3(CO)_{10}$ may be metal-centred as in the addition of $Pt(C_2H_4)_2(PR_3)$; two ethylene molecules are lost to give the product $H_2Os_3Pt(PR_3)(CO)_{10}$ (36). One could describe in this way many of the condensation reactions of neutral metal carbonyls with carbonyl anions to give higher nuclearity clusters, such as the reaction of $Os_3(CO)_{12}$ with $[Re(CO)_5]^-$ to give $HOs_3Re(CO)_{16}$ after acidification (37) or of $[Rh_6(CO)_{15}]^{2-}$ with $Rh_6(CO)_{16}$ to give $[Rh_{12}(CO)_{30}]^{2-}$ (5b,38). Other unsaturated trinuclear clusters are $[H_4Re_3(CO)_{10}]^-$ (39) and $[H_3Re_3(CO)_{10}]^{2-}$ (40), which is isoelectronic with $[H_3Os_3(CO)_{10}]^+$ (18), the protonated form of $H_2Os_3(CO)_{10}$. Nucleophilic additions at these Re_3 clusters might be expected although none has been described, and their electrophilicity would certainly be reduced by the charge they carry.

Large clusters might be expected to have more easily modified geometries but even so, simple nucleophilic additions are rare. The addition of isocyanides L to $Os_6(CO)_{18}$ to give $Os_6(CO)_{18}L_2$ leads to metal-metal bond reduction but in this case, Figure 6, the two isocyanides act as a terminal 2e-donor and a rather unusual μ^3-4e-donor respectively (41). A simple rationalisation is that three Os-Os bonds have been cleaved to accommodate the extra six electrons, but whatever the precise description the skeleton of $Os_6(CO)_{18}$ has opened out in the adduct. $Os_6(CO)_{16}L_2$ (L = Bu^{t}NC) is formally a substitution product of $Os_6(CO)_{18}$, although actually synthesised by pyrolysis of $Os_3(CO)_{11}L$ in the same way that $Os_6(CO)_{18}$ was formed by pyrolysis of $Os_3(CO)_{12}$, and as expected has the same Os_6 geometry as the parent carbonyl (42). The easy direct addition of CO (apparently four molecules of these) to $[Rh_{12}(CO)_{30}]^{2-}$

with no change of nuclearity is another important
example but the structure of the adduct is unknown and
rather intriguing (5b).

Modification of the metal-metal bonding is one
way simple nucleophilic additions may be achieved but
the presence of ligands which can change their mode
of attachment at the cluster allows an alternative
route. For example, $Fe_3(\mu^3$-alkyne$)(CO)_9$ (43) and
$Os_3(\mu^3$-alkyne$)(CO)_{10}$ (44) have different M_3C_2 geometries
and one should expect interconversions between compounds
of this sort to allow uptake (or loss) of ligands, but
addition reactions of the iron cluster do not seem to
have been established experimentally. In another
example, the SMe ligand in $HOs_3(\mu^3$-SMe$)(CO)_9$ can be
converted from a μ^3-5e-donor to a μ^2-3e-donor as a
nucleophile is added (30). Figure 4 shows the
addition of CO but other less easily coordinated
ligands such as C_2H_4 may be added directly.

2.2.2. Addition at Ligands

Nucleophilic addition of hydroxide at coordinated
CO followed by elimination of H^+ and CO_2, each reacting
with further OH^-, is an important route to carbonyl
anions, clusters and mononuclear complexes alike.

$$M_x(CO)_y + OH^- \rightarrow [M_x(CO)_{y-1}(CO_2H)] \rightarrow [M_x(CO)_{y-1}]^{2-} + CO_2 + H^+$$

Although a 2e-donating CO is converted to a 1e-donating
(hence terminal) hydroxycarbonyl ligand, the negative
charge satisfies the electron-requirements of the
cluster. CO_2H-containing clusters have not been
isolated but the reversible formation of a CO_2Me-
containing one, which may be isolated, has been

described (5a):

$$Rh_4(CO)_{12} \underset{H^+/MeOH}{\overset{MeO^-/MeOH}{\rightleftharpoons}} [Rh_4(CO)_{11}(CO)_2Me)]^-$$

Nitrogen-based nucleophiles also add at CO as in the formation of a μ^2-carb moyl complex by reaction of

$$Os_3(CO)_{12} + PhCH_2NH_2 \longrightarrow HOs_3(\mu^2-PhCH_2NHCO)(CO)_{10} + CO$$

benzylamine with $Os_3(CO)_{12}$ and this is almost certainly occurring in the formation of dibenzylurea from benzylamine catalysed by $Os_3(CO)_{12}$ (45).

Nucleophilic addition at CO has been used as an effective way of displacing CO from inert metal clusters so as to introduce other ligands under particularly mild conditions. Metal carbonyls were originally used to reduce amine oxides but the same reaction has been employed with the above objective:

$$M-CO + Et_3NO \rightarrow \left[M^- - C \underset{O-\overset{+}{N}Et_3}{\overset{O}{\diagdown}} \right] \rightarrow M-NEt_3 + CO_2$$

NEt_3 is very weakly bound and may readily be replaced by an appropriate ligand which is usually present in situ. $Os_3(CO)_{12}$ (46) and $Ir_4(CO)_{12}$ (47) have been substituted in this way and complexes such as $Os_3(CO)_{10}(CH_3CN)_2$, using acetonitrile as solvent, have been isolated. The acetonitrile itself may be readily displaced so that this complex can usefully lead to substitution or oxidative addition products.

Presumably nucleophilic addition at CO is the first step in the hydride reduction of this ligand as in the treatment of $Ru_3(CO)_{12}$ with BH_4^- followed by acidification (48). A range of Ru_3, Ru_4, Ru_5 and Ru_6 clusters is obtained containing the hydrocarbon

414.

ligands CH, CH_2, C_2H_2, CCH_3 and C_3H_4, but the reaction is far too unselective to make it synthetically useful. For Fischer-Tropsch type chemistry one would imagine that a source of hydride must be generated from H_2 which is nucleophilic enough to add at a carbonyl carbon.

Nucleophilic additions at coordinated CO are also presumably involved in the metal-catalysed water gas shift reaction ($H_2O + CO \rightleftharpoons H_2 + CO_2$) (49), and one rather exotic display of this type of chemistry is in the reaction of $[HOs_3(CO)_{11}]^-$ with $Os_6(CO)_{18}$ to give $[HOs_3(CO)_{10}(\mu-CO_2)Os_6(CO)_{17}]^-$ (27) Figure 7(a). This can be regarded as the conjugate base of the carboxylic acid $[Os_6(CO)_{17}(CO_2H)]^-$ acting as a bridging ligand as do other carboxylato-ligands in the related complexes $HOs_3(CO)_{10}(\mu^2-RCO_2)$, Figure 7(b). The mechanism of its formation is unknown but $[HOs_3(CO)_{11}]^-$ reacts with aqueous acid to give $HOs_3(CO)_{10}(OH)$ and it seems likely that an oxanion derived from this is the nucleophile that attacks $Os_6(CO)_{18}$.

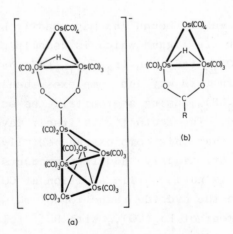

Figure 7

At this stage there seems to be no clear
evidence concerning the mechanism of formation of
cluster carbides in the pyrolysis of metal carbonyls,
but carbides can be produced directly from carbon
monoxide as well as from halocarbons. Combining the
nucleophilicity of one CO ligand (considerable for
extensively bridged CO in carbonyl anions) with the
electrophilicity of another (highest for terminal CO
in carbonyl cations) one might expect to form a C_2O_2-
bridge as shown below which would reasonably break
down into CO_2 and a reformed cluster carbide.

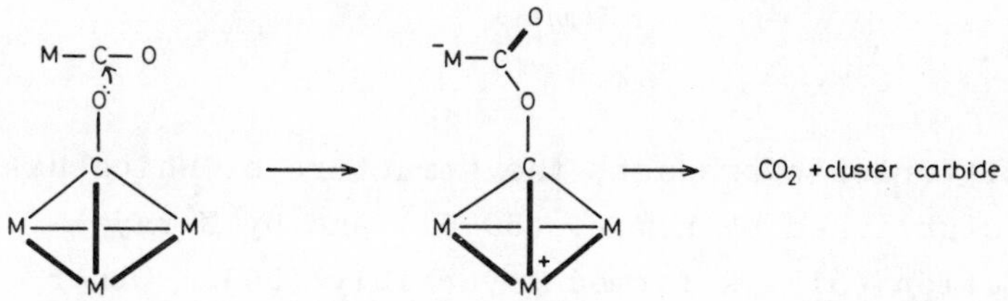

In a cluster that was particularly open geometrically
there could be an intramolecular coupling and extrusion
of CO_2, but this does not seem to be a requirement for
this chemistry. One requirement though is that electron-
rich and electron-poor components should come together
with CO-CO coupling but alternatively metal-metal bond
formation might occur as the components meet to give
higher nuclearity clusters. Indeed higher clusters
$[Os_5(CO)_{16}, Os_6(CO)_{18}, Os_7(CO)_{21}, Os_8(CO)_{23}]$ and
carbides as well at higher temperatures $[Os_5C(CO)_{15}$ and
$Os_8C(CO)_{21}]$ are formed in the sealed tube pyrolysis of
$Os_3(CO)_{12}$ (4). Carbide formation appears more facile
for ruthenium and $Ru_6C(CO)_{17}$ is formed in good yield
in the pyrolysis of $Ru_3(CO)_{12}$.

Carbonyl clusters may contain other ligands, such

as those derived from alkenes and alkynes, which are
more susceptible to nucleophilic attack than
coordinated CO. The best example is addition of
tertiary phosphine at the β-carbon of a μ^2-CH:CH$_2$
ligand, Figure 8.

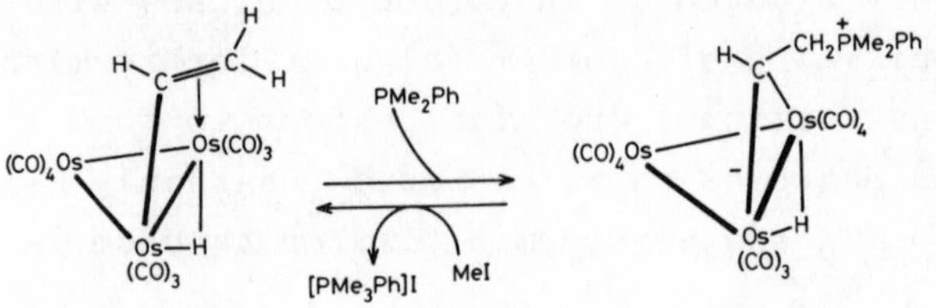

Figure 8

The zwitterionic product, the structure of which has
been established by n.m.r. (50,51) and by X-ray
diffraction (51) is formed reversibly (50). Other
hydrocarbon ligands also undergo nucleophilic attack,
for example $Os_3(\mu^3$-HC$_2$H)(CO)$_{10}$ (Figure 9) undergoes
attack by PMe$_2$Ph at carbon but this addition is
accompanied by an oxidative addition (H-transfer from
C to Os) to give an adduct of structure (A) or (B)
(50). The but-2-yne analogue does not react,
presumably because the oxidative addition is
fundamental to the reaction and the ligand has no
comparable hydrogen atom to transfer.

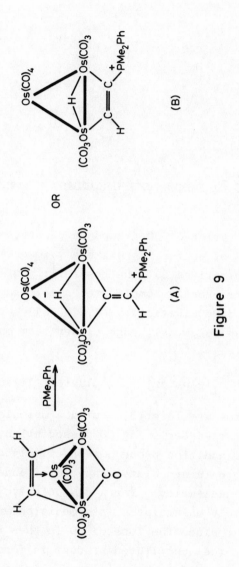

Figure 9

3. ADDITIONS AND ELIMINATIONS WITH CLUSTER REFORMING VERSUS
   SUBSTITUTION

Substitution of CO groups in clusters by other
ligands is, of course, an extremely important route to
cluster derivatives but the possibility of cluster
reforming frequently complicates simple substitution.
Systems studied kinetically (1,2,52-54) have followed
the rate law below, although either term may vanish in
particular cases:

$$\text{Rate} = k_1 \left[\text{cluster}\right] + k_2 \left[\text{cluster}\right] \left[\text{ligand}\right]$$

The two terms are identified with a unimolecular process,
usually a rate-determining CO dissociation, and a
bimolecular addition process respectively and either
could lead to intermediates capable of undergoing
changes in nuclearity. For a dissociative mechanism
coordinatively unsaturated intermediates must be
involved but since the loss of CO is then rate-
determining the substituted product is formed readily
in the presence of reacting ligand. It is only in the
absence of any suitable ligand to replace the lost CO
group that an increase in nuclearity is expected (4).
It is well known that second and third row metal
carbonyls will increase their nuclearity more readily

than first row. An alternative to cluster reforming
is that other ligands present might modify their
attachment to the metal by oxidative addition as in,
for example, the conversion of $HOs_3(OPh)(CO)_{10}$ to
$H_2Os_3(OC_6H_4)(CO)_9$ (55,56) (see Section 4) or by
employing an available electron pair for coordination
as in the conversion shown in Figure 10 (57). It is
in an associative pathway that one would expect cluster

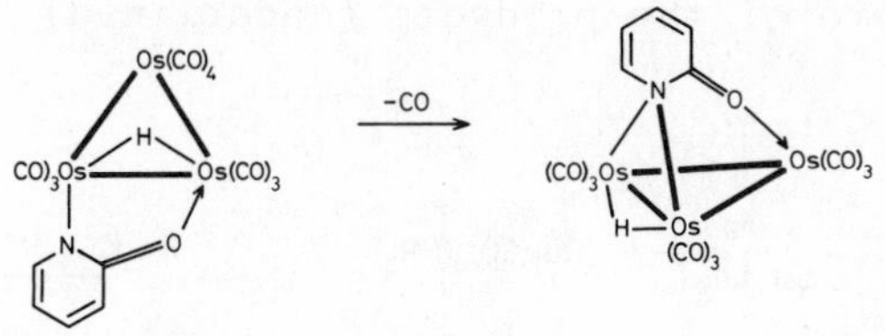

Figure 10

reforming, in this case degradation, to complete
significantly with substitution. In Section (2.2) a
few cases of stable adducts of carbonyl clusters were
cited and for $H_2Os_3(CO)_{10}$ a tertiary phosphine adduct
formed rapidly at $20^\circ C$ only slowly converts to the
substitution product in refluxing hexane (29):

$$H_2Os_3(CO)_{10} \xrightarrow[20^\circ C]{PPh_3} H_2Os_3(CO)_{10}(PPh_3) \xrightarrow[70^\circ C]{-CO} H_2Os_3(CO)_9(PPh_3)$$

    purple   yellow   purple

But this is entirely exceptional behaviour and in
general the adduct must be either close to the
transition state with bond breaking and making occurring
simultaneously or it is a short-lived intermediate which
looses CO or breaks up into compounds of lower nuclearity.
Very little of this has been established experimentally
but in the substitution of $Ru_3(CO)_{12}$ by PR_3 a linear

free energy relationship has been found relating k_2
with the phosphine basicity and indicates considerable
Ru-P bond formation in the transition state for the
bimolecular path (2). $Ru_3(CO)_{12}(PR_3)$ is, of course,
a transient species of unknown structure but one would
predict that it only contains two Ru-Ru bonds. In
this system there is a fine balance between substitution
and cluster breakdown and where $R = Bu^n$ the scheme
below has been proposed to account for the kinetics
and the nature of the products (underlined).

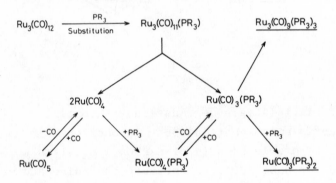

Cluster breakdown is favoured by high ligand
concentrations and temperatures and mononuclear
complexes are obtained exclusively on photolysis (58).
A PR_3-substituted cluster more readily breaks down
thermally than the parent carbonyl, certainly in the
above example, and the greater stability of third row
metal clusters is illustrated by the PR_3-substitution
of $Os_3(CO)_{12}$ can give solutions containing all four
complexes, $Os_3(CO)_{12-x}(PR_3)_x$ (x = 0, 1, 2 or 3), (59)
and mononuclear species are only formed under forcing
conditions.

It is universally observed in the substitution of

mononuclear carbonyls by PR_3 that successive
substitutions become increasingly difficult, slower
and more easily reversed, mainly it seems because of
electronic effects which lead to M-CO strengthening as
PR_3 ligands are introduced. In clusters this would
account for the substitution of only one CO per metal
atom by monodentate ligands, but an interesting effect
observed for the CO-substitution of $Ir_4(CO)_{12-x}(PR_3)_x$
(x = 0, 1, 2) is that a PR_3 ligand at one metal atom,
while presumably making CO groups at that atom less
labile, makes those at adjacent metal atoms more so
(1). Thus, $Ir_4(CO)_{12}$ leads to $Ir_4(CO)_9(PR_3)_3$ without
the intermediates $Ir_4(CO)_{11}(PR_3)$ and $Ir_4(CO)_{10}(PR_3)_2$
being observed and when these intermediates were
studied separately they also gave $Ir_4(CO)_9(PR_3)_3$
rapidly and at rates consistent with their being only
transient intermediates in the substitution of $Ir_4(CO)_{12}$.
There is a difference of rate law and a large difference
in activation parameters between the substitution of
$Ir_4(CO)_{12}$ and of $Ir_4(CO)_{12-x}(PR_3)_x$ (x = 1 or 2)
(Table 1) and this seems to be related to structural
differences. Only the parent carbonyl exists with
the non-CO-bridging structure (60). One cannot know
at this stage whether the labilising effect of
introduced PR_3 is related to structural ideosyncracies
of Ir_4 clusters or is a general phenomenon of carbonyl
clusters which undergo substitution without change of
nuclearity. High CO pressures are required to break
down Ir_4 clusters, $Ir_4(CO)_{12-x}(PR_3)_x$ (x = 3 or 4)
giving $Ir_2(CO)_{8-x}(PR_3)_x$ (x = 1 or 2) (61), and
substitution of these clusters occurs without
breakdown.

TABLE 1

Kinetic Data for the Substitution of $Ir_4(CO)_{12}$

	$\Delta H^{\ddagger}/kJ\ mol^{-1}$	$\Delta S^{\ddagger}/JK^{-1}\ mol^{-1}$
$Ir_4(CO)_{12} + PR_3 \longrightarrow Ir_4(CO)_{11}(PR_3) + CO$ Rate $= k_a[cluster][PR_3]$	86	-92
$Ir_4(CO)_{11}(PR_3) + PR_3 \longrightarrow Ir_4(CO)_{10}(PR_3)_2 + CO$ Rate $= k_b[cluster]$	133	+59
$Ir_4(CO)_{10}(PR_3)_2 + PR_3 \longrightarrow Ir_4(CO)_9(PR_3)_3 + CO$ Rate $= k_c[cluster]$	130	+71

4. OXIDATIVE ADDITION: BOND CLEAVAGE IN LIGANDS

If a molecule is reduced while entering the
coordination sphere of a metal atom which is itself
oxidised in the process, we say that the molecule has
added oxidatively. This term is usually restricted
to where a bond is cleaved in the addendum molecule
which can then occupy two coordination sites (see
for example, reference 62). In many or even most
reactions catalysed by transition metal compounds both
nucleophilic and electrophilic (oxidative) additions
are found to occur in succession and the ability of a
metal to readily interconvert between two oxidation
states differing by one or two is a usual requirement
of a homogenous catalyst. Oxidative addition has been
identified by some with dissociative chemisorption and
so a study of oxidative addition reactions of clusters
could provide a useful link between the two areas of
chemistry.

4.1. H–H Cleavage

It is clear that oxidative addition at a
coordinatively saturated complex, as in the

hydrogenation of $Os(CO)_3(PPh_3)_2$ (63), requires a loss
of ligand if both fragments of the adding molecule are
to remain coordinated, reaction (a). A reaction

$$Os(CO)_3(PPh_3)_2 + H_2 \rightarrow OsH_2(CO)_2(PPh_3)_2 + CO \qquad a$$

$$Os(CO)_3(PPh_3)_2 + Cl_2 \rightarrow \left[OsCl(CO)_3(PPh_3)_2\right]Cl \qquad b$$

without dissociation can occur if one of the fragments,
Cl^- in reaction (b), does not coordinate in the initial
step (64). Current evidence supports the idea that H_2
can only react with coordinatively unsaturated
molecules, such as $Os(CO)_2(PPh_3)_2$ formed by an initial
CO loss, so that two metal to hydrogen bonds may be
formed simultaneously to give a coordinatively
saturated product. Preliminary CO dissociation is
necessary in reaction (a) because the alternative
formation of free H^- or the formation of a
coordinatively supersaturated intermediate are both
energetically prohibitive. With chlorination no such
restriction applies because the formation of free Cl^-
is feasible, but it should be noted that subsequent
substitution of CO by Cl^- gives $OsCl_2(CO)_2(PPh_3)_2$ so
that whatever the mechanism the products of oxidative
addition may correspond, that is $OsX_2(CO)_2(PPh_3)_2$
(X = H or Cl).

Similar differences between hydrogenation (65)
and chlorination (66,59) apply to clusters as
exemplified in Figure 11. The final products
$Os_3X_2(CO)_{10}$ (X = H or Cl) are equivalent but this
should not be allowed to disguise gross mechanistic
differences. Thus chlorination of $Os_3(CO)_{12}$ does not
lead initially to CO dissociation whereas hydrogenation
does. There is evidence for preliminary adduct
formation in the chlorination, perhaps the adduct
being $\left[Os_3Cl(CO)_{12}\right]^+Cl^-$. Instead of Cl^- replacing a
CO ligand after the oxidative addition as in (b) above,

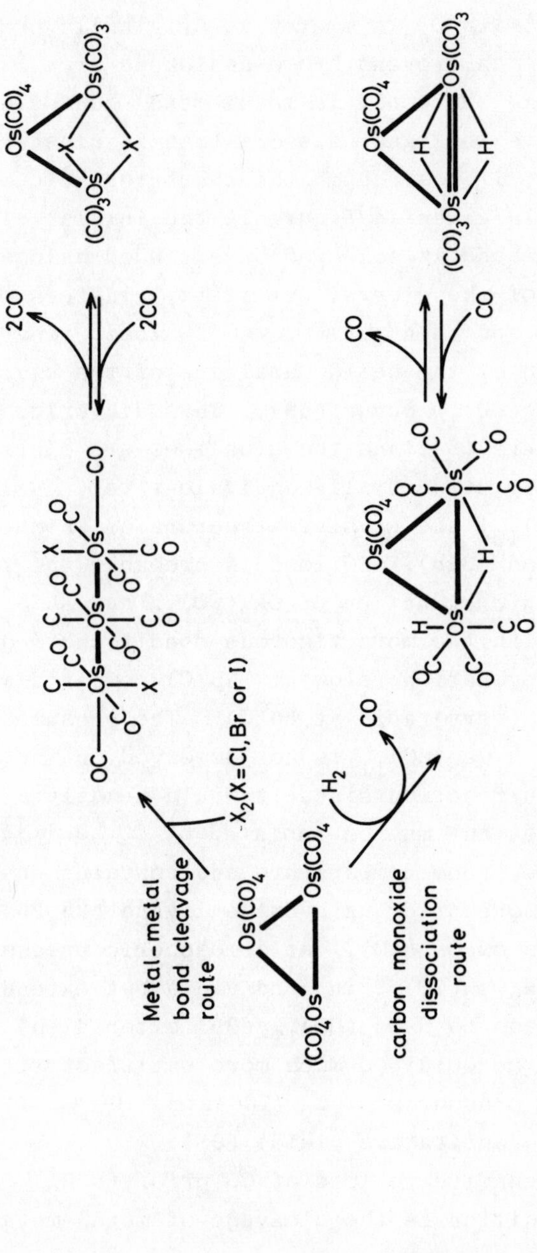

Figure 11

Os-Os bond cleavage gives the ring-opened product.
The complex $Os_3I_2(CO)_{12}$ has the chain structure shown
with cis-configurations at terminal osmium atoms (67),
and not that with D_{4h} symmetry as originally proposed
(66), and the chloro and bromo-analogues are
isostructural. Not only is metal-metal bond cleavage
an alternative to ligand dissociation in cluster
halogenation, it has become the characteristic feature.
In the example given in Figure 11 the initial cleavage
of one Os-Os bond is followed on extended halogenation
by cleavage of the others, but if the clusters $Os_3X_2(CO)_{12}$
are isolated and then thermolysed CO loss, ring closure
and formation of the halide analogue of the hydrogenation
product $H_2Os_3(CO)_{10}$ occur (59). This dihydride is of
known structure (34) and the dihalides are electronically
and presumably structurally equivalent to
$Os_3(OMe)_2(CO)_{10}$, the crystal structure of which has also
been described (34b). CO loss is probably the first
step in the hydrogenation of $Os_3(CO)_{12}$ and this is
consistent with the more vigorous conditions required
for the hydrogenation (slow at $125^{o}C$) compared with
chlorination (very rapid at $80^{o}C$). The presumed
intermediate $H_2Os_3(CO)_{11}$ is not observed in the
hydrogenation reaction since it would readily
decarbonylate, but may be isolated by CO addition to
$H_2Os_3(CO)_{10}$ at room temperature and contains one
terminal and one bridging hydride ligand (28,29).
Hydrogenation of $Os_3(CO)_{12}$ at atmospheric pressure at
$125^{o}C$ gives $H_2Os_3(CO)_{10}$ in good yield but extended
treatment leads to $H_4Os_4(CO)_{12}$ (29% after 41 h) (65),
but this may be obtained much more satisfactorily by
using higher pressures of H_2 (100 atm, $100^{o}C$, 24 h,
essentially quantitative yield) (68).

An alternative to loss of CO prior to H_2
oxidative addition is the cleavage of metal-metal bonds

and it appears that, although $Re_2(CO)_{10}$ is generally
inert to H_2, on photolysis (311 nm) rhenium hydrides
{$HRe(CO)_5$, $H_2Re_2(CO)_8$, $H_3Re_3(CO)_{12}$} are formed probably
via the intermediate $Re(CO)_5$ (69). It is clear though
that cluster breakdown should occur unless CO ligands
are lost.

Oxidative addition of H_2 at metal atoms appears
to be the only way for it to react with clusters and
in cases where one or both hydrogen atoms are located
on a ligand in an isolated product one would assume
that at some stage both had been bound to metal atoms.

4.2. C–H Cleavage in Alkenes

Fission of vinylic C-H bonds at clusters in
preference to allylic ones is a clear indication that
the combined influence of several metal atoms generate
a unique ligand chemistry for clusters. Studies of
the oxidative addition reactions of alkenes and alkynes
is still largely restricted to triosmium and
triruthenium clusters and more recently to higher
clusters of these metals although more complex sequences
of reactions occur with these. The best example is the
reaction of ethylene (1 atm) with $Os_3(CO)_{12}$ at $125^{\circ}C$
to give compound (A), Figure 12. The two presumed
intermediates may be isolated by the room temperature
reactions b, d, e and f but as these compounds are not
observed in the reaction at $125^{\circ}C$ the initial
substitution reaction a must be rate-determining. A
general feature of the reactions of
$Os_3(CO)_{12}$ is that once CO has been displaced
coordinatively unsaturated species are generated which
are very reactive towards oxidative addition. Hence

428.

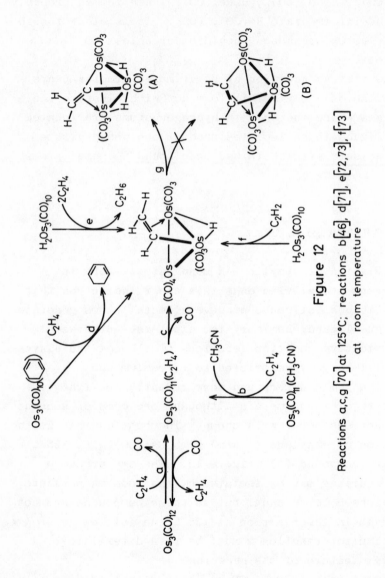

Figure 12

Reactions a,c,g [70] at 125°C; reactions b[46], d[71], e[72,73], f[73]
at room temperature

the presence of readily replaced ligands L (alkenes or CH_3CN) in the complex $Os_3(CO)_{11}L$ (46) or $Os_3(CO)_{10}L_2$ (71,74) allows either oxidative additions to occur under milder conditions or the isolation of intermediates prior to oxidative addition.

Interestingly, the vinyl group in the intermediate complex is doubly-bridging in the configuration shown in Figure 12 (74) and hence donates 3 electrons through one σ-Os-C bond and a familiar η^2-alkene to metal interaction to the other metal atom bridged whereas the corresponding product from reaction f (but using $CF_3C \equiv CCF_3$) contains the η^3-vinyl ligand $CF_3C.CHCF_3$ which donates the same number of electrons by forming three Os-C σ-bonds (75). This leads to the interesting speculation that the hydrocarbonyl vinyl ligand might transform into the μ^3-bonding mode prior to the second oxidative addition.

Although the isomeric form (B) is not obtained from the $C_2H_4/Os_3(CO)_{12}$ reaction, both (A) and (B) types are formed from the corresponding reaction with $Ru_3(CO)_{12}$ (17) and from the reaction of $CH_3CH:CH_2$ with $Os_3(CO)_{12}$ whereas cyclic alkenes only give products of type (B) as expected (70, 76-80). Cyclic alkenes and benzene are chemically very different in most respects but here benzene adds in the same way to give $H_2Os_3(\mu^3-C_6H_4)(CO)_9$ (70), as does pyrrole (NC_4H_5) to give $H_2Os_3(\mu^3-NC_4H_3)(CO)_9$ in which two C-H bonds are cleaved leaving the N-H bond intact (81).

$Os_6(CO)_{18}$ also reacts with ethylene (at 165°C) to give a green complex $Os_6(CO)_{18}(C_2H_4)_2$ (8), but this deceptively simple formulation disguises the fact that several reaction steps (including oxidative addition) must have led to it since it contains two CH_3C ligands bridging (μ^3 and μ^4) triangular and square metal faces respectively, see Chapter 2, structure 78. It would

certainly be valuable as with the Os_3 clusters
(Figure 12) to attempt the interception of intermediates by generating complexes such as $Os_6(CO)_{16}L_2$
where L might be readily displaced by ethylene.

Oxidative addition of alkenes might occur widely
for third row transition metal clusters, but is found
for only a few metals other than osmium and ruthenium.
For example, $Ir_4(CO)_{12}$ reacts with cycloocta-1,5-diene
(C_8H_{12}) to give $Ir_4(CO)_5(C_8H_{12})_2(C_8H_{10})$ as the major
product which contains the dehydrogenated ligand
cycloocta-1,5-enyne; the two alkyne carbon atoms and
the metal atoms form an Ir_4C_2 octahedron as in
$Co_4(CO)_{10}(EtC_2Et)$ (82). A minor product of this
reaction $Ir_7(CO)_{12}(\eta^4-C_8H_{12})(\mu^2-C_8H_{11})(\mu^3-C_8H_{10})$ is
something of a curiosity in that it contains alkene
ligands in three stages of dehydrogenation (83).
Although the μ-ligands were clearly generated by
oxidative addition, subsequent reactions such as
reductive elimination of H_2 or alkene hydrogenation
have led to hydrogen atom removal.

Types of cluster different from those in Figure
12 are obtained by oxidative addition of hexadienes
(C_6H_{10}) to $Ru_3(CO)_{12}$ to give two isomers of
$HRu_3(C_6H_9)(CO)_9$. Isomer (A) (Figure 13) contains a
μ^3-allenyl and the thermally derived isomer (B)
a η^3-allyl group with M-C σ-bonds to the terminal
carbon atoms (84) and for osmium this class of compound
is made accessible by oxidative addition of but-2-yne
(73,85). The mode of bonding in (B) (Figure 13)
appears to be particularly stable and is found in
products from such unlikely reactions as that of
$Ru_3(CO)_{12}$ with cyclododecatriene (86), the treatment
of $Ru_3(CO)_{12}$ with PhLi (87) and the insertion of
$Ph_2C=C=O$ into an Os-H bond of $H_2Os_3(CO)_{10}$

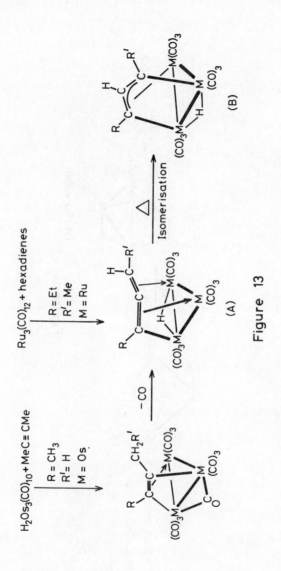

Figure 13

432.

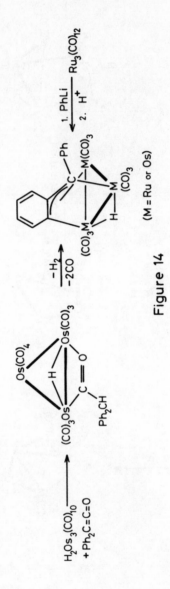

Figure 14

to give a μ^2-acyl complex which is then pyrolysed (88). The last two routes are illustrated in Figure 14.

4.3. C-H Cleavage in Alkynes

The conversion of the μ^3-but-2-yne into the μ^3-1-methylallenyl triosmium complex by oxidative addition is a clear example (Figure 13) but oxidative addition with $RC \equiv C-H$ cleavage is the normal behaviour for terminal alkynes to give μ^2- and μ^3-alkynyl groups in clusters. For example, $PhC \equiv CH$ reacts with $H_2Os_3(CO)_{10}$ to give the insertion product $HOs_3(\mu^2-CH=CHPh)(CO)_{10}$ together with an oxidative addition product $HOs_3(\mu^2-C \equiv CPh)(CO)_{10}$ (Figure 15) in which the alkynyl ligand is bound in a σ,π-manner comparable with the μ^2-$CH=CH_2$ ligand shown in Figure 12 (73). However, alkynyl ligands have two more electrons available for coordinating than do alkenyls and the μ^3-5e-donor mode

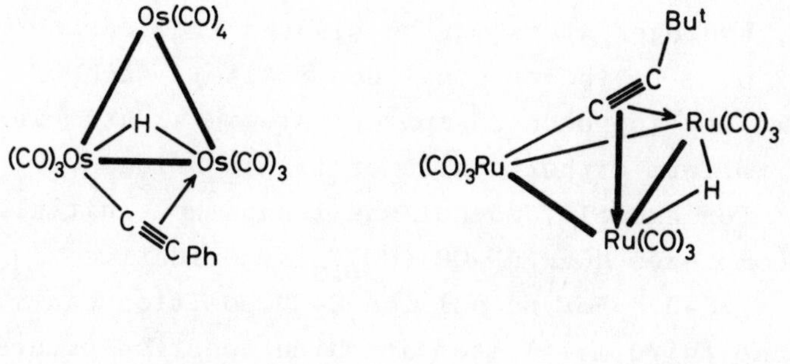

Figure 15 μ^2- and μ^3 - alkynyl ligands

should be regarded as more characteristic for clusters than the μ^2-mode. The best characterised example of this is $HRu_3(\mu^3\text{-}C\equiv CBu^t)(CO)_9$ obtained from $Ru_3(CO)_{12}$ and $Bu^tC\equiv CH$ (89,90) (Figure 15).

4.4. C-H Cleavage in Other Molecules

Accumulating results on the oxidative addition reactions of $Os_3(CO)_{12}$ show the remarkable generality of this reaction. Other clusters have been studied but it is only in the triosmium system that a wide range of substrate types have been used and some examples, mainly from our own work are collected in Figure 16. We can say with confidence that essentially all functionalised organic molecules, excluding alkanes for the present, are capable of undergoing oxidative addition. It would seem that a n- or π- donor attachment at metal is needed to position the C-H bond close to the metal atoms for long enough for reaction. In hoping to expand this chemistry, selectivity will be a problem because as seen in Figure 16, hydrogen atoms may be cleaved from carbon atoms α, β or γ to the original donor site. Each of these atoms may approach an appropriate metal atom in a cluster to form either double or triple bridges or chelates. For example, phenol and 1-naphthol initially give the complexes $HOs_3(\mu^2\text{-}OR)(CO)_{10}$ by oxidative addition of RO-H. For phenol the 2-CH position can approach the third metal atom at which reaction occurs to give $H_2Os_3(OC_6H_4)(CO)_9$ which interestingly contains the $\mu^3\text{-}OC_6H_4$ ligand in the non-aromatic dienone form as established by X-ray diffraction work on the 2-PhCH$_2$- substituted compound (99). Aniline gives the

α-Metallation

$$PEt_3 \xrightarrow{[91,85]}$$

Et–P–Et, Et, CH₃ structure with $(CO)_3Os$—$Os(CO)_3$, H—$Os(CO)_3$—H

$$\xrightarrow[]{PhCH:NMe \; [95,96]}$$

$Os(CO)_4$ / $(CO)_3Os$ —H— $Os(CO)_3$ / $C=N$ / Ph — Me

Pyridine $\xrightarrow{[93]}$ $Os(CO)_4$ / $(CO)_3Os$ —H— $Os(CO)_3$ / N-pyridyl

$$\xrightarrow[-H_2]{NMe_3 \; [96]}$$

$Os(CO)_4$ / $(CO)_3Os$ —H— $Os(CO)_3$ / C / Me—N—Me

2,2'-bipyridine $\xrightarrow{[94]}$ $Os(CO)_4$ / $(CO)_3Os$ —H— $Os(CO)_2$ / N—N bipyridyl

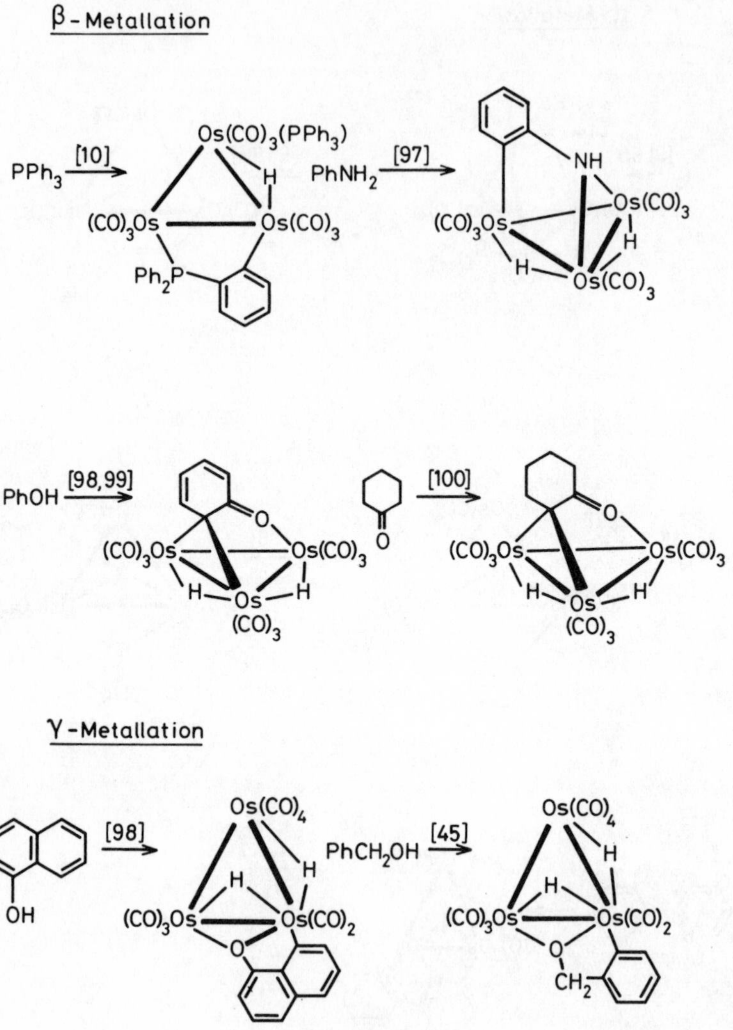
β - Metallation
PPh3
[10]
Os(CO)3(PPh3)
H
(CO)3Os
Os(CO)3
Ph2P
PhNH2
[97]
NH
Os(CO)3
H
(CO)3Os
H
Os(CO)3
PhOH
[98,99]
O
(CO)3Os
Os(CO)3
H
Os
H
(CO)3
[100]
O
(CO)3Os
Os(CO)3
H
Os
H
(CO)3
γ -Metallation
OH
[98]
Os(CO)4
H
H
(CO)3Os
Os(CO)2
O
PhCH2OH
[45]
Os(CO)4
H
H
(CO)3Os
Os(CO)2
O
CH2

Figure 16

analogous cluster $H_2Os_3(NHC_6H_4)(CO)_9$ but this should
not be regarded as the imine derived from the dienone
since an alternative aromatic configuration is adopted;
the OC_6H_4 complex was thought to contain an aromatic
ring (98) prior to X-ray diffraction work. β-Metalla-
tion is also available to 1-naphthol but unlike phenol
there is a competing reaction in which the 8-CH group
approaches an osmium atom already bonded to oxygen to
give the chelated γ-metallated product shown in
Figure 16. Similar competitive reactions are found
for benzyl alcohol which metallates at the γ-site
(Figure 16) but also at the α-site to give PhCHO.
(The site α to the donor atom is, of course, β to the
metal and the formation of PhCHO may alternatively be
regarded as a β-elimination)

As can be seen oxidative addition is an extremely
versatile route to many interesting or novel organometallic clusters containing doubly- and triply-
bridging ligands and is applicable to both saturated
and unsaturated organic molecules.

4.5. The Unique Aspects of Oxidative Addition in Clusters

The ability to cleave bonds very close to the
original point of attachment of a ligand to a
triosmium cluster is in contrast to overriding
5-membered chelate ring formation in cyclometallation
reactions of mononuclear systems (101). There are a
few cases recently of metallations close to the donor
site at single metal atoms, such as in the conversion
of $Fe(PMe_3)_5$ to $HFe(Me_2PCH_2)(PMe_3)_4$ which contains a
FePC ring (102), but this cannot be regarded as
characteristic behaviour. Almost certainly in clusters

hydrogen atom transfer from a ligand is to an adjacent
metal atom (Figure 17). Thus the first oxidative
addition of ethylene at an Os_3 cluster is likely to be
as in (A) and that of pyridine as in (B). Since the
complexes $Os_3(CO)_{10}L_2$ (L is a readily replaced ligand)
react with C_2H_4 or py to give oxidative addition
products directly with $Os_3(CO)_{10}L_2$ (L = C_2H_4 or py)
not reaching detectable concentrations whereas
$Os_3(CO)_{11}L$ (L = C_2H_4 or py) are unreactive enough to
be isolated, it must be assumed that it is the
coordinatively unsaturated complexes $Os_3(CO)_{10}L$ that
undergo oxidative addition. This is, of course,
expected. If CO loss is the first step in the
internal oxidative addition reaction of $HOs_3(\mu^2\text{-}CH{:}CH_2)$
$(CO)_{10}$ and similar compounds (reaction g, Figure 12),
one would expect that the 1,1- and 1,2-isomers [(A)
and (B) respectively, Figure 12] are formed via the
configurations (C) and (D) (Figure 17) respectively.
Configuration (C) is found for the $CH=CH_2$ and $CH=CHEt$
complexes in the solid state (74) and this almost
certainly persists in solution. An alternative
mechanism to be considered involves oxidative
addition before CO loss, in which case one would
expect a coordinatively unsaturated cluster to be
generated by the vinyl group becoming terminal and so
allowing internal oxidative addition to take place
through configurations (E) and (F) respectively.

Whatever the precise mechanistic details of these
oxidative addition reactions (there certainly needs to
be work in this direction), it would seem that
coordinatively unsaturated intermediates must be
generated. Hence it follows that an examination of
such clusters containing hydrocarbon would be extremely
relevant to oxidative addition. Recently it has been
shown that the cluster formed by an equimolar addition

439.

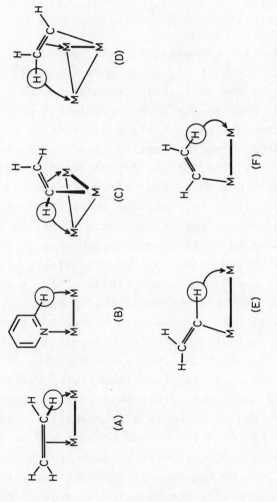

Figure 17   Some oxidative addition products of $Os_3(CO)_{12}$
The other reagent and one product only
is shown in each case.

of CH_2N_2 to $H_2Os_3(CO)_{10}$ undergoes a tautomeric equilibrium that may be regarded as an oxidative

$$HOs_3(\mu^2-CH_3)(CO)_{10} \qquad H_2Os_3(\mu^2-CH_2)(CO)_{10}$$
$$(A) \qquad\qquad\qquad (B)$$
$$\text{unsaturated} \qquad\qquad \text{saturated}$$

addition/reductive elimination equilibrium (103).
The elements of CH_4 in (A) would seem to donate 2e to the metal atoms but 4e in (B). Separate n.m.r. signals were observed for these isomers but spin saturation transfer experiments confirm their ready interconversion. Two interesting conclusions have been made from a study of the dideuterated complexes prepared from $D_2Os_3(CO)_{10}$ and CH_2N_2. Firstly, where there is an equilibrium involving transfer of hydrogen atoms between metal- and carbon-bound sites, as in this case where an equilibrium distribution is readily achieved, deuterium atoms will preferentially accumulate at the carbon atoms (104). This is a consequence of the higher vibrational frequencies associated with C-H compared with M-H or M-H-M bonds. Secondly, an analysis of the large chemical shifts between the $C\underline{H}_3$, $C\underline{H}_2D$ and $C\underline{H}D_2$ signals in the ^1H n.m.r. spectra of variously deuterium-substituted versions of (A) has shown that there are methyl bridges as shown in Figure 18 (105). There is a preferential occupancy of H rather than D in the C-H-Os bridges and so the H-atoms in the CH_3 complex on an average have a lower occupancy of the bridge than in the CHD_2 complex, and as might be expected there is quite a large chemical shift difference between bridging and non-bridging sites. The rate of hydrogen atom transfer between these sites is high.

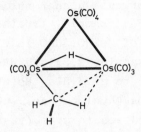

Figure 18

Thus there is evidence that electrons occupying C-H
bonding orbitals may be delocalised on to metal atoms
in coordinatively unsaturated clusters, and one
certainly suspects that similar interactions would
occur in intermediates prior to oxidative addition.
While Figure 17 represents possible hydrogen atom
movements during oxidative addition it could also
represent ground state C-H-M interactions in
coordinatively unsaturated intermediates.

4.6. Cleavage of Other Bonds

Bonds between hydrogen and heteroatoms are also
readily cleaved by oxidative addition at clusters and
there are examples of cleavage of O-H bonds in
alcohols (98), water (106) and carboxylic acids (71),
of S-H bonds in thiols (107-110) or hydrogen sulphide
(111), of N-H bonds in amines (97) (but not in pyrrole
which undergoes oxidative addition at $Os_3(CO)_{12}$ by
cleavage of two C-H bonds rather than the N-H one
(112)), of Cl-H, Br-H and I-H bonds (71) and of M-H

bonds. In the latter case higher nuclearity clusters may be generated. For example, treatment of $Os_3(CO)_{12}$ with one equivalent of $Me_3NO.2H_2O$ in CH_3CN displaces one CO as CO_2 (see section 2.2) and $HRe(CO)_5$ can then be added oxidatively to give $HOs_3[Re(CO)_5](CO)_{11}$, while adding $HRe(CO)_5$ to $Os_3(cyclooctene)_2(CO)_{10}$ allows a double oxidative addition to give $(\mu^2-H)_2Os_3|\eta^1-Re(CO_5]_2$ $(CO)_{10}$ (113). One might by analogy consider the possibility of a double H_2 oxidative addition to give the unknown cluster $H_4Os_3(CO)_{10}$ with two terminal and two bridging hydride ligands. Os_3W clusters have been similarly prepared by addition of $HW(CO)_3(\eta^5-C_5H_5)$ to the above cyclooctene osmium cluster to give $HOs_3W(CO)_{12}(\eta^5-C_5H_5)$ and $H_3Os_3W(CO)_{11}(\eta^5-C_5H_5)$ (114).

Bonds between many other pairs of atoms may be cleaved but just a few highlights will be presented here. P-C or As-C bonds may be broken as in the formation of $\mu^3-C_6H_4$ from phenyl phosphines or arsines (10,115). For example $Os_3(CO)_{11}(PMe_2Ph)$ gives $HOs_3(PMe_2)(C_6H_4)(CO)_9$ as a thermal decomposition product (115) and an analogous μ^3-acetylene cluster $HOs_3(PEt_2)(C_2H_2)(CO)_9$ was obtained as a pyrolysis product of $Os_3(CO)_{11}(PEt_3)$ but in this case H_2-elimination occurs as well as oxidative addition (92). In forming $\mu^3-C_6H_4$ and $\mu^3-C_2H_2$ in these reactions C-H cleavage in the phosphine substituent occurs first to give bridging ligands (see Figure 16); the hydrocarbon group is then separated by P-C bond breaking in the bridge. This may not be the only way P-C bonds may break since the μ^2-phenyl ligand in $Os_3(CO)_8(PPh_2)(Ph)$ $(PhPC_6H_4)$ is apparently directly cleaved from a phosphorus atom either in PPh_3 or $Ph_2PC_6H_4$ ligands of precursors (10). P-O and C-O bonds are broken in the thermolysis of $Os_3(CO)_{11}[P(OMe)_3]$ but clearly a lot of other chemistry is required to form

$Os_5C(CO)_{13}[\mu^3-OP(OMe)OP(OMe)_2]$ from this treatment (116). It is not surprising that this is not the only product from this reaction which involves such major modification of both cluster and ligand.

Another important type of bond cleavage is the triple-bond cleavage of alkynes as in the formation of $M_3(C_5H_5)_3[\mu^3-CNEt_2]_2$ (M = Co or Rh) from the reaction of $M(C_5H_5)(CO)_2$ with $Et_2NC\equiv CNEt_2$ (117). Two related trirhodium clusters formed similarly from $PhC\equiv CPh$ are $Rh_3(C_5H_5)_3(CO)(PhC\equiv CPh)$ and one without the CO group, $Rh_3(C_5H_5)_3(PhC\equiv CPh)$ (118). The CO-containing cluster has been thoroughly characterised structurally as (A) (Figure 19) (119), and this complex is analogous to $Os_3(CO)_{10}(RC\equiv CR)$, which incidentally contains a bridging CO where R = H or Me but not where R = Ph, and has structure (B) with the same attachment of the alkyne to the metal atoms as in (A). A cluster with one less CO interestingly adopts a quite different M_3C_2 framework based on a trigonal bipyramid (C), and this appears to be a particularly rare form of alkyne bonding and the only other example as established by diffraction studies seems to be $Ni_4(Bu^tNC)_4(\mu^3-PhC\equiv CPh)_3$ (92). A similar geometry was proposed for the isoelectronic cluster $Rh_3(C_5H_5)_3(PhC\equiv CPh)$ (D), but there are, however, different trigonal bipyramidal geometries available and an alternative to (D) is (E) in which the C-C bond has broken. The close similarity of the ^{13}C n.m.r. signals for the Et_2N- and Ph-acetylene derivatives shows that they both adopt geometry (E), but why the tri-iron complex has structure (C) is an open question. Thus an analysis of $RC\equiv CR$ cleavage reactions which seemingly requires such major ligand modification simply resolves into a change of the distribution of skeletal atoms in a cluster polyhedron

444.

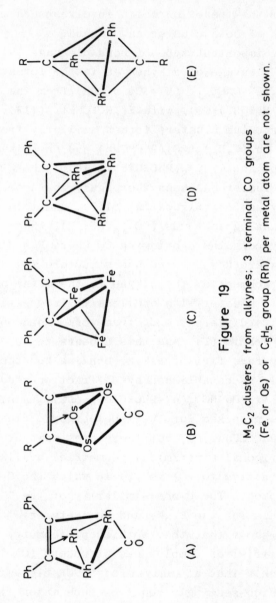

(A)   (B)   (C)   (D)   (E)

Figure 19

M_3C_2 clusters from alkynes; 3 terminal CO groups (Fe or Os) or C_5H_5 group (Rh) per metal atom are not shown.

445.

and the classification of this as an oxidative
addition is not very useful conceptually.

Another example of alkyne bisection is the
formation of $Os_6(CO)_{16}(\mu^3-CPh)(\mu^4-CPh)$ from $Os_6(CO)_{18}$
and $PhC{\equiv}CPh$ (120), the product being structurally just
like the corresponding CMe complex formed from the
reaction of ethylene with $Os_6(CO)_{18}$ (8), structure (A)
(Figure 20). Also formed in the reaction with ethylene

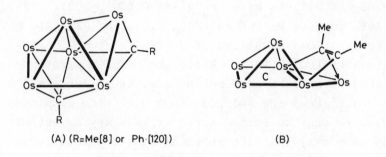

(A) (R=Me[8] or Ph·[120])   (B)

Figure 20   (CO ligands omitted)

is product (B) which contains coupled CMe ligands, so
that the scission of $RC{\equiv}CR$ and the reverse coupling
of the fragments RC are both reactions that can occur
in clusters.

Some other examples of bond cleavages to give
clusters which contain only fragments of the original
reagent are given below:

$Ru_3(CO)_{12}$ $\quad + Me_3Sn-CH_2NMe_2 \longrightarrow HRu_3(\mu^2-CNMe_2)(CO)_{10}$ (121)

$Fe_2(CO)_9$ $\quad + Me_3SiN-N_2 \quad \longrightarrow Fe_3(\mu^3-NSiMe_3)(CO)_9$ (122)

$Fe_2(C_5H_5)_2(CO)_4 + RS-S-R \quad \longrightarrow Fe_3(\mu^3-S)(\mu^3-SR)(C_5H_5)_3(CO)$ (123)

$Fe_3(CO)_{12}$ $\quad + EtNO_2 \quad \longrightarrow Fe_3(\mu^3-NEt)(CO)_{10}$ (124)

Clusters may also be derived by substitution of Cl at
various elements by transition metal centres and, for

example, CCl_4 can be used to generate $Co_3(\mu^3\text{-CCl})(CO)_9$ (125) or $[Rh_6C(CO)_{15}]^{2-}$ (126). Chlorine atoms are displaced from CCl_4 to form the μ^3-CCl ligand and the μ^3-CR ligand may be similarly generated from $RCCl_3$ and in certain cases all four chlorine atoms are displaced from CCl_4 to form carbide ligands which in the rhodium case above, lie at the centre of octahedra of rhodium atoms. Halo-compounds are useful reagents to introduce various atoms into clusters and further examples are the use of SiI_4 to give $Co_3[\mu^3\text{-Si-Co(CO)}_4](CO)_9$ (128), of AsF_3 to give $Fe_3(\mu^3\text{-As})(CO)_9$ (129), of $PRCl_2$ to give $Co_3(\mu^3\text{-PR})(CO)_9$ (130) and of $S=PCl_3$ to give $Co_3(\mu^3\text{-P=S})(CO)_9$ (131). While the metal reagent used in these reactions need not be a cluster, for example $[Co(CO)_4]^-$, once the halogen atoms have been displaced a cluster must be formed since it is a key aspect of this chemistry that the residual atom or group forms bonds to several metal atoms. This allows some clusters to be synthesised with a degree of design.

5. OTHER REACTIONS RELATED TO CATALYSIS

5.1. Oligomerisation Reactions

Cyclotrimerisation reactions of alkynes are
induced by many transition metal compounds, including
clusters. The most notable difference between
clusters and mononuclear systems is the nature of the
metal to hydrocarbon interactions. Indeed the ability
of alkynes to couple in clusters was recognised very
early by Hübel and co-workers (132) when two isomers
of $Fe_3(CO)_8(PhC\equiv CPh)_2$ (violet and black) were isolated
from the reaction of $PhC\equiv CPh$ with $Fe_3(CO)_{12}$ which were
later shown by X-ray diffraction (133) to be as
in Figure 21.

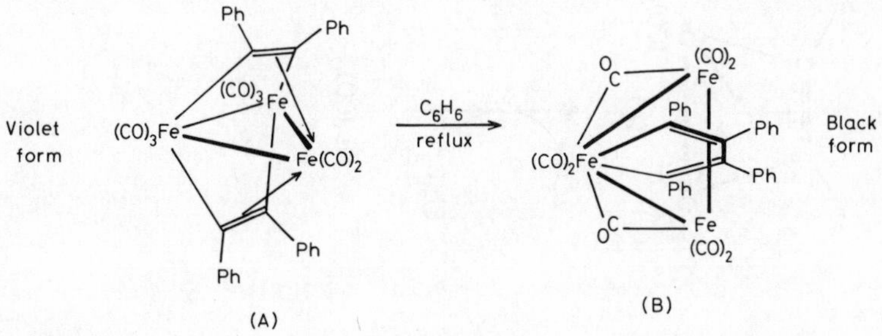

Figure 21 [132, 133]

448.

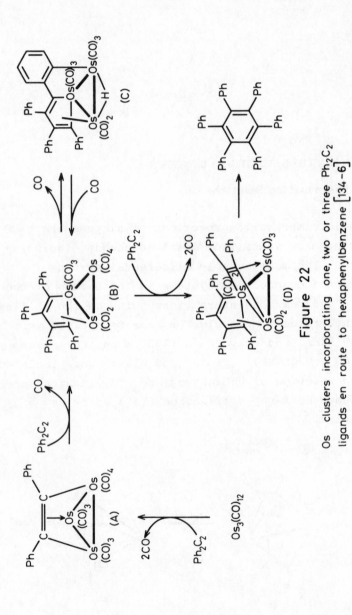

Figure 22

Os clusters incorporating one, two or three Ph_2C_2 ligands en route to hexaphenylbenzene [134-6]

Metallocyclopentadiene rings are formed in most
alkyne oligomerisations with mononuclear, dinuclear
and cluster systems, but in clusters carbon bridges
between metal atoms are present. In the tri-iron
system (Figure 21) all three metal atoms are
associated with the C_4 chain whereas in the closely
related tri-osmium clusters (Figure 22) the C_4 chain
is associated with only two metal atoms as in other
dinuclear molecules. The difference is of course
related to the different stoichiometries of compounds
(B) (Figures 21 and 22), $M_3(CO)_x(Ph_4C_4)$ [M = Fe, x = 8;
M = Os, x = 9]. In the osmium case the complex
$Os_3(CO)_8(Ph_4C_4)$ with the same stoichiometry as the
iron compound is formed reversibly from the
nonacarbonyl but interestingly this does not have a
structure like that of the iron compound but rather
two extra electrons are fed into the cluster by an
oxidative addition. A hydrogen atom is transferred
from a phenyl group as shown in (C) so that the
difference between iron and osmium relates to the
greater facility with which third row metals undergo
oxidative addition. All iron and osmium clusters with
uncoupled alkynes contain these in the common μ^3-mode
in these systems.

There are very many more varied ways alkynes can
couple in clusters than described so far and the
complexes of formulae $Fe_3(CO)_8(MeC\equiv CH)_2$,
$Fe_3(CO)_8(MeC\equiv CH)_3$, $Fe_3(CO)_8(MeC\equiv CH)_4$ (137),
$M_3(CO)_7(Bu^tC\equiv CH)_4$ (M = Ru or Os) (138),
$Os_3(CO)_{10}(EtC\equiv CH)_2$, $Os_3(CO)_9(EtC\equiv CH)_2$ (139),
$Fe_3(CO)_7(EtC\equiv CH)_4$ (140) and $Os_3(CO)_{10}(PhC\equiv CH)_2$ (141)
are a few examples which contain a fascinating array
of ligand types, none actually containing alkyne
ligands, and which can only satisfactorily be described
with diagrams. As well as coupling of alkynes, CO

groups may be coupled to alkyne carbon atoms through C atoms alone (139) or through both C and O atoms (138,141), the RC≡CH triple bond may be cleaved and the components coupled separately (140) and hydrogen atoms may be transferred from carbon to carbon (137, 138) or from carbon to metal (139). Given this scope there seems to be few limitations as to what products may be obtained and an analysis of how these are formed is a daunting prospect. Interestingly though, having established the structures of crystalline products by diffraction methods (and there seems to be no other way), all can be shown to fit the 18e-rule and the metal to ligand bonding can be understood by breaking it down into well-known types, some of which have been described already.

5.2. Hydrogen Transfer Reactions and Triple Bond Reductions

In the chemistry of organic ligands with Os_3, Os_4 and Os_6 clusters and corresponding Ru clusters hydrogen transfer reactions are extremely important. Rapid site exchange of metal-bound hydrogens is a characteristic of such clusters and oxidative addition with H-transfer from ligand to metal by oxidative addition has been discussed in section 4. Since H_2 addition at clusters is expected to occur exclusively at the metal atoms, subsequent H-transfer from metal to ligand is important in the H_2-reduction of multiple bonds of ligands in clusters. Hydrogen transfer to alkenes (142) and alkynes (73) from $H_2Os_3(CO)_{10}$ to give alkyl and vinyl products respectively are both cis as expected. The reductions of alkynes and organonitriles in clusters have been

studied. The reduction of a μ^3-C≡CR ligand in a
triosmium cluster to μ^3-C.CH$_2$R via the intermediacy of
μ^3-C=CHR (where R = H or alkyl) is shown in Figure 23,
the appropriate species being (B), (D) and (F) (143,
73). Complex (A) containing a μ^3-alkyne can loose CO
and take up H$_2$. Now this can be done in two ways
(73). Direct hydrogenation with displacement of CO
leads to (C) and under atmospheric pressure treatment
no further hydrogenation takes place and no H-atom is
transfered to the alkyne ligand. On the other hand, an
initial displacement of CO from (A) by thermolysis
gives the μ^3-alkynyl complex (B) by oxidative addition
(see Section 4.3). In this form hydrogenation proceeds
directly through to (F) presumably via (D) although
this could not be observed in this treatment. (D), can,
however, be synthesised separately as shown in Figure
12 and was shown to hydrogenate to (F) so is reasonably
the intermediate. This could be an important
conclusion: it is the asymmetric 1,1-isomer (D) that
can be hydrogenated rather than the 1,2-isomer (C).
In certain cases it is possible to duplicate or
reverse these hydrogenations by successive additions
or removals of H$^+$ and H$^-$ with ionic intermediates such
as (E). (D) has the tilted side-on arrangement of
C=CHR and the observation of hydride signals with
intensity ratio 1:2 indicates that the ligand is also
tilted in complex (E) (17). The structure of the
corresponding complex $[Co_3(CCH_2)(CO)_9]^+$ is unknown
but it probably adopts the tilted geometry rather
than a vertical configuration. N.m.r. evidence (144)
is that the positive charge is considerably
delocalised on to the metal and calculations (12)
indicate that a tilted geometry is most stable. The
"acylium" ion $[Co_3(C-CO)(CO)_9]^+$ is related and,
although the structure is unknown, we prefer

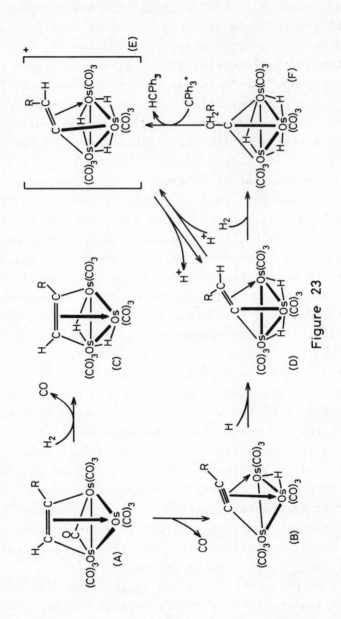

Figure 23

structure (B) (Figure 24) which is best described as
a ketenide.

The reduction of RCN in Fe_3 clusters (147) is
closely related to the reductions in Figure 23. The
two isomers (B) and (C), Figure 25, are directly
analogous to isomers (C) and (D) in Figure 23 and
significantly it is only the 1,1-isomer (C) that is
hydrogenated to (D). The different in this system is
that the 1,2-isomer readily isomerises to the 1,1-
isomer but not for the alkyne derivatives, but to
what extent this reflects the greater lability of Fe
compared with Os cannot be assessed at present.

There is a lot of interest in the reduction of
alkynes, N_2 and nitriles in their own right but also in
relation to CO reduction (148). It is not possible
to discuss the whole problem of CO reduction here but
there are significant observations that have been made
for clusters that are relevant. The initial stages
of CO reduction might be speculatively considered to
involve:

(i) The conversion of 2e- into 4- or 6e-donating CO.
4e-donating CO has been observed in $Mn_2(CO)_5(PPh_2CH_2PPh_2)_2$
(149) and $[Me_3NCH_2Ph][HFe_4(CO)_{13}]$ (150), (A) and (B)
(Figure 26). Increasing the number of C to M contacts
weakens the CO bond by increasing "back-donation" and
might favour its reduction but against this is the
build-up of electron density at CO reducing the
susceptibility to nucleophilic attack by a hydride
source, e.g. metal hydride formed by oxidative addition
of H_2. A side-on interaction such as in (A) and (B)
should lead to CO bond weakening without this build-up
of charge. By analogies with known μ^3-alkyne and
μ^3-nitrile compounds one might speculate on other
possible ways this could be achieved. Thus as shown
in Figure 26 (C) corresponds with (A) (Figure 23), (D)

454.

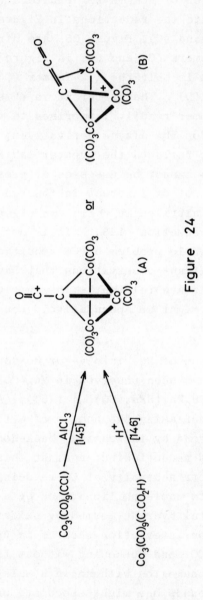

Figure 24

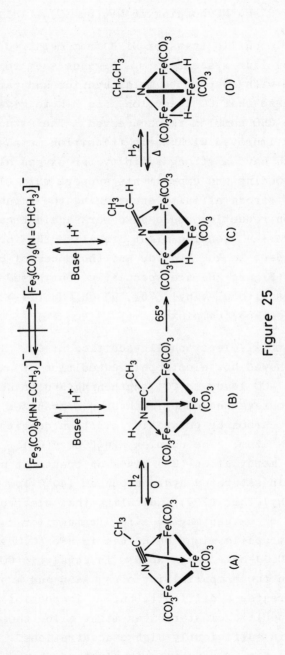

Figure 25

with (D) (Figure 23) and (E) with (B) (Figure 23) and with the μ^3-Bu^{t}NC bonding in Ni$_4$(Bu^{t}NC)$_7$ (151).

(ii) Nucleophilic transfer of H from metal to CO. In mononuclear systems formyl ligands seem to be unstable with respect to metal hydride and carbon monoxide so that CO insertion into M-H to give an isolable CHO complex is unobserved. The situation might be improved with early transition metals (e.g. Zr) which have a strong affinity for oxygen allowing η^2-CHO bonding and experiments support this (152), but this strong affinity encouraging the turnabout of the CO on reduction could also work against catalysis. In clusters μ^n-bonding might stabilise the formyl with respect to the hydride and the modes of bonding (F)-(H) (Figure 26) are speculative but based on analogies with μ^3-vinyl (75), μ^3-CH$_3$CNH (147) and μ^4-alkyne (153) complexes.

(iii) Initial electrophilic addition at CO. In section 2.1 we showed how electrophilic addition at oxygen of μ^2- or μ^3-CO leads to CO lengthening and it should also encourage nucleophilic hydrogen transfer from metal to carbon by polarising electron density away from this atom. The complex HFe$_3$(COMe)(CO)$_{10}$ formed by μ^2-CO methylation (23) gives on treatment with H$_2$ (1 atm) in saturated hydrocarbon at 140°C low yields of dimethylether (154) indicating that electrophilic addition of oxygen may be a requirement for reduction. The electrophile might be H$^+$ (as in HFe$_3$(C.OH)(CO)$_{10}$, section 2.1) or a Lewis acid. In catalytic CO reduction the compatibility of an acid and a hydride source creates a difficulty but it is possible that highly specific cluster sites might allow these to coexist in sufficiently high concentrations. A study of models for CO reduction in clusters should certainly lead to interesting new structures and patterns of behaviour.

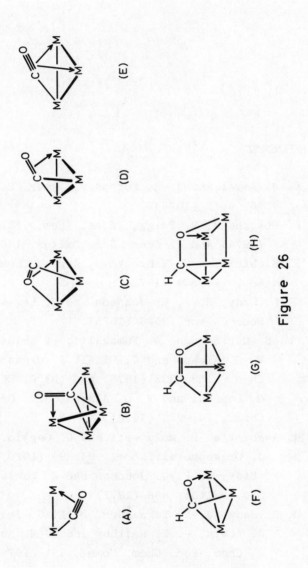

Figure 26

458.

6. REFERENCES

1. K. J. Kavel and J. R. Norton, J. Amer. Chem. Soc.,
   $\underline{96}$, 6812 (1974).
2. A. Poë and M. V. Twigg, Inorg. Chem., $\underline{13}$, 2982,
   (1974) and J. Chem. Soc. Dalton, 1860 (1974).
3. F. Richter and H. Vahrenkamp, Angew. Chem. Int.
   Ed., $\underline{17}$, 864 (1978).
4. C. R. Eady, B. F. G. Johnson and J. Lewis, J. Chem.
   Soc. Dalton, 2606 (1975).
5. (a) S. Martinengo, A. Fumagalli, P. Chini,
       V. G. Albano and G. Ciani, J. Organometal.
       Chem., $\underline{116}$, 333 (1976) and (b) P. Chini,
       G. Longoni and V. G. Albano, Adv. Organometal.
       Chem., $\underline{14}$, 285 (1976).
6. M. Angoletta, L. Malatesta and G. Gaglio,
   J. Organometal. Chem., $\underline{94}$, 99 (1975).
7. C. R. Eady, B. F. G. Johnson and J. Lewis, J. Chem.
   Soc. Dalton, 838 (1977).
8. C. R. Eady, J. M. Fernandez, B. F. G. Johnson,
   J. Lewis, P. R. Raithby and G. M. Sheldrick,
   J. Chem. Soc. Chem. Comm., 421 (1978).
9. C. R. Eady, B. F. G. Johnson and J. Lewis, J. Chem.
   Comm., 302 (1976) and M. McPartlin, C. R. Eady,
   B. F. G. Jonson and J. Lewis, J. Chem. Soc.
   Chem. Comm., 883 (1976).

10. C. W. Bradford, R. S. Nyholm, G. J. Gainsford,
    J. M. Guss, P. R. Ireland and R. Mason,
    J. Chem. Soc. Chem. Comm., 87 (1972);
    G. J. Gainsford, J. M. Guss, P. R. Ireland,
    R. Mason, C. W. Bradford and R. S. Nyholm,
    J. Organometal. Chem., 40, C70 (1972) and
    C. W. Bradford and R. S. Nyholm, J. Chem.
    Soc. Dalton, 529 (1973).

11. D. Seyferth, G. H. Williams, A. T. Wehman and
    M. O. Nestle, J. Amer. Chem. Soc., 97, 2107
    (1975).

12. E. R. Schilling and R. Hoffmann, J. Amer. Chem.
    Soc., 100, 6274 (1978) and references
    therein.

13. H. D. Kaesz, Chem. Brit., 9, 344 (1973).

14. J. Knight and M. J. Mays, J. Chem. Soc. (A), 711
    (1970).

15. A. J. Deeming, B. F. G. Johnson and J. Lewis,
    J. Chem. Soc. (A), 2967 (1970).

16. A. J. Deeming, B. F. G. Johnson and J. Lewis,
    J. Chem. Soc. (A), 2517 (1970).

17. A. J. Deeming, S. Hasso, M. Underhill, A. J. Canty,
    B. F. G. Johnson, W. G. Jackson, J. Lewis
    and T. M. Matheson, J. Chem. Soc. Chem. Comm.,
    807 (1974).

18. E. G. Bryan, W. G. Jackson, B. F. G. Johnson,
    J. W. Kelland, J. Lewis and K. T. Schorpp,
    J. Organometal. Chem., 108, 385 (1976).

19. J. S. Kristoff and D. F. Shriver, Inorg. Chem.,
    13, 499 (1974).

20. G. Schmid, V. Bätzel and B. Stutte, J. Organometal.
    Chem., 113, 67 (1976).

21. B. Stutte, V. Bätzel, R. Boese and G. Schmid,
    Chem. Ber., 111, 1603 (1978).

22. M. F. Lappert, D. S. Patil and J. B. Pedley,
    J. Chem. Soc. Chem. Comm., 830 (1975).

460.

23. D. F. Shriver, D. Lehman and D. Strope, J. Amer.
    Chem. Soc., 97, 1594 (1975).

24. H. A. Hodali, D. F. Shriver and C. A. Ammlung,
    J. Amer. Chem. Soc., 100, 5239 (1978).

25. V. Bätzel and G. Schmid, Chem. Ber., 109, 3339,
    (1976).

26. M. R. Churchill and B. G. DeBoer, Inorg. Chem.,
    16, 878 (1977).

27. C. R. Eady, B. F. G. Johnson, J. Lewis and
    M. C. Malatesta, J. Chem. Soc. Dalton, 1358,
    (1978).

28. J. R. Shapley, J. B. Keister, M. R. Churchill
    and B. G. DeBoer, J. Amer. Chem. Soc., 97,
    4145 (1975).

29. A. J. Deeming and S. Hasso, J. Organometal. Chem.,
    88, C21 (1975) and 114, 313 (1976).

30. B. F. G. Johnson, J. Lewis, D. Pippard and
    P. R. Raithby, J. Chem. Soc. Chem. Comm.,
    551 (1978).

31. R. Bau, B. Don, R. Greatrex, R. J. Haines,
    R. A. Love and R. D. Wilson, Inorg. Chem.,
    14, 3021 (1975).

32. F. Richter and H. Vahrenkamp, Angew. Chem. Int.
    Ed., 17, 444 (1978).

33. B. F. G. Johnson, J. Lewis, P. R. Raithby,
    G. M. Sheldrick, K. Wong and M. McPartlin,
    J. Chem. Soc. Dalton, 673 (1978).

34. (a) M. R. Churchill, F. J. Hollander and
    J. P. Hutchinson, Inorg. Chem., 16, 2697
    (1977), (b) V. F. Allen, R. Mason and
    P. B. Hitchcock, J. Organometal. Chem., 140,
    297 (1977) and (c) R. W. Broach and
    J. M. Williams, Inorg. Chem., 18, 314 (1979),
    (d) A. G. Orpen, A. V. Rivera, E. G. Bryan,
    D. Pippard, G. M. Sheldrick and K. D. Rouse,
    J. Chem. Soc. Chem. Comm., 723 (1978).

35. M. R. Churchill and B. G. DeBoer, Inorg. Chem.,
    16, 2397 (1977).

36. L. J. Farrugia, J. A. K. Howard, P. Mitrpachachon,
    J. L. Spencer, F. G. A. Stone and P. Howard,
    J. Chem. Soc., Chem. Comm., 260 (1978).

37. J. Knight and M. J. Mays, J. Chem. Soc. Dalton,
    1022 (1972).

38. V. G. Albano, A. Ceriotti, P. Chini, G. Ciani,
    S. Martinengo and M. Anker, J. Chem. Soc.
    Chem. Comm., 859 (1975) and references
    therein.

39. G. Ciani, G. D'Alfonso, M. Freni, P. Romiti,
    A. Sironi and A. Albinati, J. Organometal.
    Chem., 136, C49 (1977).

40. A. Bertolucci, M. Freni, P. Romiti, G. Ciani,
    A. Sironi and V. G. Albano, J. Organometal.
    Chem., 113, C61 (1976).

41. A. V. Rivera, G. M. Sheldrick and M. B. Hursthouse,
    Acta Cryst., B34, 1985 (1978).

42. M. J. Mays and P. D. Gavens, J. Organometal. Chem.,
    124, C37 (1977) and A. G. Orpen and
    G. M. Sheldrick, Acta Cryst., B34, 1989
    (1978).

43. J. F. Blount, L. F. Dahl, C. Hoogzand and
    W. Hübel, J. Amer. Chem. Soc., 88, 292 (1966).

44. C. G. Pierpont, Inorg. Chem., 16, 636 (1977).

45. K. A. Azam, C. Choo Yin and A. J. Deeming,
    J. Chem. Soc. Dalton, 1201 (1978).

46. B. F. G. Johnson, J. Lewis and D. Pippard,
    J. Organometal. Chem., 145, C4 (1978).

47. J. R. Shapley, M. Tachikawa, G. F. Stuntz and
    J. B. Keister, 173rd ACS National Meeting,
    New Orleans, 1977, INOR53.

48. C. R. Eady, B. F. G. Johnson and J. Lewis, J. Chem.
    Soc. Dalton, 477 (1977).

462.

49. R. M. Laine, R. G. Rinker and P. C. Ford, J. Amer.
     Chem. Soc., 99, 252 (1977) and H. Kang,
     C. H. Mauldin, T. Cole, W. Slegeir, K. Cann
     and R. Pettit, J. Amer. Chem. Soc., 99, 8323
     (1977).

50. A. J. Deeming and S. Hasso, J. Organometal Chem.,
     112, C39 (1976).

51. M. R. Churchill, B. G. DeBoer, J. R. Shapley and
     J. B. Keister, J. Amer. Chem. Soc., 98, 2357
     (1976) and M. R. Churchill and B. G. DeBoer,
     Inorg. Chem., 16, 1141 (1977).

52. S. K. Malik and A. Poë, Inorg. Chem., 17, 1484
     (1978).

53. R. Rossetti, P. L. Stanghellini, O. Gambino and
     G. Cetini, Inorg. Chim. Acta, 6, 205 (1972).

54. R. Rossetti, G. Gervasio and P. L. Stanghellini,
     J. Chem. Soc. Dalton, 222 (1978).

55. K. A. Azam, A. J. Deeming, R. E. Kimber and
     P. R. Shukla, J. Chem. Soc. Dalton, 1853
     (1976).

56. K. A. Azam, A. J. Deeming, I. P. Rothwell,
     M. B. Hursthouse and L. New, J. Chem. Soc.
     Chem. Comm., 1086 (1978).

57. A. J. Deeming and R. Peters, unpublished work.

58. B. F. G. Johnson, J. Lewis and M. V. Twigg,
     J. Organometal. Chem., 67, C75 (1974) and
     J. Chem. Soc. Dalton, 1876 (1975).

59. A. J. Deeming, B. F. G. Johnson and J. Lewis,
     J. Chem. Soc. (A), 897 (1970).

60. G. F. Stuntz and J. R. Shapley, J. Amer. Chem.
     Soc., 99, 607 (1977) and references
     therein.

61. A. J. Drakesmith and R. Whyman, J. Chem. Soc.
     Dalton, 362 (1973).

62. A. J. Deeming, MTP Reviews of Science, Inorganic
    Chemistry, Series One, 9, 117 (1972).

63. F. L'Eplattenier and F. Calderazzo, Inorg. Chem.,
    7, 1290 (1968).

64. J. P. Collman and W. R. Roper, J. Amer. Chem.
    Soc., 88, 3504 (1966).

65. S. A. R. Knox, J. W. Koepke, M. A. Andrews and
    H. D. Kaesz, J. Amer. Chem. Soc., 97, 3942
    (1975).

66. G. R. Crooks, B. F. G. Johnson, J. Lewis and
    I. G. Williams, J. Chem. Soc. (A), 2761
    (1969).

67. N. Cook, L. Smart, and P. Woodward, J. Chem. Soc.,
    Dalton, 1744 (1977).

68. B. F. G. Johnson, J. Lewis and S. Bhaduri,
    private communication.

69. J. L. Hughey, C. R. Bock and T. L. Meyer, J. Amer.
    Chem. Soc., 97, 4440 (1975).

70. A. J. Deeming and M. Underhill, J. C. S. Dalton,
    1415 (1974).

71. E. G. Bryan, B. F. G. Johnson and J. Lewis,
    J. Chem. Soc. Dalton, 1328 (1977).

72. J. B. Keister and J. R. Shapley, J. Organometal.
    Chem., 85, C29 (1975).

73. A. J. Deeming, S. Hasso and M. Underhill,
    J. Chem. Soc. Dalton, 1614 (1975).

74. A. G. Orpen, D. Pippard, G. M. Sheldrick and
    K. D. Rouse, Acta Cryst., B34, 2466 (1978) and
    J. J. Guy, B. E. Reichert and G. M. Sheldrick,
    Acta Cryst., B32, 3319 (1976).

75. M. Laing, P. Sommerville, Z. Dawoodi, M. J. Mays
    and P. J. Wheatley, J. Chem. Soc. Chem. Comm.,
    1035 (1978).

76. M. Tachikawa and J. R. Shapley, J. Organometal.
    Chem., 124, C19 (1977).

464.

77.　A. P. Humphries and S. A. R. Knox, J. Chem. Soc.
　　　　Dalton, 1710 (1975).

78.　J. Evans, B. F. G. Johnson, J. Lewis and
　　　　T. W. Matheson, J. Organometal Chem., 97,
　　　　C16 (1975).

79.　A. J. Canty, A. J. P. Domingos, B. F. G. Johnson
　　　　and J. Lewis, J. Chem. Soc. Dalton, 2056
　　　　(1973).

80.　A. J. Deeming, J. Organometal Chem., 150, 123 (1978).

81.　C. Choo Yin and A. J. Deeming, unpublished work.

82.　G. F. Stuntz, J. R. Shapley and C. G. Pierpont,
　　　　Inorg. Chem., 17, 2596 (1978).

83.　C. G. Pierpont, G. F. Stuntz and J. R. Shapley,
　　　　J. Amer. Chem. Soc., 100, 616 (1978).

84.　G. Gervasio, D. Osella and M. Valle, Inorg. Chem.,
　　　　15, 1221 (1976) and M. Evans, M. Hursthouse,
　　　　E. W. Randall, E. Rosenberg, L. Milone and
　　　　M. Valle, J. Chem. Soc. Chem. Comm., 545
　　　　(1972).

85.　A. J. Deeming, J. Organometal. Chem., 128, 63,
　　　　(1977).

86.　A. Cox and P. Woodward, J. Chem. Soc. (A), 3599
　　　　(1971) and M. I. Bruce, M. A. Cairns and
　　　　M. Green, J. Chem. Soc. Dalton, 1293 (1972).

87.　A. W. Parkins, E. O. Fischer, G. Huttner and
　　　　D. Regler, Angew. Chem. Int. Ed., 9, 633
　　　　(1970).

88.　K. A. Azam and A. J. Deeming, J. Chem. Soc. Chem.
　　　　Comm., 472 (1977) and J. Mol. Cat., 3, 207
　　　　(1977).

89.　E. Sappa, O. Gambino, L. Milone and G. Cetini,
　　　　J. Organometal. Chem., 39, 169 (1972).

90.　G. Gervasio and G. Ferraris, Cryst. Struct. Comm.,
　　　　3, 447 (1973) and M. Catti, G. Gervasio and
　　　　S. A. Mason, J. Chem. Soc. Dalton, 2260 (1977).

91. A. J. Deeming and M. Underhill, J. Chem. Soc.
       Dalton, 2727 (1973).

92. M. G. Thomas, E. L. Muetterties, R. O. Day and
       V. W. Day, J. Amer. Chem. Soc., 98, 4645
       (1976).

93. C. Choo Yin and A. J. Deeming, J. Chem. Soc.
       Dalton, 2091 (1975).

94. A. J. Deeming, R. Peters, M. B. Hursthouse and
       J. D. J. Backer-Dirks, unpublished results.

95. R. D. Adams and N. M. Golembeski, Inorg. Chem.,
       17, 1969 (1978).

96. C. Choo Yin and A. J. Deeming, J. Organometal.
       Chem., 133, 123 (1977).

97. C. Choo Yin and A. J. Deeming, J. Chem. Soc.
       Dalton, 1013 (1974).

98. K. A. Azam, A. J. Deeming, R. E. Kimber and
       P. R. Shukla, J. Chem. Soc. Dalton, 1853
       (1976).

99. K. A. Azam, A. J. Deeming, I. P. Rothwell,
       M. B. Hursthouse and L. New, J. Chem. Soc.
       Chem. Comm., 1086 (1978).

100. I. P. Rothwell and A. J. Deeming, unpublished
       results.

101. G. W. Parshall, Accounts Chem. Res., 3, 139 (1970)
       and 8, 113 (1975).

102. J. W. Rathke and E. L. Muetterties, J. Amer. Chem.
       Soc., 97, 3272 (1975).

103. R. B. Calvert and J. R. Shapley, J. Amer. Chem.
       Soc., 99, 5225 (1977).

104. R. B. Calvert, J. R. Shapley, A. J. Shultz,
       J. M. Williams, S. L. Suib and G. D. Stucky,
       J. Amer. Chem. Soc., 100, 6240 (1978).

105. R. B. Calvert and J. R. Shapley, J. Amer. Chem.
       Soc., 100, 7726 (1978).

106. C. R. Eady, B. F. G. Johnson and J. Lewis,
     J. Organometal. Chem., 57, C84 (1973).

107. G. R. Crooks, B. F. G. Johnson, J. Lewis and
     I. G. Williams, J. Chem. Soc. (A), 797 (1969).

108. A. J. Deeming, R. Ettorre, B. F. G. Johnson and
     J. Lewis, J. Chem. Soc. (A), 1797 and 2701
     (1971).

109. J. A. DeBeer and R. J. Haines, J. Organometal.
     Chem., 4, 757 (1970).

110. S. Jeannin, Y. Jeannin and G. Lavigne, Inorg.
     Chem., 17, 2103 (1978).

111. A. J. Deeming and M. Underhill, J. Organometal.
     Chem., 42, C60 (1972) and J. L. Vidal,
     F. A. Fiato, L. A. Cosby and R. L. Pruett,
     Inorg. Chem., 17, 2574 (1978).

112. C. Choo Yin and A. J. Deeming, unpublished result.

113. J. R. Shapley, G. A. Pearson, M. Tachikawa,
     G. E. Schmidt, M. R. Churchill and
     F. J. Hollander, J. Amer. Chem. Soc., 99,
     8064 (1977) and M. R. Churchill and
     F. J. Hollander, Inorg. Chem., 17, 3546
     (1978).

114. M. R. Churchill, F. J. Hollander, J. R. Shapley
     and D. S. Foose, J. Chem. Soc. Chem. Comm.,
     534 (1978).

115. A. J. Deeming, R. E. Kimber and M. Underhill,
     J. Chem. Soc. Dalton, 2589 (1973).

116. A. G. Orpen and G. M. Sheldrick, Acta Cryst.,
     B34, 1992 (1978).

117. R. B. King and C. A. Harmon, Inorg. Chem., 15,
     879 (1976).

118. S. A. Gardner, P. S. Andrews and M. D. Rausch,
     Inorg. Chem., 12, 2396, (1973).

119. Trinh-Toan, R. W. Broach, S. A. Gardner and
     M. D. Rauch, Inorg. Chem., 16, 279 (1977).

120. J. M. Fernandez, B. F. G. Johnson, J. Lewis and
     P. R. Raithby, Acta Cryst., B34, 3086 (1978).

121. M. R. Churchill, B. G. DeBoer, F. J. Rotella,
     E. W. Abel and R. J. Rowley, J. Amer. Chem.
     Soc., 97, 7158 (1975) and M. R. Churchill,
     B. G. DeBoer and F. J. Rotella, Inorg. Chem.,
     15, 1843 (1976).

122. E. Koerner von Gustorf, R. Wagner, B. L. Barnett
     and C. Krüger, Angew. Chem. Int. Ed., 10,
     910 (1971).

123. R. J. Haines, J. A. DeBeer and R. Greatrex,
     J. Organometal. Chem., 55, C30 (1973).

124. S. Aime, G. Gervasio, L. Milone, R. Rossetti and
     P. L. Stranghellini, J. Chem. Soc. Dalton,
     534 (1978) and references therein.

125. B. L. Booth, G. C. Casey and R. N. Haszeldine,
     J. Chem. Soc. Dalton, 1850 (1975) and
     references therein.

126. V. G. Albano, P. Chini, S. Martinengo, M. Sansoni
     and D. Strumolo, J. Chem. Soc. Dalton, 459
     (1978) and references therein.

127. G. Schmid, V. Bätzel and G. Etzrodt, J. Organo-
     metal. Chem., 112, 345 (1976).

128. K. E. Schwarzhans and H. Steiger, Angew. Chem.
     Int. Ed., 11, 535 (1972).

129. L. T. J. Delbaere, L. J. Kruczynski and
     D. W. McBride, J. Chem. Soc. Dalton, 307
     (1973).

130. L. Markó and B. Markó, Inorg. Chim. Acta, 14,
     L39 (1975).

131. A. Vizi-Orosz, G. Pályi and L. Markó, J. Organo-
     metal. Chem., 60, C25 (1973).

132. W. Hübel, Organic Synthesis via Metal Carbonyls,
     Vol. 1, Interscience, 1968, ed. by Wender
     and Pino, 273.

468.

133. R. P. Dodge and V. Schomaker, J. Organometal.
        Chem., 3, 274 (1965).

134. G. A. Vaglio, O. Gambino, R. P. Ferrari and
        G. Cetini, Inorg. Chim. Acta, 7, 193 (1973).

135. R. P. Ferrari, G. A. Vaglio, O. Gambino and
        G. Cetini, J. Chem. Soc. (A), 1998 (1972).

136. M. Tachikawa, J. R. Shapley and C. G. Pierpont,
        J. Amer. Chem. Soc., 92, 7172 (1975).

137. S. Sappa, L. Milone and A. Tiripicchio, J. Chem.
        Soc. Dalton, 1843 (1976) and S. Aime,
        L. Milone, E. Sappa and A. Tiripicchio,
        J. Chem. Soc. Dalton, 227 (1977).

138. G. Gervasio, S. Aime, L. Milone, E. Sappa and
        M. Franchini-Angela, Transition Met. Chem.,
        1, 96 (1976) and Inorg. Chim. Acta, 27,
        145 (1978).

139. M. R. Churchill and R. A. Lashewycz, Inorg. Chem.,
        17, 1291 (1978) and M. R. Churchill,
        R. A. Lashewycz, M. Tachikawa and
        J. R. Shapley, J. Chem. Soc. Chem. Comm.,
        699 (1977).

140. E. Sappa, A. Tiripicchio and A. M. M. Lanfredi,
        J. Chem. Soc. Dalton, 552 (1978).

141. G. Gervasio, J. Chem. Soc. Chem. Comm., 25 (1976).

142. J. B. Keister and J. R. Shapley, J. Amer. Chem.
        Soc., 98, 1056 (1976).

143. A. J. Deeming and M. Underhill, J. Chem. Soc.
        Chem. Comm., 277 (1973) and A. J. Deeming,
        S. Hasso and M. Underhill, XVI Conf. Coord.
        Chem., Dublin, 4.35 (1974).

144. D. Seyferth, G. H. Williams and D. D. Traficante,
        J. Amer. Chem. Soc., 96, 604 (1974).

145. D. Seyferth, G. H. Williams and C. L. Nivert,
        Inorg. Chem., 16, 758 (1977).

146. J. E. Hallgren, C. S. Eschbach and
          D. Seyferth, J. Amer. Chem. Soc., 94, 2547,
          (1972).

147. M. A. Andrews and H. D. Kaesz, J. Amer. Chem.
          Soc., 99, 6763 (1977).

148. E. L. Muetterties, Bull. Soc. Chim. Belg., 85,
          451 (1976).

149. C. J. Commons and B. F. Hoskins, Aust. J. Chem.,
          28, 1663 (1975).

150. M. Manassero, M. Sansoni and G. Longoni, J. Chem.
          Soc. Chem. Comm., 919 (1976).

151. V. W. Day, R. O. Day, J. S. Kristoff,
          F. J. Hirsekorn and E. L. Muetterties,
          J. Amer. Chem. Soc., 97, 2571 (1975).

152. J. M. Manriquez, D. R. McAlister, R. D. Sanner
          and J. E. Bercaw, J. Amer. Chem. Soc., 100,
          2716 (1978) and references therein.

153. P. F. Heveldt, B. F. G. Johnson, J. Lewis,
          P. R. Raithby and G. M. Sheldrick, J. Chem.
          Soc. Chem. Comm., 340 (1978).

154. A. J. Deeming, unpublished result.

Transition Metal Clusters
Edited by Brian F.G. Johnson
© 1980 John Wiley & Sons Ltd.

CHAPTER 7

Ligand Mobility in Clusters

B. F. G. Johnson and R. E. Benfield

University Chemical Laboratory,
Cambridge, U.K.

472.

1. INTRODUCTION

In recent years the migration of atoms or small
molecules over or within polymetal aggregates which
contain metal-metal bonds has gained widespread
attention, especially with regard to the possible
relationship of such migrations to those occurring
during the chemisorption of substrates to a metal
surface. An understanding of the basic pathways by
which movement of these coordinated groups occurs
relative to the framework of the metallic cluster is
therefore of some importance. However, the analogy
between a small metal cluster contained in a sheath of
ligands - a three-dimensional system - and the bulk
surface - essentially a two-dimensional system - should
not be pressed too far. There can be no doubt that the
pathways followed by the migrating group will be
governed not only by the geometry and type of metallic
cluster but also by the number and type of other
ligands also present; a situation which, in general,
will differ considerably from the bulk surface.

Fluxionality of the metal cluster and ligand
mobility in a $M_m L_n$ system indicates that there are a
number of possible cluster structures with energies
similar to that of the ground-state that can be
excited by thermal energies. Thus a classical

474.

(localised) bond description of these systems may be
inappropriate since the ground-state structure does
not sit in a well-defined (deep) potential well in the
potential energy surface governing nuclear motion (1).

A relatively minor reorientation of the ligands
about the central M_m cluster can often bring about an
apparently major change in the overall structure. This
is the case with dodecacarbonyltri-iron, $Fe_3(CO)_{12}$
(Fig. 1). In the ground-state structure of this
molecule of C_{2v}-symmetry, there are ten terminal and
two edge-bridging groups. The three iron atoms form
an isosceles triangle and the twelve carbonyl ligands
an icosahedron (2). Simple rotation of the Fe_3
triangle (about a C_2 axis) within the icosahedron
generates a new form of D_3-symmetry in which the twelve
CO-ligands still define an icosahedron but which now
contains only terminally bound CO-groups (3). Thus an
apparently major structural change in fact corresponds
to a moderately trivial reorganisation of CO ligands
about the iron triangle.

Figure 1a

The C_{2v} structure of $Fe_3(CO)_{12}$ as a
metal atom triangle within an
icosahedron of ligands

Figure 1b

The possible D_3 structure of $Fe_3(CO)_{12}$

A third, closely related, form of $Fe_3(CO)_{12}$ of D_{3h}-symmetry is also possible. In this the triangle sits within an anti-cubeoctahedron of twelve CO groups. Again all CO-groups are terminally-bound but in this case a more substantial reorganisation has occurred. The ^{13}C nmr data for the carbonyl mobility in $Fe_3(CO)_{12}$ has been explained by involving interconversion between these various forms (3,4).

It has become customary to regard cluster M_mL_n as constructed of well defined $ML_{n/m}$ fragments and to investigate what happens when the fragments are put together. Use of this approach (using a semi-empirical M.O. Theory (5)) has led to some understanding of cluster geometries (1). To some extent ligand mobility has been treated similarly and the migration of ligands over a cluster surface considered in terms of localised ligand movements. In effect, the problem of ligand mobility or fluxionality has been broken down to give a convenient, often pictorial representation of the fluxional process. Consider again, for example, dodecarbonyltri-iron, $Fe_3(CO)_{12}$, a highly fluxional molecule exhibiting total ligand mobility at very low temperatures and for which an activation energy of <20 kJ mole^{-1} has been estimated (6). A convenient mechanistic pathway for this carbonyl mobility corresponds to a series of simple localised bridge opening- bridge closing operations of the type:-

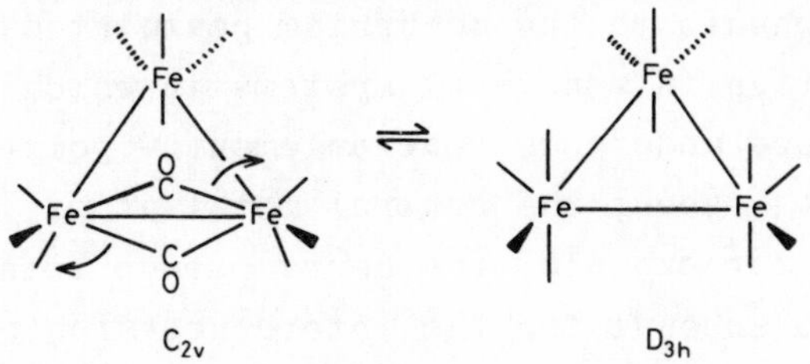

Figure 2

476.

In reality, the single bridge-opening operation must involve the concerted movement of all twelve CO ligands (see above) and a scissoring motion of the Fe_3 triangle:-

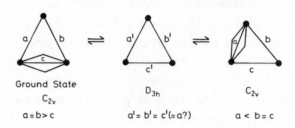

Figure 3

This concerted movement of ligands must be general to all systems even though ligand mobility may appear to correspond to (and may be interpreted as) localised migrations. Any apparently localised movement of ligands about the M_m cluster requires movement of all other ligands although they may retain their individual identities. This is not necessarily the case for molecules chemisorbed on the surface.

This is not to say that this view of localised scrambling is without merit. It provides the most convenient and easily understood method of classifying fluxional behaviour and will be used as such throughout this Chapter. But an alternative, more general, approach must also be considered (7). This bears a close relationship to the mechanism postulated for ligand migration in monometal systems in which the ligand polytope undergoes rearrangement - polytopal rearrangement - about the central metal atom. Classically, for example, the Berry pseudo rotation may be invoked to account for the interconversion of a trigonal bipyramidal ML_5 species (8). In cluster

systems the problem is clearly more complicated.
First, the cluster unit M_m usually exhibits higher
coordination numbers - twelve in $M_3(CO)_{12}$ - for
example. Rearrangements of polytopes with twelve or
more vertices are less well characterised than for
smaller numbers. Also, because of the (generally)
lower symmetry of the M_mL_n species compared to the
isolated L_n polytope the degeneracy of the modes of
interconversion of the L_n polytope may be removed.

   To illustrate this, consider the simple
hypothetical, example of an M_2 unit within an O_h-octa-
hedron of ligands, M_2L_6. Ligand mobility in a
monometal octahedral complex may be considered to
occur _via_ a trigonal prismatic excited state which is
achieved by rotation of any one of the eight triangular
faces (about a C_3 axis), which in an ML_6 system, are
equivalent.

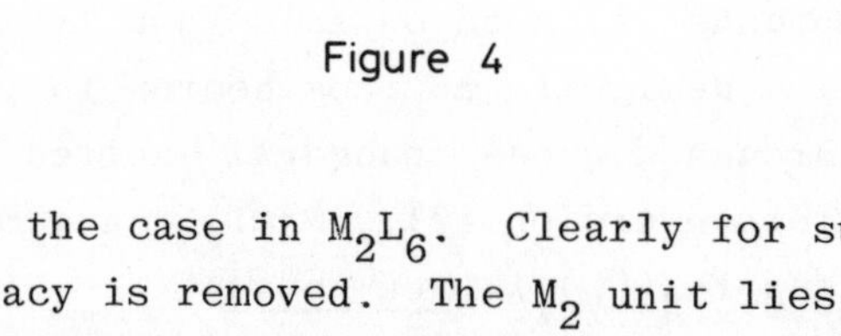

Figure 4

This is not the case in M_2L_6. Clearly for such a species
the degeneracy is removed. The M_2 unit lies along one
C_3 axis and rotation about this axis will bring about
a different result to rotation about any one of the
three remaining C_3 axes. Rotation about the unique C_3
axis would correspond to local ligand rearrangement
about one metal atom, whereas rotation of the second
type would involve migration from one metal atom to

478.

another.

Localised Exchange   Delocalised Exchange

Figure 5

Possible Ligand Mobility in M_2L_6

    The barriers to these two processes will be different but the overall fluxional process(es) may be rationalised by the mechanism of polytopal rearrangement of the ligand shell and as such provides a general mechanistic pathway (or rationalisation) which may be applied to all species M_mL_n.

    An alternative description of CO-mobility in $M_4(CO)_{12}$ compounds is based on this idea. The twelve CO-groups which designate an icosahedron in the ground-state pass through a cube-octahedral excited state to enable CO interconversion (7). Similar arguments may also be applied $Fe_3(CO)_{12}$ (see above).

    In this account we have chosen to subdivide our discussion into three major parts. In Section 2 - types of stereodynamic behaviour - an assessment of the various possible fluxional processes is given. Section 3 - ligand types - is concerned largely with the most commonly observed types of ligand mobility viz. CO and

H-migrations and finally in Section 4, a survey of
stereodynamic behaviour in metal carbonyl clusters and
related systems is given. This section, we hope, will
serve as a guide to the information currently available
in this area. We would draw the reader's attention to
other reviews of this subject currently available
(references 9).

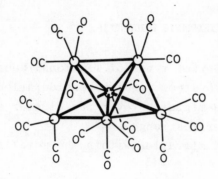

Figure 6

$Os_6(CO)_{18}$

2. TYPES OF STEREODYNAMIC BEHAVIOUR

Three important types of fluxional behaviour within cluster systems require consideration:-
1. Ligand migration over the cluster surface
2. Cluster rearrangement
3. Ligand migration within the metallic cluster

2.1. Ligand Migration Over the Surface of the Cluster

Ligand migration over the cluster surface is the most commonly observed type of fluxional behaviour and may be further subdivided into:-

2.1.1. Localised Migration About a Single Atom

Most processes available to monometal systems will be available to single metals within the cluster unit. Intramolecular exchange within $M(CO)_3$ and $M(CO)_4$ units, have for example, been observed, e.g. the temperature dependent spectra of $Os_6(CO)_{18}$ (Figure 6) from -123° to

$100^{\circ}C$ show (10) that the interchange of the carbonyl groups is localised and that at each type of osmium atom Os_1, Os_2 and Os_3 an independent exchange process is taking place with a different activation energy but no evidence of intermetal CO migration within that temperature range.

In $Ru_3(CO)_{10}(NO)_2$ the $Ru(CO)_3$ units are rigid in the temperature range -50° to 40° whereas the $Ru(CO)_4$ unit is stereochemically non-rigid via axial equatorial exchange (11).

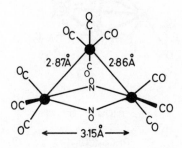

Figure 7

Structure of $Ru_3(CO)_{10}(NO)_2$

2.1.2. Localised Migration Over Two or More Metal Atoms

Behaviour of this type has been observed with the mixed cluster $RhCo_3(CO)_{12}$ (Fig. 8). The proposed mechanism of carbonyl exchange in this molecule has been said to involve two separate steps (12). In the first (low-temperature) process ten of the twelve CO-

482.

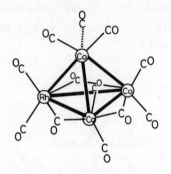

Figure 8

$RhCo_3(CO)_{12}$

ligands undergo migration such that the two terminal
carbonyls on the rhodium atom remain distinct (but
see below, Section 4.3).

2.1.3. Delocalised Migration Over the Whole Cluster Unit

This commonly observed phenomenon leads to the
total equilibration of all ligands. Dodecacarbonyl-
tetrarhodium, $Rh_4(CO)_{12}$, provides an ideal example.
At low temperature ($-65^{\circ}C$) the ^{13}C nmr. spectrum is
in accord with the solid-state structure. At higher
temperature a quintet is observed indicating that all
twelve CO-groups are in association with four
equivalent rhodium nuclei and that the total CO
scrambling must be occurring (13). The mechanism of
this process is discussed in detail below.

Irrespective of whether (CO) migration between
metal atoms is localised or delocalised, it must
involve the formation of bridged-intermediates: μ_2, μ_3
or $\mu_4(?)$. The carbonyl and hydrido-ligands are
special in that in general they can donate only two or
one electron respectively to the overall bonding scheme
independent of the bonding mode μ_1(terminal), μ_2(edge-
bridge) or μ_3(three centre-bridge) (see below). In
effect this means that the metal-metal bond(s) is
retained irrespective of the bonding mode of the
ligand, as shown in Fig. 9.

It would appear from bond enthalpy measurements
that $\Delta E_{M-CO(\mu_1)} \simeq 2\Delta E_{M-CO(\mu_2)}$ (see Chapter 5). A
similar relationship must hold for metal-hydride
systems.

In contrast, for halogeno-clusters the number of
electrons donated to the overall bonding scheme by the

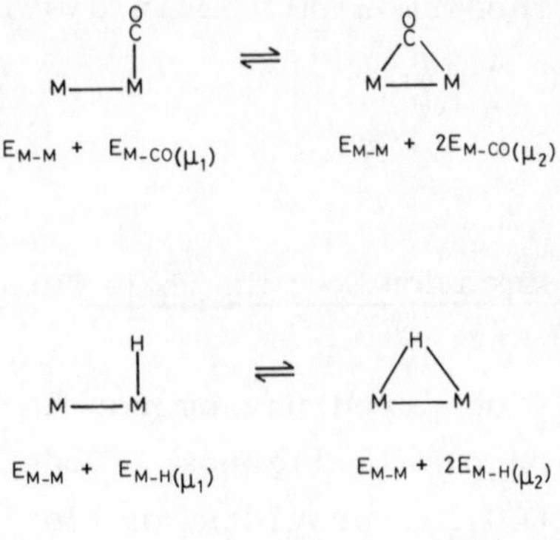

$E_{M-M} + E_{M-CO(\mu_1)}$ $E_{M-M} + 2E_{M-CO(\mu_2)}$

$E_{M-M} + E_{M-H(\mu_1)}$ $E_{M-M} + 2E_{M-H(\mu_2)}$

Figure 9

halide atom is normally dependent on the bonding mode adopted. In this case the formation of an X-bridge must be at the expense of metal-metal bond(s) or ligand ejection.

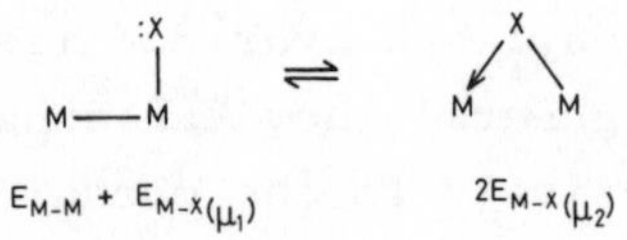

$E_{M-M} + E_{M-X(\mu_1)}$ $2E_{M-X(\mu_2)}$

Figure 10

Appropriate bond enthalpies are not available for such systems but it would be reasonable to assume that $\Delta E_{M-X} > \Delta E_{M-M}$ and halogeno-systems might not therefore exhibit the same degree of fluxional behaviour.

Terminal and bridging modes have also been established for isocyanide, alkyl, aryl and nitrosyl ligands but apart from isocyanide little data are available to indicate ligand mobility.

Unlike CO isocyanide exhibits a wide range of bridging modes in addition to the simple $\overset{\displaystyle CNR}{\underset{\displaystyle M}{\diagup}}$ link and their structures and consequently their fluxional behaviour are expected to be more complicated. The alkyl and aryl ligands are similar to CO in that irrespective of their bonding pattern μ_1(terminal) or μ_2(bridging), they are able to donate only one electron to the bonding scheme and the M-M bond will be retained. Finally the NO ligand is rather like halide in that it usually (although not necessarily) forms bridges (μ_2) at the expense of the M-M bond.

2.1.4. Hidden Processes

At this stage it is worth considering the so-called "hidden-process". This may involve any one of the three processes, (a), (b) and (c), or any combination of them, and is a process that cannot be detected by direct spectroscopic methods. Consider, for example, the possible mechanisms of CO exchange in $Os_3(CO)_{12}$. Two singlet resonances are observed in the ^{13}C nmr. spectrum at low temperature for the axial and equatorial CO groups (11). However, although this spectrum is in accord with the ground-state structure, localised scrambling of the six equatorial CO-groups via a triply bridged intermediate (Fig. 11) cannot be excluded. This merry-go-round behaviour is an example of a "hidden-process".

Dodecacarbonyltetrairidium, $Ir_4(CO)_{12}$, provides a second example. In this molecule, which has T -symmetry, all CO-groups are equivalent (14). A hidden fluxional

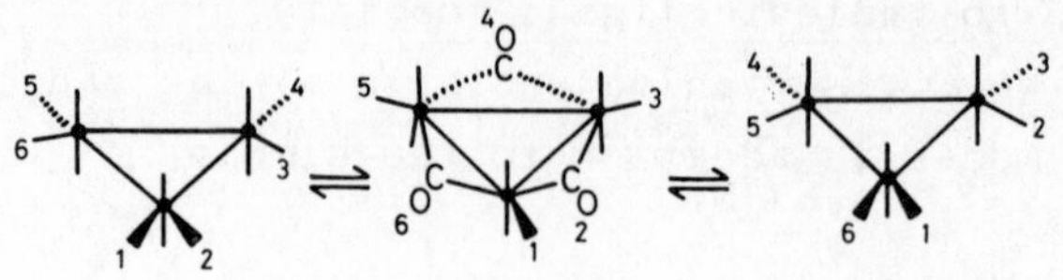

486.

Figure 11

Possible equatorial carbonyl exchange
in $Os_3(CO)_{12}$

process involving, for example, the interconversion of
a cube-octahedron of CO-ligands into an icosahedron and
vice versa (7) would not be detectable by conventional
methods.

2.2. Cluster Rearrangement

Only a few well established examples of this
class are known. Such examples fall within two groups,
(a) rearrangement with retention of the ligand
arrangement and (b) rearrangement in concert with
ligand mobility.

2.2.1. Rearrangement With Retention of the Ligand Arrangement

An excellent example is the dianion $[Pt_9(CO)_{18}]^{2-}$
(Fig. 12). The structure of this species in the solid
consists of three stacked $Pt_3(\mu_2-CO)_3(\mu_1-CO)_3$
triangulated units having a more or less eclipsed
configuration (15).

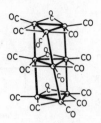

Figure 12

$\left[Pt_9(CO)_{18}\right]^{2-}$

In solution, these $Pt_3(\mu_2\text{-}CO)_3(\mu_1\text{-}CO)_3$ units apparently rotate but retain the identity of the individual CO-groups (viz. three terminal and three edge-bridges) (16).

2.2.2. Rearrangement in Concert With Ligand Mobility

Consider as an example the $\left[H_3Ru_4(CO)_{12}\right]^{-}$ anion. From the ^1H nmr. spectrum at low temperature it has been shown that two structural isomers exist, one having C_2- (or C_{2v}) symmetry and the other C_{3v}-symmetry. Each undergoes fluxional behaviour and apparently interconvert at higher temperatures (17).

X-ray analysis has shown that two structural isomers exist in the solid with metal-hydrogen geometries approximating to C_2(isomer I) and C_{3v} (isomer II) symmetry (18) (Fig. 13).

In isomer I three tetrahedral edges, Ru(1)-Ru(2), Ru(1)-Ru(3) and Ru(2)-Ru(4) are significantly longer (mean 2.923(1) Å) than the others (mean 2.803(1) Å) and the CO ligands appear pushed back from these edges.

488.

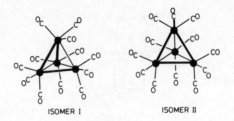

Figure 13
The two isomers of $\left[H_3Ru_4(CO)_{12}\right]^{\ominus}$
Hydride ligands bridge the solid Ru-Ru bonds

This is consistent with the three H-ligands bridging
these longer edges (see Section 2.5 of Chapter 2).
The isomer II has three Ru-Ru bonds, in this case
forming the triangle face Ru(1), Ru(2), Ru(3) which
are significantly longer than the others (mean
2.937(1) $\overset{o}{A}$).

Thus, in solution, when undergoing ligand
migration which interconverts isomers I and II it is
necessary not only for the H-ligands to migrate from edge
to edge but also the R_4-tetrahedron to undergo stretching
and bending of the metal framework. This process is
hidden in the sense that it cannot be separated from
ligand migration about the cluster framework.

A similar process occurs in the exchange process
postulated for $H_2FeRuOs_2(CO)_{13}$ (19) (Fig. 14). Here
again carbonyl and hydride migration must occur in
concert with the deformation of the quasi-tetrahedral
metal framework.

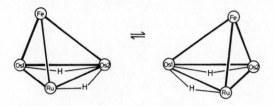

Figure 14

Skeletal rearrangement

in $H_2FeRuOs_2(CO)_{13}$

2.3. Ligand Migration Within the Metallic Cluster

Apart from hydrogen, few other ligands are able
to occupy interstitial sites within small metallic
clusters. In general interstitial ligand migration
will be observable only with body centred polyhedra
containing several interstitial sites. The possibility
that the H-ligand may reside along the inside edges or
faces of simple non body-centred polyhedra must exist
but the migration of such ligands would be difficult
to define (see Section 3.2).

3. LIGAND TYPES

Studies of the migration of ligands over the cluster surface have, for obvious reasons, been confined largely to three main types of ligand:-

1. Carbon monoxide
2. Hydrogen (or hydride)
3. Simple organic ligands including isocyanides.

3.1. Carbon Monoxide

The binary carbonyls form the largest class of binary molecular clusters within the transition-series. The advent of Fourier transform n.m.r. methods have largely overcome the difficulties encountered in the observation of ^{13}C resonances but in studies of the metal carbonyls ^{13}CO enrichment is normally essential. In addition, because of the long spin relaxation times T_1 for carbon atoms not bound to hydrogen, an inert, paramagnetic, shiftless relaxation reagent such as chromium trisacetylacetonate, $Cr(acac)_3$, is added. This gives a considerable enhancement in the signal to noise ratio.

3.1.1. Common Types of Bonding Mode

Three common types of CO-linkage have been identified:-

(i) Two-centre/two-electron, <u>terminal</u> M-C-O (μ_1).

(ii) Three-centre/two-electron, <u>edge-bridge</u> (μ_2). This is almost always found in association with a metal-metal bond and may by symmetric or asymmetric (Fig. 15).

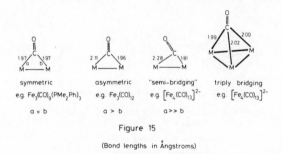

Figure 15

(Bond lengths in Ångstroms)

(iii) Four-centre/two-electron, <u>face-bridge</u> (μ_3). Always observed in association with an M_3-triangular face and may show varying degrees of asymmetry.

3.1.2. Other Points of Note

(iv) All bonding modes (i), (ii) and (iii) are implicated in the chemisorbed state.

(v) Irrespective of the bonding mode (i), (ii)

or (iii), the CO can donate only two electrons.

(vi) Clusters with greater than four-centre interactions, _e.g._ $M_4(\mu_4-CO)$, have not been observed - although M_4CO, M_5CO and even M_6CO have been considered for surfaces.

(vii) No cluster carbonyl $M_m(CO)_n$ has m>n; $[Pt_{19}(CO)_{22}]^{4-}$ comes closest to breaking this rule (20).

(viii) Two, as yet unique, bonding modes have also been observed. In the structure of the anion $[Fe_4(CO)_{13}H]^-$ one CO is bound to a "butterfly" arrangement of four iron atoms and donates a total of four electrons, two from carbon and two from oxygen (21). In $Mn_2(CO)_5(Ph_2PCH_2PPh_2)_2$ one CO is bent and serves as a four-electron donor (22), (Fig. 16).

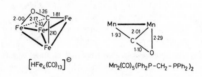

$$[HFe_4(CO)_{13}]^{\ominus} \qquad Mn_2(CO)_5(Ph_2P-CH_2-PPh_2)_2$$

Figure 16

Unusual carbonyl bonding modes

(Bond lengths in Ångstroms)

In general, interconversion of bonding modes (i), (ii) or (iii) has formed the basis of our understanding of CO-mobility in clusters. Other bonding modes, e.g. M_4CO, M_5CO and M_6CO, are not considered as possible intermediates. This has clear ramifications in considerations of $M_m(CO)_n$ clusters in which the metal polytope has, e.g. square faces as in $[Rh_6C(CO)_{15}]^{2-}$ (23).

The barrier to carbon monoxide mobility in the cluster appears to be lowered as the negative charge on the cluster is increased:- (a) by an increase in anionic charge or (b) by the substitution of ligands

which are better donors e.g. R_3P. Few experiments
have been conducted to test these hypotheses but
certainly $[Rh_{13}(CO)_{24}H_2]^{3-}$ is stereochemically non-
rigid at -80° whereas the closely related species
$[Rh_{13}(CO)_{24}H_3]^{2-}$ only undergoes CO-migration at 25^oC
and above (24), and the barrier to axial-equatorial CO-
exchange in the monosubstituted complex $Os_3(CO)_{11}PEt_3$
is significantly less than that for $Os_3(CO)_{12}$ (25).

The argument is based on the idea that a bridging
CO group is a better π-acceptor and therefore a better
electron acceptor than a terminal one. This being the
case, then for any process involving the formation of
a bridged-intermediate will be favoured by increase in
charge. Certainly the increase in electron density
often changes the distribution of bonding modes (i),
(ii) and (iii) within a cluster. This is apparent
from the following series of cobalt carbonyl complexes
(Table 1).

3.1.3. Factors Influencing the Distribution of Bonding Modes

The distribution of bonding modes (i), (ii) or
(iii) apparently depends on a number of factors:-
1. The metal, 2. The negative charge on the metal
framework - this may be increased either by addition
of charge to form anionic species or substitution of
CO by ligands which are better σ-donors, e.g. tertiary
phosphines. The effect of charge may be seen from the
following series of cobalt carbonyl complexes, (see
Table 1). The number of bridges increasing with
increasing negative charge.

The effect of substitution on this ground-state

TABLE 1

Cluster	CO-Types			Reference
	μ_1(terminal)	μ_2(edge-bridge)	μ_3(face-bridge)	
$Co_6(CO)_{16}$	12	0	4	26
$Co_6(CO)_{15}^{2-}$	9	3	3	27
$Co_6(CO)_{14}^{4-}$	6	0	8	28

geometry is apparent from structural observations on $Ir_4(CO)_{12}$ and its derivatives $Ir_4(CO)_{n-x}L_x$. The parent carbonyl has a T_d-ground-state with no CO-bridges (14). Replacement of CO to give $Ir_4(CO)_{11}(RNC)$ by isocyanide (RNC), which is electronically similar to CO, causes no CO-bridges to be formed (29). In contrast, substitution by tri-phenylphosphine, a better σ-donor than CO, produces derivatives which have a quasi-$Co_4(CO)_{12}$ structure, possessing three CO-bridges (30). It follows that for any fluxional process which involves the formation of a bridged intermediate will be aided by the introduction of a phosphine ligand. The effect of the substitution of CO by phosphine has been investigated (25) in $Os_3(CO)_{12-x}(PEt_3)_x$ (x = 1, 2 or 3). In the mono-substituted derivative n = 1, the observed spectra are consistent with dynamic behaviour in which only CO_a, CO_c, CO_d and CO_g are exchanging via a bridged intermediate (Fig. 17).

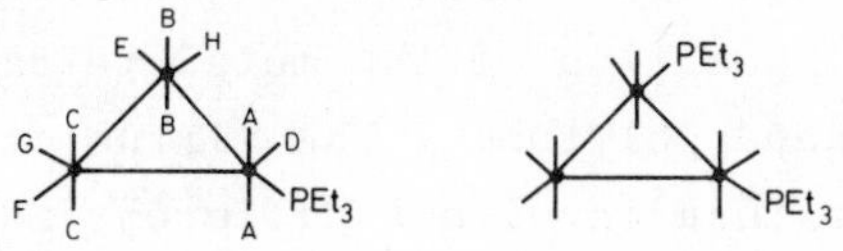

Figure 17

Structures of $Os_3(CO)_{11}(PEt_3)$
and $Os_3(CO)_{10}(PEt_3)_2$

The activation energy for this process is considerably smaller than that calculated for ligand mobility in the parent carbonyl and phosphine substitution clearly promotes axial equatorial CO exchange.

Although these "ideas" are attractive they are by

no means general. Many anionic carbonyls show no
tendency to form CO-bridges and the introduction of
phosphine ligands in $Os_3(CO)_{12}$ shows no evidence for
CO-bridge formation in the ground-state structure.

An alternative explanation may be offered. It
has been argued that the numbers and types of CO ligand
in a polynuclear species $M_m(CO)_n$ reflect (i) the polyhedral arrangement of the n-carbonyl groups and
(ii) the orientation of the M_m unit within this
polyhedron. The structural difference between
$Co_4(CO)_{12}(C_{3v})$ and $Ir_4(CO)_{12}(T_d)$, for example, merely
reflecting the two different CO-polyhedra, being
icosahedral in $Co_4(CO)_{12}$ and cubeoctahedral in $Ir_4(CO)_{12}$.
It has been suggested that the fluxional behaviour of
$Co_4(CO)_{12}$ may be interpreted in terms of the polytopal
rearrangement of the twelve CO ligands. Thus, the
initial arrangement of ligands rearranges along a well-
defined reaction coordinate via a cubeoctahedral
transition-state; the existence of different patterns
of CO-scrambling then arises as a consequence of the
geometrical disposition of the metal tetrahedron
within the ligand polytope. The virtue of the scheme
is that rather than involve a different ad hoc mechanism
to rationalise each observed fluxional process in
cluster carbonyls, this single type of carbonyl polytopal rearrangement involves the concerted motion of
all the CO groups (and possibly metal atoms) in ways
that are well-defined in terms of molecule geometry.
A more detailed consideration of this approach to
M_2L_6 systems has been given above.

3.2. Hydrogen

The mobility of the H-ligand in transition-metal clusters is a well established phenomenon but no simple, binary cluster hydrides are known at present and all studies of H-mobility have been of mixed ligand systems. Any migration of the H-ligand must by necessity therefore be governed by the requirements of the other ligands (usually CO).

Recent studies may be taken to indicate that the hydrido-ligand is similar in size to CO and may occupy a single coordination site about the central cluster unit. In addition, and in contrast to most other common ligands, hydride is special in that it may occupy an interstitial site within the metal polytope. This introduces an additional variant on ligand mobility, viz. the ability of the ligand to migrate from one interstice to another within the cluster unit (see below). The possibility that the H-ligand may also migrate from the outside (regular coordination site) to the inside (interstitial site occupancy) of the cluster unit must also be considered.

Three common types of external hydride-bonding mode have been identified (Fig. 18).

(i) Two-centre/two-electron, terminal (μ_1).

(ii) Three-centre/two-electron, edge-bridge (μ_2). This is always found in association with a metal-metal bond.

(iii) Four-centre/two-electron, face-bridge (μ_3). Always observed in association with an M_3 triangular-face.

498.

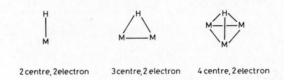

2 centre, 2 electron   3 centre, 2 electron   4 centre, 2 electron

Figure 18

Bonding in metal hydrides

Interconversion of these three types of H-bond offers a method by which H-mobility may occur about a cluster surface.

Two additional points arise:-

(a) unlike CO (see above) the hydrido-ligand apparently prefers to adopt bonding modes (ii) or (iii). The terminal mode (i) is not as commonly observed.

(b) in general adoption of modes (ii) or (iii) causes the metal-metal bonds to be longer than those not supporting bridges although this effect may be masked by other structural effects such as a carbonyl bridging (and shortening) the same M-M bond (see Chapter 2).

An exception is $H_2Os_3(CO)_{10}$. In this molecule which has a bi-μ_2-bridged form, the Os-Os distance supporting the H-bridges is short (31) (Fig. 19).

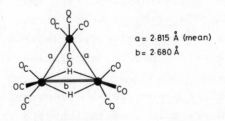

a = 2·815 Å (mean)
b = 2·680 Å

Figure 19

$H_2Os_3(CO)_{10}$

It follows, therefore, that all migrations of the hydride-ligand about the cluster surface must also involve rearrangement of the metal polytope. This aspect is covered above.

The characterisation of species containing interstitial H-atoms has been a highlight of more recent structural studies of hydrido-complexes. Such species are of two types. Firstly, those which contain H in a single interstitial site. Examples are $[HRu_6(CO)_{18}]^-$ (32) and $[HCo_6(CO)_{15}]^-$ (33). Secondly, those which occupy one or more of several available sites. This type will only exist for body-centred clusters such as Rh_{13} (34) (Fig. 20).

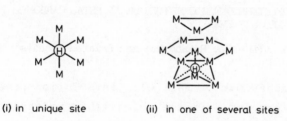

(i) in unique site   (ii) in one of several sites

Figure 20

Interstitial hydride ligands

There is no evidence as yet available that hydride can migrate from the outside to the inside of the metal polytope (or *vice versa*) but clear evidence has been provided for migration from interstitial site to interstitial site. This has been demonstrated for the Rh_{13} cluster anion $[H_3Rh_{13}(CO)_{24}]^{2-}$ (24) (Section 4.5.4).

4. SURVEY OF STEREODYNAMIC BEHAVIOUR IN METAL CARBONYL CLUSTERS AND RELATED SYSTEMS

4.1. $M_3(CO)_{12}$ (M = Fe, Ru and Os) and Related Species

The behaviour of $Fe_3(CO)_{12}$ in solution poses a difficult problem. In the solid, this molecule has the doubly bridged, C_{2v} structure (2) (Fig. 21), but

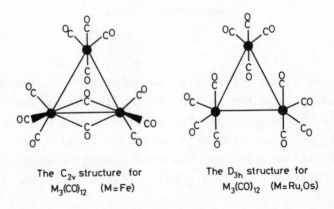

The C_{2v} structure for
$M_3(CO)_{12}$ (M = Fe)

The D_{3h} structure for
$M_3(CO)_{12}$ (M = Ru, Os)

Figure 21

in solution, ^{13}C nmr. shows that all carbonyls are
equivalent at -150°C (6). Thus the activation energy
for total scrambling is less than 20 KJ/mole. No
further information is obtainable directly because of
solubility problems at such low temperatures. In
addition, $Fe_3(CO)_{12}$ has an anomalous infra-red
solution spectrum which shows fewer than the expected
number of bands and a weak, broad absorption in the
bridging carbonyl region (35). The heteronuclear
species $RuFe_2(CO)_{12}$ and $[MnFe_2(CO)_{12}]^-$, which have
similar doubly bridged structures, also exhibit a
single ^{13}CO signal, which shows that their carbonyls
are mobile over all three metal atoms (11).

Cotton proposed (2) that in $Fe_3(CO)_{12}$ fluxionality
occurs by concerted bridge-terminal pairwise exchange,
involving a D_{3h}(unbridged) intermediate (Fig. 22).
This is appealing because the homologous $Ru_3(CO)_{12}$ and
$Os_3(CO)_{12}$ have D_{3h} crystal structures (36,37).

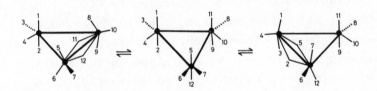

Figure 22

If in solution the $Fe_3(CO)_{12}$ molecules populate the
entire range of structures attained during this
fluxional process, then the anomalous i.r. spectrum is
understandable as a convolution of those of the different
structures. This mechanism, operative about all three
edges of the Fe_3 triangle, necessarily involves a

'scissoring' of this triangle since in the ground state
it is isosceles with Fe-Fe bond lengths of 2.558(1),
2.677(2), 2.683(1) Å (cf. Section 2.2.2). This lability
of the metal skeleton is of particular interest in
connection with the problem of 'bond lengths versus
bond strengths' in metal clusters (38).

Cotton's proposal is by no means the only possible
explanation for the nmr. and i.r. behaviour of $Fe_3(CO)_{12}$.
The solid state structure comprises a metal atom
triangle enveloped by a flattened icosahedron of carbonyl groups (4). If in solution the ligands were to
maintain a fixed spatial relationship while the metal
triangle relaxes into a slightly different orientation,
a structure of D_3 symmetry with no bridging carbonyls
would be attained (3) (Fig. 1). This D_3 symmetry
would be consistent with the i.r. spectrum, and
repeated interchange of C_{2v} and D_3 forms would exchange
all the carbonyls in fewer steps than the Cotton
mechanism.

Recent study by ^{13}C nmr. of a number of phosphine-
and isocyanide-substituted molecules $Fe_3(CO)_{11}(PR_3)$ and
$Fe_3(CO)_{11}(CNR)$ has shown that, over an intermediate
temperature range, there occurs a pattern of carbonyl
scrambling which cannot be explained on the basis of
either of the mechanisms outlined above (39). At
higher temperatures, total carbonyl scrambling is
observed; in no case was the ground state structure
of one of these molecules 'frozen out', again
illustrating the very low activation energy for
fluxionality in the Fe_3 system. The observed spectra
may be interpreted on the basis of a concerted
mechanism in which the icosahedral ligand envelope
rearranges via a cuboctahedral transition state, in
identical fashion to the process recently postulated
for tetranuclear dodecacarbonyls and discussed in

Section 4.3. In addition, rotation of unsubstituted $M(CO)_3$ units appears to occur, a process observed in many other carbonyl clusters. However, experimental data offers conflicting evidence and further investigation into these systems is in progress (39).

$Ru_3(CO)_{12}$ (36) and $Os_3(CO)_{12}$ (37) possess D_{3h} crystal structures. $Ru_3(CO)_{12}$ exhibits a single ^{13}CO nmr. signal at $-100^{\circ}C$ (40), while $Os_3(CO)_{12}$ shows a single signal at high temperature and two peaks of equal intensity below $60^{\circ}C$ (11). It is tempting to interpret the low temperature spectrum of $Os_3(CO)_{12}$ in terms of the ground state structure with all fluxional processes 'frozen out'. However, a variety of 'hidden' carbonyl scrambling mechanisms could be operating, provided that axial-equatorial interchange does not occur (41) (see Section 2.1.4); in particular, an in-plane cyclic permutation of the six equatorial carbonyls; there is strong circumstantial evidence for this, as discussed below. Additionally, scrambling of axial ligands might occur; a similar process has been observed in one isomer of $Rh_3Cp_3(CO)_3$ (42).

The mechanism of interchange of axial and equatorial carbonyls in $Os_3(CO)_{12}$ above $60^{\circ}C$ is unclear. By analogy with Cotton's proposal for $Fe_3(CO)_{12}$ we may envisage carbonyl scrambling about edges of the Os_3 triangle, via a transition state of C_{2v} symmetry. However, it is not necessary for <u>internuclear</u> scrambling to take place; a polytopal rearrangement which exchanges the axial and equatorial carbonyls of each individual $Os(CO)_4$ unit would account for the observed spectra. It is impossible to ascertain whether internuclear scrambling does occur in $M_3(CO)_{12}$ (M = Ru, Os) because of the absence of any magnetic isotope of these metals with natural abundance greater than 0.5%. We may consider a good deal of circumstantial evidence from

504.

substituted derivatives.

4.1.1. $D_{3h} \rightleftharpoons C_{2v} \rightleftharpoons D_{3h}$ Mechanism

The equatorially phosphine-substituted molecules $Os_3(CO)_{11}(PEt_3)$, $Os_3(CO)_{10}(PEt_3)_2$ (25), $Ru_3(CO)_{10}[Ph_2CH_2-PPh_2]$ (43) (Fig. 23) and $Os_3(CO)_{10}(PPh_2-CH_2-CH_2-PPh_2)$ (44) all exhibit this mechanism.

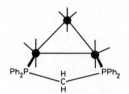

Figure 23
The structure of
$Ru_3(CO)_{10}(Ph_2P-CH_2-PPh_2)$

Coalescence of ^{13}C nmr. signals corresponding to six carbonyls about an edge of the M_3 triangle is clearly observed. The less mobile phosphine ligands restrict this exchange to a single edge, although it might be expected to operate about two edges in the monosubstituted $Os_3(CO)_{11}PEt_3$ (Fig. 17). In $Os_3(CO)_{10}(Ph_2P-CH_2-CH_2-PPh_2)$ a second process occurs at higher temperatures (see paragraph 4.1.3 below).

The behaviour of $Os_3(CO)_{12-n}(PEt_3)_n$ (n = 0, 1, 2) shows that successive phosphine substitution lowers the activation energy for carbonyl scrambling. In the trisubstituted $Os_3(CO)_9(PEt_3)_3$, no carbonyl scrambling about an edge of the Os_3 triangle is possible if the phosphines are to remain immobile, and fluxionality takes place at higher temperature than in the mono- and di-substituted species. In all these substituted compounds, complete scrambling, appearing to involve a breakdown in the rigidity of the phosphines, occurs at $40^{\circ}C$ (25). This temperature is still lower than that at which axial-equatorial exchange occurs in $Os_3(CO)_{12}$.

4.1.2. Polytopal Rearrangement of $M(CO)_4$ Units

In $Ru_3(CO)_{10}(NO)_2$ (Fig. 7), the nitrosyl groups act as 3-electron donors and occupy bridging positions in a structure of C_{2v} symmetry (45). ^{13}C n.m.r. indicates that the $Ru(CO)_4$ unit undergoes localised scrambling with its two axial and two equatorial ligands exchanging at $-50^{\circ}C$ (11).

Similarly, in a range of molecules $HOs_3(CO)_{10}X$ (X = 3-electron donor) (Fig. 24), the same pattern of

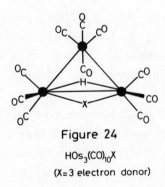

Figure 24

$HOs_3(CO)_{10}X$

(X=3 electron donor)

carbonyl scrambling on the unique metal atom is observed (46). In $HOs_3(CO)_{10}(COOEt)$ the coalescence of the signals

from this $Os(CO)_4$ unit is a two-step process, (47);
three carbonyls exchange at $80^{o}C$ while the fourth,
axial, one remains distinct until $90^{o}C$. We may
conclude that the $Os(CO)_4$ rearrangement occurs by a
trigonal twist mechanism.

In $Os_3(CO)_{10}(C_6H_8)$ the cyclohexa-1,3-diene ligand
occupies one equatorial and one axial site on a single
osmium atom (48) (Fig. 25). In the lowest energy

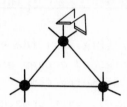

Figure 25

$Os_3(CO)_{10}(C_{10}H_8)$

fluxional process the $Os(C_6H_8)(CO)_2$ group undergoes a
rotational motion which, by analogy, could also cause
site exchange in an $Os(CO)_4$ unit. This observation
has recently been confirmed for several other $Os_3(CO)_{10}$
(diene) species (44).

Thus polytopal rearrangement of $M(CO)_4$ units in
the absence of internuclear scrambling has been
demonstrated in a variety of molecules, although those
with non-labile bridging ligands may be of doubtful
relevance to the unbridged $Ru_3(CO)_{12}$ and $Os_3(CO)_{12}$.

4.1.3. Equatorial In-plane Scrambling

In $Os_3(CO)_{10}(C_7H_8)$ the norbornadiene ligand also occupies one axial and one equatorial position on a single osmium atom (49). In contrast to $Os_3(CO)_{10}(C_6H_8)$ a different fluxional process is observed. The norbornadiene group 'swings round' on its osmium atom while the equatorial carbonyls undergo a limited cyclic exchange via a triply bridged transition state. The process is restricted by the inability of the diene to transfer from one metal atom to another, but the similarity to the possible 'hidden process' of in-plane equatorial scrambling in $Os_3(CO)_{12}$ is clear (49).

In $Os_3(CO)_{10}(C_4H_4N_2)$ (Fig. 26) the bidentate

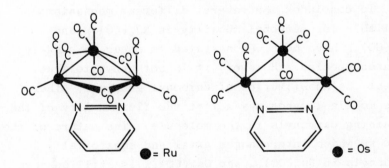

Figure 26

Structures of

$Ru_3(CO)_{10}(C_4H_4N_2)$ and $Os_3(CO)_{10}(C_4H_4N_2)$

diazine ligand occupies two axial coordination sites, bridging an Os-Os bond (50). The six equatorial carbonyls became equivalent at $7^{\circ}C$, showing that in-plane scrambling takes place. The two axial carbonyls on the unsubstituted osmium atom also exchange with these six carbonyls, in a simultaneous $Os(CO)_4$ rearrangement. The ruthenium analogue $Ru_3(CO)_{10}(C_4H_4N_2)$ (51) possesses a crystal structure with three bridging carbonyls (Fig. 26), in contrast to the unbridged osmium structure. Thus a form proposed as a transition state for a fluxional process in one molecule is the ground state structure for a second, closely related molecule. In $Ru_3(CO)_{10}(C_4H_4N_2)$ the six equatorial carbonyls exchange with the two axial carbonyls on the unsubstituted metal atom; one of the latter two carbonyls joins in this scrambling at a higher temperature than the other, providing further evidence for the trigonal twist mechanism for $M(CO)_4$ site exchange. This conclusion is reinforced by ^{13}C n.m.r. line-shape analysis of a higher temperature fluxional process in $Os_3(CO)_{10}(Ph_2P-CH_2-CH_2-PPh_2)$ (44).

To conclude, the several different mechanisms available for carbonyl mobility in $Ru_3(CO)_{12}$ and $Os_3(CO)_{12}$ have been demonstrated to occur in closely related species. However, it is not known to what extent the substitution of carbonyl groups by other, less mobile ligands may affect the fluxionality of the remaining carbonyls in the molecule. The nature of the process which interchanges axial and equatorial carbonyls in $Ru_3(CO)_{12}$ and $Os_3(CO)_{12}$ is still unknown.

4.2. Trinuclear Carbonyl Hydrides of Fe, Ru and Os

Several species of this class have been
investigated. In $[HFe_3(CO)_{11}]^-$ (Fig. 27), a low
temperature ^{13}C n.m.r. spectrum with all fluxional

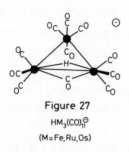

Figure 27

$HM_3(CO)_{11}^{\ominus}$

(M = Fe, Ru, Os)

processes frozen out has been obtained at $-107^{O}C$ (52).
At $-20^{O}C$ there is a signal of very low intensity whose
chemical shift indicates it to represent an average of
the bridging with one terminal carbonyl. The other
nine carbonyls also exchange at this temperature and
this spectrum has been interpreted in terms of a
mechanism involving a D_{3h} type transition state with a
terminal hydride ligand. At higher temperatures,
complete carbonyl scrambling occurs; only at the low
temperature limit is ^1H coupling observed in the ^{13}C
spectrum (52).

In contrast, in the isostructural $[HRu_3(CO)_{11}]^-$
(53), which exhibits a low temperature limiting n.m.r.
at $-116^{O}C$, there is exchange of the bridging (CO) with

a number of pairs of equivalent terminal carbonyls; the unique carbonyls remain distinct. It was concluded that the hydride remains rigid in its bridging position; $^1H-^{13}C$ coupling is observed over a wide temperature range. At $32^{\circ}C$, total carbonyl scrambling occurs.

The crystal structure of $[HOs_3(CO)_{11}]^-$ has not been determined, but at $-105^{\circ}C$ the ^{13}C n.m.r. spectrum is consistent with a rigid structure identical to those of the iron and ruthenium analogues (54). At $-85^{\circ}C$, the several resonances collapse into two broad peaks and at $-50^{\circ}C$ a single resonance is observed. Chemical shift values indicate that the bridging carbonyl might not be among those contributing to this single resonance, and the mechanism of carbonyl scrambling is unclear.

In this series of three anions there does not seem to be any underlying pattern of fluxionality, despite the fact that they are isostructural.

The μ-(CO)-alkylated derivatives $HOs_3(CO)_{10}(COEt)$ (referred to in Section 4.1) and $HFe_3(CO)_{10}(COMe)$ undergo fast flipping of their (COR) groups, thought to involve inversion at the oxygen atom rather than rotation about the C-O bond (47). They exhibit $M(CO)_4$ polytopal rearrangement, followed at higher temperature by site exchange on $M(CO)_3$ units. Internuclear (CO) exchange is not observed, and alkyl migration has been shown not to occur in the iron species (47).

$H_2Os_3(CO)_{10}$ (Fig. 19), has a low temperature ^{13}C n.m.r. spectrum consistent with its crystal structure (30,55). At $70^{\circ}C$ the six carbonyls on the two bridged Os atoms are equivalent; this may indicate $Os(CO)_3$ rotation, although it is not known if the hydride ligands are mobile. At the decomposition temperature of $120^{\circ}C$ there is only slight broadening of the ^{13}C resonances corresponding to the $Os(CO)_4$ unit.

The molecules, $H_2Os_3(CO)_{10}L$ (L = CNMe, PMe_2Ph, $AsMe_2Ph$), have been studied by [1]H n.m.r. (56). At $-60^\circ C$ the two hydrides have separate resonances in agreement with the crystal structure of $H_2Os_3(CO)_{11}$ (Fig. 28), which has one hydride bridging and one

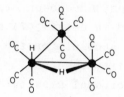

Figure 28

$H_2Os_3(CO)_{11}$

terminal (disordered with a carbonyl) (37). By $50^\circ C$ they exchange, and observation of separate ^{1}H signals for coordinated and extra free ligand L shows that ligand dissociation does not occur. It was proposed that the hydrides exchange via a transition state in which they are both on one osmium atom, but no ^{13}C study to investigate simultaneous rearrangement of the carbonyls has been carried out.

Simple mobility over the cluster surface is not the only fluxional behaviour exhibited by the hydride ligand. [1]H n.m.r. shows that a reversible equilibrium exists in solution between $H_2Os_3(CO)_{10}(CH_2)$ and $HOs_3(CO)_{10}(CH_3)$ (57,58). Here the hydride interacts directly with another ligand, a process often invoked in reaction schemes.

512.

4.3. Tetranuclear $M_4(CO)_{12}$ (M = Co, Rh or Ir) Clusters

Several clusters, $M_4(CO)_{12}$ (M = Co, Rh, Ir) and
substituted derivatives, have been studied; they have
recently been reviewed in detail (7) and a single
mechanism proposed to account for the different
patterns of (CO) mobility observed.

$Co_4(CO)_{12}$ has a crystal structure of C_{3v} symmetry
(Fig. 29) (59), but ^{13}C n.m.r. in solution shows three

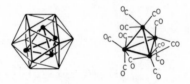

The C_{3v} structure of $M_4(CO)_{12}$ (M = Co, Rh)
as a tetrahedron within an icosahedron of
ligands

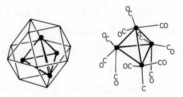

The Td structure of $M_4(CO)_{12}$ (M=Ir)
as a tetrahedron within a cuboctahedron of
ligands

Figure 29

signals of equal intensity instead of the expected four
(60). This, together with infra-red spectral anomalies
(61,62), led to long controversy over the solution
structure, (63), although ^{59}Co n.m.r. indicates a C_{3v}

solution structure (64). The ^{13}C n.m.r. result seems
to arise from quadrupole effects, even in the presence
of a relaxation agent, although $Co_4(CO)_{11}[P(OMe)_3]$ (64)
and $RhCo_3(CO)_{12}$ (12) show no intensity anomalies.
Until very recently, sample decomposition prevented
the high temperature ^{13}C spectrum of $Co_4(CO)_{12}$ being
recorded. Now, a single broad signal has been observed
in viscous solution at $61^{o}C$ (65); the chemical shift
value of this average peak shows that, in the low
temperature spectrum, one of the iron-bridging
carbonyl signals should be given double weighting.
Together with further infra-red studies (66), this
appears to confirm the C_{3v} structure of $Co_4(CO)_{12}$ in
solution.

$Rh_4(CO)_{12}$ has a low temperature ^{13}C n.m.r.
spectrum (67) entirely consistent with its C_{3v} crystal
structure (68); assignments are aided by $^{103}Rh-^{13}C$
coupling (^{103}Rh, I = ½, 100% natural abundance). As
the temperature is raised, the signals broaden and
coalesce. At $50^{o}C$ the spectrum is a binomial quintet,
unequivocally showing that the carbonyls are mobile
over the whole Rh_4 cluster (13).

The heteronuclear $RhCo_3(CO)_{12}$ (Fig. 8) exhibits
a more restricted pattern of carbonyl mobility (12).
The low temperature ^{13}C n.m.r. spectrum is consistent
with the ground state structure if the three apical
carbonyls give a single resonance; the $Co(CO)_3$ unit
appears to be rotating even at $-85^{o}C$. Over an
intermediate temperature range, there is a 10:2
pattern of carbonyl equivalence with two distinct (CO)s
bonded to the rhodium atom, to which coupling is
maintained.

The substituted derivative $Ir_4(CO)_{11}(PPh_2Me)$
(Fig. 30) has three bridging carbonyls, (30), in
contrast to the unbridged $Ir_4(CO)_{12}$ (14); the phosphine

514.

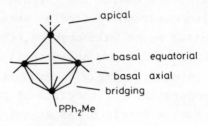

Figure 30

$Ir_4(CO)_{11}(PPh_2Me)$

ligand is in a basal axial position. The fluxional
process of lowest activation energy is a cyclic
permutation of the three bridging and three terminal
equatorial carbonyls about the basal plane. At higher
temperatures, all carbonyls exchange except one, which
$^{31}P-^{13}C$ coupling shows to be antipodal to the phosphine.
At still higher temperatures, all (CO)s in this molecule
become equivalent.

The apparently dissimilar patterns of carbonyl
scrambling in these clusters may be explained in terms
of a single type of process (7). The twelve ligands
in the ground state structure of each molecule define
an icosahedron. This polyhedron may undergo concerted
distortion into a cuboctahedron, (Fig. 31); further
motion along the same distortion coordinate regenerates
an icosahedron, with vertices permuted in a certain
pattern. For an isolated icosahedron, this distortion may
be carried out in five equivalent ways. When there is
a tetrahedron of metal atoms within the icosahedron of
ligands, these five 'modes' are no longer degenerate;
the permutations produced by them correspond to the
observed carbonyl scrambling patterns in the $M_4(CO)_{12}$
species. The only additional process which has to be

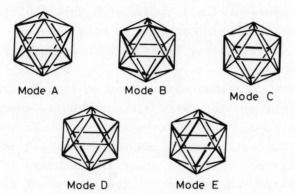

(a) The five modes of distortion of an icosahedron
into a cuboctahedron; the thick lines denote
edges that are being stretched.

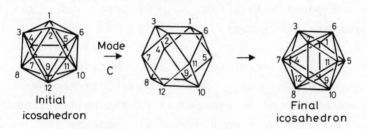

(b) The permutation of vertices in the icosahedron
by complete operation of one of these
distortion modes.

Figure   31

invoked is rotation of the apical $M(CO)_3$ group, which takes place at low temperature in $RhCo_3(CO)_{12}$ but only at high temperature in $Ir_4(CO)_{11}(PPh_2Me)$. Rotation of $M(CO)_3$ units has been observed in many cluster carbonyls, particularly those of high nuclearity (Section 4.6).

$Ir_4(CO)_{11}(CN-tBu)$ retains the unbridged structure of the parent cluster, with a cuboctahedral envelope of ligands (29). A 10:1 pattern of carbonyls shows that the (CO) antipodal to the isocyanide ligand remains distinct while the other ten exchange. This may be interpreted in terms of the inverse of the mechanism outlined above, i.e. rearrangement of the cuboctahedron via an icosahedral transition state. At higher temperatures, rotation of $Ir(CO)_3$ units causes antipodal relationships to be lost, and all carbonyls are equivalent.

As mentioned in Section 4.1, it has also proved possible to apply this model to the substituted derivatives of $Fe_3(CO)_{12}$ (39). However, it appears that such concerted rearrangement of the entire ligand sphere can occur only when very high symmetry is present, such as the I_h symmetry of the icosahedron, and that clusters of lower symmetry are restricted to more localised scrambling patterns. This seems particularly apparent in higher nuclearity clusters (Section 4.6).

4.4. Tetranuclear Carbonyl Hydrides of Fe, Ru and Os

The cluster $H_4Os_4(CO)_{12}$ has a structure of D_{2d} symmetry (69), with hydrides bridging four edges of the metal atom tetrahedron (Fig. 32). It exhibits a

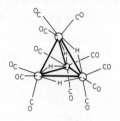

Figure 32

Structure of $H_4M_4(CO)_{12}$

$(M = Ru, Os)$

single ^{13}C n.m.r. signal over a wide temperature range (69), showing that carbonyl scrambling is occurring. The isostructural $H_4Ru_4(CO)_{12}$ (70) has not been studied by ^{13}C n.m.r.; all hydrides are equivalent in the crystal structure. The heterometal derivative $H_4FeRu_3(CO)_{12}$ shows only one ^1H signal at $-130^\circ C$ (71).

The substituted species $H_4Ru_4(CO)_{12-n}[P(OMe)_3]_n$ have been studied by ^1H n.m.r. for n = 1, 2, 3, 4 (72). Successive substitution occurs at different ruthenium atoms, and the simplicity of the i.r. spectra shows that only one of the several possible isomers is formed for each compound. Each species exhibits only a single chemical shift for all hydrides; each hydride appears to couple equally to all the ^{31}P nuclei. In the tetrasubstituted molecule, this was so even at $-100^\circ C$. Rotation of $M(CO)_2L$ groups was thought unlikely because of non-observance of $^{31}P-^{31}P$ coupling, and it was concluded that the hydrides are mobile over the Ru_4 tetrahedron (72).

In $H_4Ru_4(CO)_{10}(Ph_2P-CH_2-CH_2-PPh_2)$ (73), the ^1H n.m.r. at $-50^\circ C$ corresponds to the crystal structure in which the diphosphine ligand chelates a single ruthenium atom. The hydrides are disposed to give C_s

518.

type symmetry, with the two unbridged Ru-Ru bonds adjacent (Fig. 33), in contrast to the D_{2d} symmetry

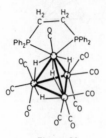

Figure 33

$H_4Ru_4(CO)_{10}(Ph_2P-CH_2-CH_2-PPh_2)$

of the parent cluster. Their n.m.r. resonances at
-50°C are multiplets because of ^{31}P-^1H coupling. By
+25°C, certain hydride resonances broaden and collapse.
A computer-generated simulation of the n.m.r.-differen-
tiable permutations in this molecule led to the
conclusion that only one could broaden the multiplets
in the observed way, over a temperature range of -44°C
to -31°C. This corresponds to a certain pattern of
exchange of hydrides between edges of the Ru_4 tetrahedron, involving a transition state with terminal
hydrides. Above -31°C another process becomes
operative, and spectra to 0°C are accurately simulated
by mixing in more and more of a second permutation
corresponding to a different pattern of hydride
exchange (73).

The anion $[H_3Ru_4(CO)_{12}]^-$ was shown by ^1H n.m.r.
to exist in solution as a mixture of two isomers (17).
This has been confirmed by their separate crystallographic characterisation (18) (Fig. 13). The C_{3v}
isomer has all hydrides equivalent, and the C_s isomer
gives a 2:1 pattern. All hydrides become equivalent

above $-5^{\circ}C$ and as discussed in Section 2.2 their mobility must involve lengthening and contraction of Ru-Ru bonds.

In $[H_2Ru_4(CO)_{12}]^{2-}$ the carbonyls adopt a bridged configuration (Fig. 34) (74). A cyclic exchange of

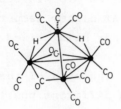

Figure 34

$[H_2Ru_4(CO)_{12}]^{2-}$

bridging and terminal carbonyls about the basal plane occurs at $-50^{\circ}C$; at higher temperatures, total averaging of carbonyl resonances is observed.

In $H_2FeRu_3(CO)_{13}$ (Fig. 35) several distinct

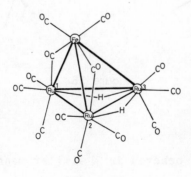

Figure 35

$H_2FeRu_3(CO)_{13}$

fluxional processes have been observed (19):

(i) The four carbonyls bonded to the iron atom become equivalent. In contrast to simple polytopal rearrangement of $M(CO)_4$ units discussed in Section 4.1, two of the (CO)s involved bridge Fe-Ru bonds in the ground state structure.

(ii) A cyclic permutation of six carbonyls about the $FeRu_1Ru_2$ plane occurs at higher temperature. Two further (CO)s are already being mixed in by process (i).

(iii) The iron atom at the apex of the asymmetric tetrahedron moves to a different position over the Ru_3 triangle, with a concomitant shift of the hydride ligands (Fig. 36).

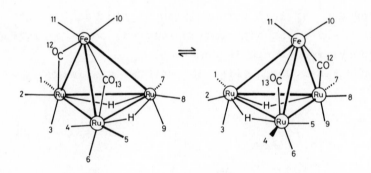

Figure 36

$H_2Ru_4(CO)_{13}$ behaves in a similar manner, but one carbonyl on the apical ruthenium atom remains distinct (75).

In $H_2FeRuOs_2(CO)_{13}$, two isomers are present,

corresponding to different positions of the iron atom over the $RuOs_2$ triangle (19). Processes (i) and (ii) are observed, and process (iii) interconverts the two isomers if the iron atom shifts in a suitable manner (Fig. 36). [13]C n.m.r. shows process (iii) to take place above $40^{\circ}C$ which is confirmed by [1]H n.m.r. as the temperature above which isomer interconversion does occur (19).

Thus in these molecules the metal cluster undergoes significant rearrangement (Section 2.2); rearrangement of atoms is common on metal surfaces that bear chemisorped species (76).

4.5. Other Tetranuclear Clusters

$HFeCo_3(CO)_{12}$ (Fig. 37) shows anomalous [13]C n.m.r. peak intensities (65), in a similar manner to $Co_4(CO)_{12}$.

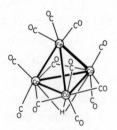

Figure 37

$HFeCo_3(CO)_{12}$

Broadening of certain signals above $26^{\circ}C$ indicates that the carbonyls bonded to cobalt atoms are mobile, but the $Fe(CO)_3$ unit remains rigid. It should be noted that the model of fluxional behaviour proposed for C_{3v} $M_4(CO)_{12}$ species (7) does not apply in this case, because despite overall C_{3v} symmetry there is a 13-ligand envelope which cannot undergo the distortions available to the 12-vertex icosahedron.

Phosphine ligands substitute $HFeCo_3(CO)_{12}$ stepwise on different cobalt atoms in axial positions (65). ^{13}C n.m.r. intensities for the substituted derivatives are not anomalous, which offers an interesting parallel to the $Co_4(CO)_2$ system; exchange of cobalt-bonded carbonyls again appears to occur. In contrast to the $Os_3(CO)_{12}$ derivatives (25), phosphine substituents increase the energy barrier for fluxionality in $HFeCo_3(CO)_{12}$; this may be because phosphines stabilise a (CO)-bridged structure (Section 3.1), which in this case is the ground state of the molecule.

$[Rh_4(CO)_{11}]^{2-}$ is totally fluxional at $-70^{\circ}C$; its ^{13}C n.m.r. shows a binomial quintet (77). In the solid state this anion has four terminal and seven edge-bridging carbonyls (78). In $[Rh_4(CO)_{11}(COOMe)]^-$, the crystal structure of which is unknown, all carbonyls are equivalent at $-70^{\circ}C$. The ^{13}C n.m.r. quintet shows that they are mobile over all four rhodium atoms (77). The $^{13}COOMe$ signal appears as a doublet, indicating that this ligand remains bonded to a single rhodium atom. This molecule may belong in the $M_4(CO)_{11}L$ class (Section 4.3); if the ligand envelope is icosahedral, no antipodal relationship to the COOMe ligand is maintained.

4.6. Higher Nuclearity Clusters

Carbonyl fluxionality in larger transition metal cluster molecules is possibly of greater relevance to mobility on metal surfaces. Recently, a good deal of information on fluxionality in such clusters has been obtained. However, the great difficulty of planned synthesis hinders work in this field, and an overall pattern is slow to emerge.

In general, carbonyl mobility in higher nuclearity clusters is found to be restricted. Some delocalised exchange of terminal and bridging carbonyls has been observed, but more usually, $M(CO)_3$ exchange localised on individual metal atoms takes place. Several large clusters are stereochemically rigid at the highest experimentally attainable temperatures.

4.6.1. Os_5 Clusters

A series of Os_5 clusters has been studied. $[Os_5(CO)_{15}]^{2-}$, the crystal structure of which is unknown, displays two ^{13}C n.m.r. signals of relative intensity 3:2 from $-119^{\circ}C$ to $0^{\circ}C$ (79). Electron counting and i.r. indicate a trigonal bipyramidal cluster structure and if there are three terminal carbonyls per osmium atom, the equatorial $Os(CO)_3$ units must be rotating to give the observed spectrum.

In the crystal structure of $[HOs_5(CO)_{15}]^-$ (80), the hydride bridges an equatorial edge of the trigonal bipyramid of osmium atoms; there are three terminal carbonyls on each osmium (Fig. 38). At $-108^{\circ}C$ the ^{13}C n.m.r. (79) indicates that $Os(CO)_3$ rotation occurs on the unique equatorial osmium atom; at $-50^{\circ}C$ two more $Os(CO)_3$ groups rotate, shown by $^1H-^{13}C$ coupling to be the other two equatorial groups. At $+55^{\circ}C$ a 3:2 ^{13}C n.m.r. spectrum shows that the hydride ligand is migrating round the cluster; the axial $Os(CO)_3$ groups

524.

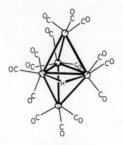

Figure 38

$[HOs_5(CO)_{15}]^{\ominus}$

may or may not be rotating.

The crystal structure of the neutral $H_2Os_5(CO)_{15}$ is unknown. Infra-red spectra show it to have no bridging carbonyls. The ^1H n.m.r. shows a sharp singlet down to $-70^\circ C$; the hydrides are either equivalent by symmetry, or exchanging at this temperature (79).

4.6.2. Ru$_6$ and Os$_6$ Clusters

In $Os_6(CO)_{18}$ (Fig. 6) the Os_6 cluster is a C_{2v} bicapped tetrahedron (81). There are three different types of osmium atom, and each has three terminal carbonyls. The ^{13}C n.m.r. spectrum shows that $Os(CO)_3$ rotation occurs with slightly different activation energy at each type of Os centre (10). Even at $100^\circ C$ there is no delocalised carbonyl scrambling, or skeletal rearrangement of the Os_6 unit.

On reduction to $[Os_6(CO)_{18}]^{2-}$, the metal cluster rearranges to form an octahedron (82). Because of the overall cluster symmetry, the three terminal carbonyls

on each osmium atom are not all equivalent in the ground
state structure. Observation of a single ^{13}CO signal
at -113°C shows that the carbonyls are exchanging; this
could be either a localised or delocalised scrambling.

Protonation yields $[HOs_6(CO)_{18}]^-$, the crystal
structure of which has the hydrogen atom located on one
face of the Os_6 octahedron, to give C_{3v} symmetry (82)
(Fig. 39). In solution at -112°C, ^{13}C n.m.r. shows two
singlets of equal intensity (83); this is consistent
with the solid state structure if $Os(CO)_3$ rotation

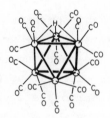

Figure 39

$[HOs_6(CO)_{18}]^{\ominus}$

takes place at all six osmium atoms. At O^{O}C, all
carbonyls are equivalent; either delocalised carbonyl
scrambling takes place, or alternatively the hydride
moves over all faces of the metal cluster to average
the carbonyl environments.

The neutral molecule $H_2Os_6(CO)_{18}$ has an unexpected
structure based on a capped square pyramid of osmium
atoms (82) (Fig. 40). The hydrides have not been
unambiguously located crystallographically but
consideration of (CO) positions indicates that they
bridge two edges of the square face of the pyramid (84).
At -85°C the ^{13}C n.m.r. (83) is consistent with this
structure, in which the two hydrides lie on a plane of

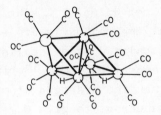

Figure 40

$H_2Os_6(CO)_{18}$

symmetry in dissimilar environments. At 100°C three
^{13}CO resonances of equal intensity are observed and
the interpretation of this spectrum remains unclear.
It was proposed (83) that as well as $Os(CO)_3$ rotation,
delocalised exchange of carbonyls occurs between the
apical and capping Os atoms.

The corresponding ruthenium compounds behave
differently. $[Ru_6(CO)_{18}]^{2-}$ (Fig. 41) possesses both
edge- and face-bridging carbonyls; ^{13}C n.m.r. shows
that all carbonyls are equivalent at -100°C (85). In

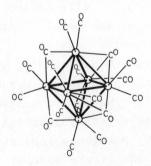

Figure 41

$[Ru_6(CO)_{18}]^{2-}$

$[HRu_6(CO)_{18}]^-$ (Fig. 42) all carbonyls are equivalent at $-104^{\circ}C$, and no $^1H-^{13}C$ coupling is observed (86).

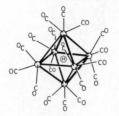

Figure 42

$[HRu_6(CO)_{18}]^{\ominus}$

Together with crystallographic work (86), this was taken as evidence that the hydride occupies an interstitial position within the cluster, a controversial proposal which has now been confirmed by a neutron diffraction study (32). $H_2Ru_6(CO)_{18}$, in contrast to its osmium analogue, has an octahedral metal atom framework (87); it has not been studied by n.m.r.

4.6.3. Larger Clusters of Os and Pt

$Os_7(CO)_{21}$ is the only larger osmium cluster to have been studied by n.m.r. Its crystal structure is based on a capped octahedron, with three terminal carbonyls per osmium atom (88) (Fig. 43). ^{13}C n.m.r. shows a 1:3:3 carbonyl pattern (88), indicating that this C_{3v} structure persists in solution with $Os(CO)_3$ rotations occurring on at least the six Os atoms of the octahedron (but not necessarily the apical one).

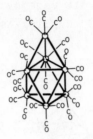

Figure 43

$Os_7(CO)_{21}$

At $-110^{\circ}C$ these $Os(CO)_3$ units begin to show evidence
of static behaviour. Even at $90^{\circ}C$ there is no
evidence of internuclear (CO)-scrambling, except
possibly within successive layers of the osmium atom
skeleton.

The platinum clusters $[Pt_n(CO)_{2n}]^{2-}$ have been
studied by ^{195}Pt n.m.r. for n = 6, 9 (16). $[Pt_6(CO)_{12}]^{2-}$
has a structure based on a trigonal prism of platinum
atoms (15); all Pt atoms are equivalent and accordingly
only one signal is seen. In $[Pt_9(CO)_{18}]^{2-}$ (Fig. 12)
the Pt_9 skeleton is made up of stacked trigonal prisms
(15); a complex ^{195}Pt spectrum arising from a statistical
distribution of platinum isotopes within the outer and
middle Pt_3 triangles is expected. Observation of a
much simpler spectrum at $-85^{\circ}C$ leads to the conclusion
that the outer Pt_3 triangles are rapidly rotating
about the pseudo-threefold axis with respect to the
central one (16). At $25^{\circ}C$ the species $[Pt_9(CO)_{18}]^{2-}$
and $[Pt_{12}(CO)_{24}]^{2-}$ were shown to exchange $Pt_3(CO)_6$
units in solution (16).

4.6.4. Rhodium Clusters

A large series of high nuclearity rhodium clusters has been investigated; the 100% natural abudance of ^{103}Rh ($I = \frac{1}{2}$) makes these ^{13}C n.m.r. studies particularly informative because of coupling to the rhodium nuclei.

$Rh_6(CO)_{16}$ (Fig. 44) is stereochemically rigid at $70^{\circ}C$; the twelve equivalent terminal carbonyls appear as a doublet in the ^{13}C n.m.r. spectrum, and the four

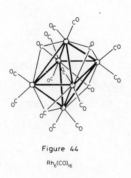

Figure 44
$Rh_6(CO)_{16}$

face-bridging carbonyls as a quartet (89,90). In view of the high symmetry of the ligand polyhedron (a tetra-capped truncated tetrahedron) in this molecule, its non-fluxionality is perhaps surprising as a concerted fluxional process might be expected by analogy with the $M_4(CO)_{12}$ system (Section 4.3). This rigidity may be connected with the stability of face-bridging carbonyls, or alternatively, the activation energy for carbonyl mobility may be high for electronic reasons.

530.

The iodo derivative $[Rh_6(CO)_{15}I]^-$ is also rigid to +30°C (91). Similarly, the ion $[Rh_{12}(CO)_{30}]^{2-}$, whose structure is based on two linked octahedra, is not fluxional at +20°C (92).

In contrast, the ion $[Rh_6(CO)_{15}]^{2-}$ (Fig. 45) is

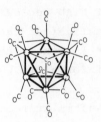

Figure 45
Probable Structure
of
$[Rh_6(CO)_{15}]^{2-}$

totally fluxional at -70°C, the ^{13}C n.m.r. showing a binomial septet (90). There is correlation between increasing negative charge on a cluster and decreasing activation energy for carbonyl scrambling (93). However, the carbido species $[Rh_6(CO)_{15}C]^{2-}$ (Fig. 46),

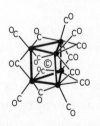

Figure 46
$[Rh_6C(CO)_{15}]^{2-}$
(bonds to interstitial carbide
omitted for clarity)

which has a trigonal prismatic metal atom skeleton
with an interstitial carbon atom (94), is stereochemically rigid at $25^{\circ}C$, despite its double negative
charge (23). This may be connected with the presence
of rectangular faces in the metal atom polyhedron
(see Section 3.1). It is clear that factors
influencing activation energies for carbonyl mobility
are not fully understood.

$[Rh_7(CO)_{16}]^{3-}$ has a C_{3v} crystal structure with a
capped octahedron of rhodium atoms (95) (Fig. 47).

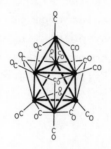

Figure 47
$[Rh_7(CO)_{16}]^{3-}$

There are three face-bridging, six edge-bridging and
seven terminal carbonyls. At $-70^{\circ}C$ the ^{13}C n.m.r. is
entirely consistent with this structure (90,96). At
$15^{\circ}C$ some site exchange occurs; three bridging and
three terminal carbonyls become equivalent, and
selective decoupling experiments show that there is a
cyclic exchange of six carbonyls about the basal Rh_3
triangle of the capped octahedron (96). The iodo
derivative $[Rh_7(CO)_{16}I]^-$ has a noteworthy bridging
iodide ligand (97). At $-31^{\circ}C$ this ion is totally
fluxional with the ^{13}C n.m.r. displaying an octet (93).

Rhodium carbonyl clusters may also be studied by
means of the magnetic ^{103}Rh nucleus. This cannot be

532.

done directly because of very low sensitivity, but may
be achieved by the method of INDOR (INternuclear
DOuble Resonance) (98). Monitoring of the frequency of
one nucleus while sweeping a second, low powered, radio
frequency irradiation through the frequency of a second
nucleus coupled to it yields a spectrum corresponding
to that of the second nucleus. Thus no receiver for
^{103}Rh frequencies is required in the spectrometer. To
work in Fourier Transform mode, the ^{103}Rh frequency is
swept in steps, and the series of spectra obtained are
analysed for changes in the signal of interest.
Different types of rhodium atom environment in high
nuclearity clusters may be distinguished by this method,
and skeletal rearrangements detected.

In $[Rh_9(CO)_{21}P]^{2-}$ (Fig. 48), which has in the solid

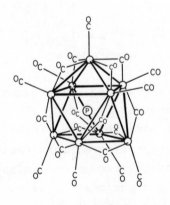

Figure 48
$[Rh_9(CO)_{21}P]^{2-}$

state a monocapped square antiprism of rhodium atoms
with an interstitial phosphorus, only one broad ^{13}CO

signal is seen from -40°C to $+45^{\circ}$C (99). ^{31}P n.m.r. at 40°C shows that all nine ^{103}Rh nuclei are coupled equivalently to the ^{31}P nucleus. The Rh_9 cluster is rearranging at this temperature; fluxional processes available to nine-vertex molecular polyhedra have been extensively studied (100). In contrast, $[Rh_5Pt(CO)_{15}]^-$ (Fig. 49) possesses a rigid octahedral metal atom

Figure 49

$[Rh_5Pt(CO)_{15}]^{\ominus}$

skeleton at 20°C, with two types of ^{103}Rh nucleus in a 1:4 ratio producing a doublet of quintets in the ^{195}Pt n.m.r. pattern (101).

The carbonyl hydrides $[H_2Rh_{13}(CO)_{24}]^{3-}$ and $[H_3Rh_{13}(CO)_{24}]^{2-}$ have been investigated (24). The metal skeleton in each case is an anticuboctahedron with the thirteenth rhodium atom interstitial; this structure represents a fragment of hexagonal close packing. There are twelve terminal and twelve edge-bridging carbonyls (34) (Fig. 50). The different ^{103}Rh nucleus environments in these clusters have been demonstrated by INDOR (24). ^{13}C n.m.r. shows that certain bridging and terminal carbonyls takes place in a concerted manner about alternate bridged and unbridged cluster edges; the lowest energy scrambling process is restricted by the need to regenerate a

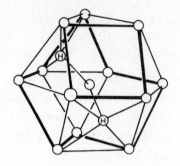

Figure 50

$$[H_2Rh_{13}(CO)_{24}]^{3-}$$

Thick Rh-Rh bonds have (CO)-bridges;

there is one terminal carbonyl on each

rhodium atom.

rearranged structure of the ground state form (24).

The ^{1}H n.m.r. of these ions reveals another type
of fluxional behaviour. The hydrides migrate about
the whole cluster, and couple to all thirteen rhodium
nuclei. The simultaneous carbonyl scrambling causes
the ^{103}Rh nuclei to be of three types, of relative
population 1:6:6, so that the ^{1}H spectrum is essentially
a doublet of septets of septets. Spin decoupling of
the outer twelve ^{103}Rh nuclei leads to a simple doublet.
It is therefore evident that the hydrides are bonded
to the central rhodium atom and migrate within the
cluster. In doing this they must occupy both tetrahedral and octahedral type interstices in the cluster.
Low temperature work has succeeded in freezing out the
hydride fluxionality in the dianion; however, in the
trianion the hydrides are still mobile at -100°C, unless
the hydrogen is replaced by deuterium (24). This
mobility of hydrogen atoms within a cluster strongly

resembles their mobility through metallic lattices (102).

To conclude this section it may be noted that in $[Rh_{17}(CO)_{32}S_2]^{3-}$, one of the largest clusters characterised, the carbonyls are rigid at $+40^{o}C$ (103).

5. REFERENCES

1. R. G. Woolley, this volume.
2. F. A. Cotton and J. M. Troup, J. Amer. Chem. Soc., 96, 4155 (1974).
3. B. F. G. Johnson, J. Chem. Soc. Chem. Comm., 703 (1976).
4. B. F. G. Johnson, J. Chem. Soc. Chem. Comm., 211 (1976).
5. M. Elian and R. Hoffmann, Inorg. Chem., 14, 1058 (1975).
6. F. A. Cotton and D. L. Hunter, Inorg. Chim. Acta, 11, L9 (1974).
7. R. E. Benfield and B. F. G. Johnson, J. Chem. Soc. Dalton, 1554 (1978).
8. See, e.g., E. L. Muetterties, Rec. Chem. Progr., 31, 51 (1970).
9. a) J. Evans, Adv. Organomet. Chem., 16, 319 (1977).
   b) E. Band and E. L. Muetterties, Chem. Rev., 78, 639 (1978).
10. C. R. Eady, W. G. Jackson, B. F. G. Johnson, J. Lewis and T. W. Matheson, J. Chem. Soc. Chem. Comm., 958 (1975).
11. A. Forster, B. F. G. Johnson, J. Lewis, T. W. Matheson, B. H. Robinson and

W. G. Jackson, J. Chem. Soc. Chem. Comm.,
1042 (1974).

12. B. F. G. Johnson, J. Lewis and T. W. Matheson,
J. Chem. Soc. Chem. Comm., 441 (1974).

13. F. A. Cotton, L. Kruczynski, B. L. Shapiro and
L. F. Johnson, J. Amer. Chem. Soc., 94,
6191 (1972).

14. M. R. Churchill and J. P. Hutchinson, Inorg. Chem.,
17, 3528 (1978).

15. J. C. Calabrese, L. F. Dahl, P. Chini, G. Longoni
and S. Martinengo, J. Amer. Chem. Soc., 96,
2614 (1974).

16. C. Brown, B. T. Heaton, P. Chini, A. Fumagalli and
G. Longoni, J. Chem. Soc. Chem. Comm., 309
(1977).

17. J. W. Koepke, J. R. Johnson, S. A. R. Knox and
H. D. Kaesz, J. Amer. Chem. Soc., 97, 3947
(1975).

18. P. F. Jackson, B. F. G. Jackson, J. Lewis,
M. McPartlin and W. J. H. Nelson, J. Chem.
Soc. Chem. Comm., 920 (1978).

19. G. L. Geoffroy and W. L. Gladfelter, J. Amer. Chem.
Soc., 99, 6775 (1977).

20. P. Chini, personal communication.

21. M. Manassero, M. Sansoni and G. Longoni, J. Chem.
Soc., 919 (1976).

22. C. J. Commons and B. F. Hoskins, Aust. J. Chem.,
28, 1663 (1975).

23. V. G. Albano, P. Chini, S. Martinengo,
D. J. A. McCaffrey, D. Strumolo and
B. T. Heaton, J. Amer. Chem. Soc., 96, 8106
(1974).

24. S. Martinengo, B. T. Heaton, R. J. Goodfellow
and P. Chini, J. Chem. Soc. Chem. Comm., 39
(1977).

538.

25. B. F. G. Johnson, J. Lewis, B. E. Reichert and
    K. T. Schorpp, J. Chem. Soc. Dalton, 1403
    (1976).

26. V. Albano, P. Chini and V. Scatturin, J. Chem. Soc.
    163 (1968).

27. V. Albano, P. Chini and V. Scatturin, J. Organomet.
    Chem., 15, 423 (1968).

28. V. Albano, P. L. Bellon, P. Chini and V. Scatturin,
    J. Organomet. Chem., 16, 461 (1969).

29. J. R. Shapley, G. F. Stuntz, M. R. Churchill and
    J. P. Hutchinson, J. Chem. Soc. Chem. Comm.,
    219 (1979).

30. G. F. Stuntz and J. R. Shapley, J. Amer. Chem.
    Soc., 99, 607 (1977).

31. A. G. Orpen, A. V. Rivera, E. G. Bryan, D. Pippard,
    G. M. Sheldrick and K. R. Rouse, J. Chem.
    Soc. Chem. Comm., 723 (1978).

32. P. F. Jackson, B. F. G. Johnson, J. Lewis,
    P. R. Raithby, S. A. Mason, W. J. H. Nelson,
    M. McPartlin and K. D. Rouse, unpublished
    results.

33. D. W. Hart, R. G. Tetler, C. -Y. Wei, R. Bau,
    G. Longoni, S. Campanella, P. Chini and
    T. F. Koetzle, Angew. Chem. Intl., 18, 80
    (1979).

34. V. G. Albano, G. Ciani, S. Martinengo and
    A. Sironi, J. Chem. Soc. Dalton, 978 (1979).

35. J. Knight and M. J. Mays, J. Chem. Soc. Chem. Comm.,
    1006 (1970).

36. M. R. Churchill, F. J. Hollander and J. P. Hutchinson,
    Inorg. Chem., 16, 2655 (1977).

37. M. R. Churchill and B. G. DeBoer, Inorg. Chem., 16,
    878 (1977).

38. C. E. Housecroft, K. Wade and B. C. Smith, J. Chem.
    Soc. Chem. Comm., 765 (1978).

39. B. F. G. Johnson, J. Lewis, M. J. Mays,
P. D. Gavens, R. E. Benfield, L. Milone
and S. Aime, submitted to J. Chem. Soc.
Dalton.

40. S. Aime, O. Gambino, L. Milone, E. Sappa and
E. Rosenberg, Inorg. Chim. Acta, 15, 53 (1975).

41. F. A. Cotton and J. D. Jamerson, J. Amer. Chem.
Soc., 98, 5396 (1976).

42. R. J. Lawson and J. R. Shapley, J. Amer. Chem.
Soc., 98, 7433 (1976).

43. F. A. Cotton and B. E. Hanson, Inorg. Chem., 16,
3369 (1977).

44. M. Tachikawa, Ph. D. Thesis, University of
Illinois, (1977).

45. J. R. Norton, J. P. Collman, G. Dolcetti and
W. T. Robinson, Inorg. Chem., 11, 382 (1972).

46. E. G. Bryan, A. Forster, B. F. G. Johnson,
J. Lewis and T. W. Matheson, J. Chem. Soc.
Dalton, 196 (1978).

47. P. D. Gavens and M. J. Mays, J. Organomet. Chem.,
162, 389 (1976).

48. E. G. Bryan, B. F. G. Johnson and J. Lewis,
J. Chem. Soc. Dalton, 144 (1977).

49. M. Tachikawa, S. I. Richter and J. R. Shapley,
J. Organomet. Chem., 128, C9 (1977).

50. F. A. Cotton and B. E. Hanson, Inorg. Chem., 16,
2820 (1977).

51. F. A. Cotton, B. E. Hanson and J. D. Jamerson,
J. Amer. Chem. Soc., 99, 6588 (1977).

52. J. R. Wilkinson and L. J. Todd, J. Organomet. Chem.,
118, 199 (1976).

53. B. F. G. Johnson, J. Lewis, P. R. Raithby and
G. Suss, J. Chem. Soc. Dalton, in the press.

54. C. R. Eady, B. F. G. Johnson and J. Lewis, J. Chem.
Soc. Dalton, 1358 (1978).

540.

55. S. Aime and L. Milone, Prog. N.M.R. Spect., 11,
    183 (1977).

56. J. R. Shapley, J. B. Keister, M. R. Churchill and
    B. G. DeBoer, J. Amer. Chem. Soc., 97, 4145
    (1975).

57. R. B. Calvert and J. R. Shapley, J. Amer. Chem.
    Soc., 99, 5225 (1977).

58. R. B. Calvert and J. R. Shapley, J. Amer. Chem.
    Soc., 100, 7726 (1978).

59. F. H. Carre, F. A. Cotton and B. A. Frenz, Inorg.
    Chem., 15, 380 (1976).

60. J. Evans, B. F. G. Johnson, J. Lewis and
    T. W. Matheson, J. Amer. Chem. Soc., 97, 1245
    (1975).

61. J. W. Cable and R. K. Sheline, Chem. Rev., 56, 1
    (1956).

62. F. A. Cotton and R. R. Monchamp, J. Chem. Soc.,
    1882 (1960).

63. D. L. Smith, J. Chem. Phys., 42, 1460 (1965).

64. M. A. Cohen, D. R. Kidd and T. L. Brown, J. Amer.
    Chem. Soc., 97, 4408 (1975).

65. S. Aime, L. Milone, D. Osella and A. Poli, Inorg.
    Chim. Acta, 30, 45 (1978).

66. G. Bor, G. Sbrignadello and K. Noack, Helv. Chim.
    Acta, 58, 815 (1975).

67. J. Evans, B. F. G. Johnson, J. Lewis and
    J. R. Norton, J. Chem. Soc. Chem. Comm., 807
    (1973).

68. C. H. Wei, Inorg. Chem., 11, 2384 (1969).

69. B. F. G. Johnson, J. Lewis, K. Wong, P. R. Raithby
    and C. Zuccaro, unpublished results.

70. R. D. Wilson, S. M. Wu, R. A. Love and R. Bau,
    Inorg. Chem., 17, 1271 (1978).

71. S. A. R. Knox, J. W. Koepke, M. A. Andrews and
    H. D. Kaesz, J. Amer. Chem. Soc., 97, 3942 (1975).

72. S. A. R. Knox and H. D. Kaesz, J. Amer. Chem.
       Soc., 93, 4594 (1971).

73. J. R. Shapley, S. I. Richter, M. R. Churchill
       and R. A. Lashewycz, J. Amer. Chem. Soc.,
       99, 7384 (1977).

74. K. E. Inkrott and S. G. Shore, J. Amer. Chem.
       Soc., 100, 3954 (1978).

75. S. Aime, L. Milone, D. Osella and E. Sappa,
       Inorg. Chim. Acta, 29, L211 (1978).

76. J. W. May, Adv. Catalysis, 21, 151 (1970).

77. P. Chini and B. T. Heaton, Top. Curr. Chem., 71,
       1 (1977).

78. V. G. Albano, G. Ciani, A. Fumagalli, S. Martinengo
       and W. M. Anker, J. Organomet. Chem., 116,
       343 (1976).

79. C. R. Eady, B. F. G. Johnson, J. Lewis,
       M. C. Malatesta and G. M. Sheldrick, J. Chem.
       Soc. Chem. Comm., 807 (1976).

80. J. J. Guy and G. M. Sheldrick, Acta Cryst., 34B,
       1722 (1978).

81. R. Mason, K. M. Thomas and D. M. P. Mingos,
       J. Amer. Chem. Soc., 95, 3802 (1973).

82. M. McPartlin, C. R. Eady, B. F. G. Johnson and
       J. Lewis, J. Chem. Soc. Chem. Comm., 883
       (1976).

83. C. R. Eady, B. F. G. Johnson and J. Lewis,
       J. Chem. Soc. Chem. Comm., 302 (1976).

84. A. G. Orpen, J. Organomet. Chem., 159, C1 (1978).

85. M. McPartlin, W. J. H. Nelson, P. F. Jackson,
       B. F. G. Johnson and J. Lewis, unpublished
       results.

86. C. R. Eady, B. F. G. Johnson, J. Lewis,
       M. C. Malatesta, P. Machin and M. McPartlin,
       J. Chem. Soc. Chem. Comm., 945 (1976.

542.

87. M. R. Churchill and J. Wormald, J. Amer. Chem.
        Soc., 93, 5670 (1971).
88. C. R. Eady, B. F. G. Johnson, J. Lewis, R. Mason,
        P. B. Hitchcock and K. M. Thomas, J. Chem.
        Soc. Chem. Comm., 385 (1977).
89. E. R. Corey, L. F. Dahl and W. Beck, J. Amer.
        Chem. Soc., 85, 1202 (1963).
90. B. T. Heaton, A. D. C. Towl, P. Chini, A. Fumagalli,
        D. J. McCaffrey and S. Martinengo, J. Chem.
        Soc. Chem. Comm., 523 (1975).
91. B. T. Heaton, P. Chini, S. Martinengo,
        D. J. McCaffrey and A. Fumagalli,
        unpublished results.
92. P. Chini, S. Martinengo, D. J. McCaffrey and
        B. T. Heaton, J. Chem. Soc. Chem. Comm.,
        310 (1974).
93. P. Chini, G. Longini and V. G. Albano, Adv.
        Organomet. Chem., 14, 286 (1976).
94. V. G. Albano, M. Sansoni, P. Chini and
        S. Martinengo, J. Chem. Soc. Dalton, 651
        (1973).
95. V. G. Albano, P. L. Bellon and G. Ciani, J. Chem.
        Soc. Chem. Comm., 1024 (1974).
96. C. Brown, B. T. Heaton, L. Longhetti, D. O. Smith,
        P. Chini and S. Martinengo, J. Organomet.
        Chem., 169, 309 (1979).
97. V. G. Albano, G. Ciani, S. Martinengo, P. Chini
        and G. Giodano, J. Organomet. Chem., 88,
        381 (1975).
98. D. Shaw, Fourier Transform N.M.R. Spectroscopy,
        Elsevier, Amsterdam, p. 279 (1976).
99. J. L. Vidal, W. E. Walker, R. L. Pruett and
        R. C. Schoening, Inorg. Chem., 18, 129 (1979).
100. L. J. Guggenberger and E. L. Muetterties, J. Amer.
        Chem. Soc., 98, 7221 (1976).

101.   A. Fumagalli, S. Martinengo, P. Chini,
            A. Albanati, S. Bruckner and B. T. Heaton,
            J. Chem. Soc. Chem. Comm., 195 (1978).
102.   F. A. Lewis, The Palladium-hydrogen System,
            Academic Press, London, Chapter 7, (1967).
103.   J. L. Vidal, R. A. Fiato, L. A. Cosby and
            R. L. Pruett, Inorg. Chem., 17, 2574 (1978).

Transition Metal Clusters
Edited by Brian F.G. Johnson
© 1980 John Wiley & Sons Ltd.

CHAPTER 8

Metal Clusters in Catalysis

R. Whyman

ICI Corporate Laboratory,
Runcorn, Cheshire, U.K.

1. INTRODUCTION

Earlier chapters in this book, and recent reviews
(1,2) serve as ample illustrations of the vast amount
of work which has been carried out on synthetic and
structural aspects of metal clusters during the last
5-10 years. One of the principal justifications for
much of this work is the expectation that metal
clusters will find uses in catalysis, directly in the
form of novel homogeneous catalysts (or catalyst
precursors) and indirectly by serving as models for
chemisorption processes in heterogeneous catalysis
(3,4). These and related aspects such as supported
metal crystallites and bimetallic clusters have been
reviewed (5-10). Catalytic applications are not of
course the only potential uses and other areas such as
photographic processes (11) and the manufacture of
thin films for electrical devices (12) may also
benefit from research on metal clusters.

The aim of this chapter is to determine whether
the above expectations are realistic by reviewing the
current state of the art with respect to both
homogeneous and heterogeneous catalysis by metal
clusters. An attempt will be made to identify those
areas where metal clusters might find applications,

with emphasis on problem areas such as the present
inadequacies of physical techniques. It is hoped that
this chapter will stimulate a shift of emphasis
towards the investigation of catalytic aspects rather
than just the preparation and structural characterisation of an ever-widening series of cluster compounds.

At the outset it should be made clear that, to
the author's knowledge, there is no really unequivocal
example of catalysis by a cluster compound - as will
become evident the problems in this area are severe
and definitive proof is extremely difficult to obtain.
At present one can at best make use of circumstantial
evidence.

2. CATALYSIS

In approaching a discussion of the potential applications of metal clusters in catalysis it is appropriate to initially consider some generalisations on the classical branches of catalysis, as applied to transition metals. Some important and contrasting features of homogeneous and heterogeneous catalysis are summarised in Table 1.

TABLE 1

Homogeneous vs. Heterogeneous Catalysis

	Homogeneous	Heterogeneous
Form	Soluble, usually mononuclear, metal complexes	Metals, usually supported, or metal oxides
Phase	Liquid	Gas/solid
Temperature	Low	High
Activity	Low	High
Selectivity	High	Low

Thus heterogeneous catalysts comprise metals or metal oxides and the metals are usually dispersed on inorganic oxide supports in order to both increase the effective surface area per gram of metal component and to dilute the catalyst for better temperature control. Heterogeneously catalysed reactions are usually operated at relatively high temperatures ($250-550^{\circ}C$) by passing reactants in the vapour phase over solid catalysts. In contrast, homogeneously catalysed reactions are carried out in the liquid phase at considerably lower temperatures ($<200^{\circ}C$) in the presence of soluble metal complexes. These reactions are frequently run under pressure in order to increase the effective concentration of gaseous reactant(s) in solution and hence usually the reaction rate. An inverse relationship between catalytic activity and selectivity usually applies to both forms of catalysis.

In terms of reaction types and mechanistic studies, some generalisations relevant to homogeneous and heterogeneous catalysis are summarised in Table 2.

Many reactions are common to both forms of catalysis but the energetically more demanding transformations, such as the activation of C-C and C-H bonds in saturated hydrocarbons and the hydrogenation of the triple bonds in carbon monoxide and nitrogen remain relatively unknown in homogeneous catalysis. However, homogeneous catalysts of very low activity have been described for some of these transformations (13,14) as of course have stoichiometric reactions involving the reduction of carbon monoxide to methanol (15) and nitrogen to hydrazine and ammonia (16).

The depth of understanding of homogeneously-catalysed reactions is much more advanced than with heterogeneous counterparts principally because the former are more selective and therefore kinetically simpler, and more amenable to direct study by

TABLE 2

Reactions and Mechanisms in Catalysis

	Homogeneous	Heterogeneous
Reactions	Hydrogenation, isomerisation, carbonylation, dimerisation, oxidation, dismutation and polymerisation of unsaturated hydrocarbons.	As for homogeneous, plus i) activation of C-H and C-C bonds in saturated hydrocarbons. ii) hydrogenation of CO to methanol, methane and Fischer-Tropsch products. iii) hydrogenation of N_2.
Mechanisms	Reasonably well understood at molecular level.	Very few properly understood. Obvious difficulties in assessing intimate mechanisms of adsorption and reaction on a surface, especially under catalytic reaction conditions.

physical techniques such as infrared and n.m.r.
spectroscopy even under actual reaction conditions.
Mechanistic studies in homogeneous catalysis have made
significant contributions to heterogeneous catalysis
by defining probable surface intermediates, many of
which were previously a matter of conjecture. For
example, the great diversity in the bonding patterns
and rearrangements of hydrocarbon fragments in cluster
compounds, discussed in preceding chapters, will
clearly provide further model configurations for
hydrocarbons adsorbed on polymetallic surfaces.

Industrially, heterogeneous catalysts are
preferred, principally on the grounds of considerably
higher activity, but also in terms of chemical
engineering requirements such as the ease of separation
of products from catalyst and reactants, and
robustness, i.e. the ability of the catalyst to
withstand vigorous working conditions.

Parshall (17) has recently reviewed the industrial
applications of homogeneous catalysis. So far as
catalysis by transition metals is concerned the only
homogeneous processes currently of major significance
are the hydroformylation of olefins to aldehydes and
alcohols (18), the carbonylation of methanol to acetic
acid (19), and the hydrocyanation of butadiene to
adiponitrile (20). In oxidation processes such as
the oxidation of p-xylene to terephthalic acid the
metal ions may function simply as initiators for the
reaction rather than playing an active role in
catalysis (17). Finally, the application of
asymmetric hydrogenation in the synthesis of L-dopa
(21) provides an elegant demonstration of selectivity
in homogeneous catalysis.

3. CLUSTERS IN CATALYSTS

Having introduced some generalisations on the subject of catalysis we should now consider which position(s) metal clusters are likely to occupy in this scenario.

Bearing in mind the apparent inverse relationship between activity and selectivity in many examples of both homogeneous and heterogeneous catalysis, the industrial catalyst designer would clearly prefer to produce catalysts which have high intrinsic activity but which also may be tailored to produce specific products with high selectivity. All this has to be achieved with minimum capital cost requirements and the expenditure of minimum amounts of energy!

The most obvious potential of metal clusters lies in using the combinations of metal atoms to bridge the gap between conventional homogeneous and hetero-geneous catalysis and hopefully combining the high selectivity achieved in the homogeneous systems with the high activity associated with heterogeneous catalysts. In order to investigate these possibilities two approaches may be considered, starting from either side of the borderline. First the use of cluster compounds in the homogeneous phase either to promote

transformations which are unknown with mononuclear
catalysts or to improve the activity of known mononuclear systems. The alternative approach is to use
cluster compounds as a source of heterogeneous
catalysts by depositing them on supports either by
physical adsorption or chemical reaction. In this
way "heterogenised" homogeous catalysts would be
produced. Extending this process one stage further,
to removal of the ligands from the supported complexes,
would in principle provide a source of highly dispersed
metal particles (<15 Å crystallite size) whose
properties may differ from conventionally prepared
supported metal catalysts. The latter generally
consist of larger metal particles (>20 Å) or bulk
metal. Let us now consider these two approaches as
they are reflected in the literature.

4. HOMOGENEOUS CATALYSTS

In recent years the literature has contained many examples of homogeneously catalysed reactions using metal cluster compounds as catalyst precursors. These have spanned classes of reaction as diverse as hydrogenation, isomerisation, hydroformylation, water-gas shift, cyclisation and oxidation. In addition to the hydrogenation of simple unsaturated carbon-carbon bonds, those less susceptible to reduction e.g. $C\equiv O$, $-C\equiv N$ and $-NO_2$, have also been hydrogenated. Examples of these reactions are summarised in Tables 3-6.

It should be emphasised that although the metal cluster compound is present in solution at the start of the reaction there is in most cases no evidence to suggest that the integrity of the cluster is maintained throughout the reaction. Indeed in several cases e.g. the hydroformylation of olefins with $Rh_4(CO)_{12}$ and $Rh_6(CO)_{16}$ and the hydrogenation and isomerisation of olefins with $[Co(CO)_2PR_3]_3$, the reverse is specifically thought to apply, in that the cluster compound breaks down under reaction conditions to highly reactive mononuclear entities. After cooling and depressurising at the end of the reaction the starting cluster may be frequently recovered. This may turn

TABLE 3

Hydrogenation of $>C = C<$, $-C \equiv C-$ and $>C = O$ Groups in the Presence of Metal Cluster Compounds

Reaction	Cluster Precursor	Comments	Reference
Reduction of $>C = C<$ and $-C \equiv C-$			
$\diagup\!\!\!\diagdown$ + H$_2$ $\xrightarrow[66^o]{15 \text{ atm}}$ $\diagup\!\!\diagdown$ $\bigcirc$ + H$_2$ $\xrightarrow[66^o]{15 \text{ atm}}$ $\bigcirc$	$[Co(CO)_2 PBu^n_3]_3$	Cluster thought to break down under reaction conditions. Also an isomerisation catalyst	(22)
$RCH=CHCOR' + CO + H_2O \xrightarrow[130^o]{100 \text{ atm}} RCH_2CH_2COR' + CO_2$	$Rh_6(CO)_{16}$	R=Ph,R'=H,Me,Ph;COR'=CN H is derived from H_2O	(23)
$RC \equiv CR' + H_2 \xrightarrow[20^o]{3 \text{ atm}} \begin{smallmatrix} R \\ H \end{smallmatrix} C=C \begin{smallmatrix} R' \\ H \end{smallmatrix}$	$Ni_4(CNR)_7$	R = C(CH$_3$)$_3$, _cis_ olefins selectively produced, no alkane observed	(24)
$\bigcirc\!\!-CH=CH_2$ + H$_2$ $\xrightarrow[20^o]{2 \text{ atm}}$ $\bigcirc\!\!-CH_2CH_3$	$Co_2Rh_2(CO)_{12}$ $Co_3Rh(CO)_{12}$	P(OMe)$_3$ enhances the activity $Co_2Rh_2(CO)_{12}$ breaks down during reaction	(25)
$\begin{smallmatrix} Me \\ \end{smallmatrix}$ CO$_2$H ... + H$_2$ → $\begin{smallmatrix} Me \\ \end{smallmatrix}$ CO$_2$H ... CO$_2$H	$H_4Ru_4(CO)_8(DIOP)_2$	(-)(S)-methylsuccinic acid and γ-lactones produced from citraconic and mesaconic acid	(26)
Reduction of $>C = O$			
$RCHO + H_2 \xrightarrow[160^o]{80 \text{ atm}} RCH_2OH$	$Rh_6(CO)_{16}$	Active catalyst thought to be mononuclear HRh(CO)$_3$	(27)
$\xrightarrow[Et]{\text{CHO}}$ + H$_2$ $\xrightarrow[80^o]{150 \text{ atm}}$ $\xrightarrow[Et]{\text{CHO}}$ + $\xrightarrow[Et]{\text{CH}_2\text{OH}}$	$[Co(CO)_2 PBu^n_3]_3$		(22)
$\bigcirc\!\!=O$ + H$_2$ $\xrightarrow[100^o]{100 \text{ atm}}$ $\bigcirc\!\!-OH$	$H_4Ru_4(CO)_{12}$ or $H_4Ru_4(CO)_{12-x}L_x$	Cluster recovered unchanged, overall activity decreased in the presence of phosphines L = PPh$_3$, PBu$^n_3$, $\frac{1}{2}$(-)DIOP	(28)

TABLE 4

Hydrogenation of $-C\equiv O$, $-N\equiv C$, $-C\equiv N$ and $-NO_2$ Groups in the Presence of Metal Cluster Compounds

Reaction	Cluster Precursor	Comments	Reference
Reduction of $-C\equiv O$			
$2CO + 3H_2 \xrightarrow[200-240^O]{1000-3000 \text{ atm}} \begin{array}{c} CH_2OH \\ \vert \\ CH_2OH \end{array}$	$M_2\left[Rh_{12}(CO)_{30}\right]$	Many rhodium cluster carbonyl anions effective, catalysis by clusters likely	(29)
$CO + H_2 \xrightarrow[140^O]{2 \text{ atm}} CH_4$	$Os_3(CO)_{12},$ $Ir_4(CO)_{12}$	No alcohols or alkenes produced 1% conversion in 3 days, rate enhanced by $P(OMe)_3$.	(30)
$CO + H_2 \xrightarrow[180^O]{1-2 \text{ atm}} CH_4 + C_2H_6$	$Ir_4(CO)_{12}/$ $NaCl.2AlCl_3$	C_2/C_1 initially $\sim 7/1$ Ir metal inactive	(31)
$CO + H_2 \xrightarrow[300^O]{100 \text{ atm}} C_nH_{2n+2}$ $(n=1-30)$	$Ru_3(CO)_{12},$ $Os_3(CO)_{12},$ $Ir_4(CO)_{12}$	Activity order Ru>Os>>Ir, catalysts deactivate during reaction	(32)
Reaction of $-N\equiv C$ and $-C\equiv N$			
$RNC + H_2 \xrightarrow[90^O]{1-3 \text{ atm}} RNHCH_3 + RNH_2$ (trace)	$Ni_4(CNR)_7/$ $Ni(CNR)_4$ (1:10)	$R = C(CH_3)_3$. Turnover rate very low (~ 0.1/hr).	(33)
$CH_3CN + H_2 \xrightarrow{900} CH_3CH_2NH_2 + RNHCH_3$	$Ni_4(CNR)_7$	Very low reaction rate	(33)
Reduction of $-NO_2$			
$PhNO_2 + 2CO + H_2 \xrightarrow[160^O]{200 \text{ atm}} PhNH_2 + 2CO_2$	$Ru_3(CO)_{12}$	$Ru(CO)_5$, $Ru(acac)_3$ also effective, diphenylurea also formed at $CO/H_2 > 1$	(34)
$PhNO_2 + 3CO + H_2O \xrightarrow[50^O \text{ NMP}]{50 \text{ atm}} PhNH_2 + 3CO_2$	$Rh_6(CO)_{16}$	NMP = N-methylpyrrolidine. Strongly basic solvents preferred	(35)
$PhNO_2 + 3CO + H_2O \xrightarrow[125-180^O]{35 \text{ atm}} PhNH_2 + 3CO_2$ Et_3N	$Rh_6(CO)_{16},$ $Ir_4(CO)_{12},$ $Os_3(CO)_{12},$ $H_2Os_3(CO)_{10},$ $H_4Os_4(CO)_{12}$	Molecular hydrogen probably does not play an important role	(36)

TABLE 5

Hydroformylation and Water Gas Shift Reactions in the Presence of Metal Cluster Compounds

Reaction	Cluster Precursor	Comments	Reference
Hydroformylation			
(cyclohexene) + CO + H_2 $\xrightarrow[150^\circ]{400\ atm}$ (cyclohexyl–CHO)	$Rh_6(CO)_{16}$	Mononuclear $HRh(CO)_3$ thought to be active species	(37)
$RCH=CH_2 + CO + H_2 \xrightarrow[75^\circ]{100\ atm} RCH_2CH_2CHO$ $+$ $RCH_2\overset{\underset{CH_3}{\mid}}{C}HCHO$	$Rh_4(CO)_{12}$	R various, n/i $\sim$ 1, no olefin isomerisation observed	(38)
$RCH=CH_2 + CO + H_2 \xrightarrow[150^\circ]{100-150\ atm} RCH_2CH_2CHO$ $+$ RCH_2CH_3 $\overset{\mid}{CHO}$	$Ru_3(CO)_{12}$	R various, significant parallel hydrogenation of olefin to paraffin	(39)
(alkene) + CO+H_2 $\xrightarrow[110-130^\circ]{30-80atm}$ (branched–CHO) + (linear–CHO)	$Co_3(CO)_9CPh$ $Co_4(CO)_{10}(PPh)_2$	Catalysis by clusters likely	(40,41)
(alkene) + CO+H_2O $\xrightarrow[125^\circ]{30\ atm}$ (iso–CHO) + (n–CHO), Et_3N	$Rh_6(CO)_{16}$ $Ir_4(CO)_{12}$	Uses CO/H_2O rather than CO/H_2	(42)
(alkene) + CO+H_2O $\xrightarrow[150^\circ]{50\ atm}$ (n–CHO), KOH	$Ru_3(CO)_{12}$ $H_4Ru_4(CO)_{12}$ $Rh_6(CO)_{16}$	With Ru n/i ratio very high (32/1) Rh also hydrogenates aldehydes to alcohols	(43)
Water Gas Shift			
$CO + H_2O \xrightarrow[100^\circ]{1\ atm} CO_2 + H_2$	$Ru_3(CO)_{12}/KOH$ $H_2FeRu_3(CO)_{13}/$ KOH	Very low reaction rates; mixed metal system more active than either metal carbonyl alone	(44)
$CO + H_2O \xrightarrow[100-150^\circ]{25\ atm} CO_2 + H_2$	$Rh_6(CO)_{16}$, $Ru_3(CO)_{12}$, $H_4Ru_4(CO)_{12}$	Et_3N used as base	(42)
$CO + H_2O \xrightarrow[135-150^\circ]{50\ atm} CO_2 + H_2$	$Ru_3(CO)_{12}/KOH$ $Rh_6(CO)_{16}/KOH$	$[H_3Ru_4(CO)_{12}]^-$ detected as major component of Ru catalyst solution	(43)

TABLE 6

Isomerisation, Cyclisation and Oxidation in the Presence of Metal Cluster Compounds

Reaction	Cluster Precursor	Comments	Reference
Isomerisation			
$\xrightarrow[66^o]{60\ atm\ H_2}$	$[Co(CO)_2PBu^n_3]_3$	cis- and trans-pent-2-ene produced. c/t = 0.26, isomerisation rate much lower under N_2	(22)
$\xrightarrow[50^o]{N_2}$	$Fe_3(CO)_{12}$	c/t = 0.35. π-allylic intermediates involved in reaction mechanism	(45)
$\xrightarrow[33^o]{N_2}$	$H_2Os_3(CO)_{10}$		(46)
$\xrightarrow[110-120^o]{N_2}$	$Os_3(CO)_{12}$	c/t = 0.4 - 0.5	(47)
$\xrightarrow{N_2}$ 70°	$Ru_3(CO)_{12}$	c/t = 0.3 - 0.4	(48)
$\xrightarrow{70^o}$	$H_4Ru_4(CO)_{12}$	c/t = 0.9 - 1.0. Catalyst more active than $Ru_3(CO)_{12}$.	(49)
$CH_2=CHCH(CH_3)OH \xrightarrow{CO} CH_3CH_2CH_2COCH_3$	$Rh_6(CO)_{16}$		(23)
$CH_2=CHCH_2OH \xrightarrow[33^o]{N_2} CH_3CH_2CHO$	$H_2Os_3(CO)_{10}$		(46)
Cyclisation			
$2HC \equiv CH + 2CO + H_2 \xrightarrow[200^o]{\substack{120\ atm\ CO \\ 10\ atm\ H_2}}$	$Ru_3(CO)_{12}$	58% yield. Higher yields (65%) obtained with CO/H_2O instead of CO/H_2	(50)
$HC \equiv CH \xrightarrow[25^o]{1\ atm}$	$Ni_4(CNR)_7$	R = $C(CH_3)_3$. No cyclooctatetraene produced	(24,51)
$\xrightarrow[20^o]{1\ atm}$	$Ni_4(CNR)_7$	Small amount of 1,3-isomer but no vinylcyclohexene or linear octatriene produced.	(24,51)
Oxidation			
$CO + O_2 \xrightarrow[100^o]{34\ atm} CO_2$	$Rh_6(CO)_{16}$	No reaction in pure O_2 because the cluster decomposes. Mononuclear catalysts also effective	(52)
$=O+CO+O_2 \xrightarrow[100^o]{30\ atm} HO_2C(CH_2)_4CO_2H$			
$CO + O_2 \xrightarrow[25^o]{1\ atm} CO_2$	$Rh_6(CO)_{16}$	Mild conditions specific to DMF as solvent	(53)

out to be a general occurrence.

Some of the reactions listed in Tables 3-6 do not appear to be of special significance, e.g., simple hydrogenations, water-gas shift, isomerisation and oxidation, in so far that they are also known to proceed with mononuclear catalysts and the cluster systems do not appear to provide any superior activities or selectivities over their mononuclear counterparts. Of greater potential significance are the examples of "difficult" hydrogenations, e.g. those involving triple bonds other than acetylenes (54), and those reactions involving reductions with carbon monoxide and water rather than hydrogen.

As an aid in highlighting problem areas let us now select some of the more interesting examples from Tables 3-6 for further discussion.

The reactivity of the tetranuclear nickel cluster $Ni_4[CNC(CH_3)_3]_7$ is remarkable in terms of promoting specific cyclisation reactions (51), which are not observed with the nearest mononuclear analogue $Ni[CNC(CH_3)_3]_4$, in selectively hydrogenating acetylenes to cis-olefins (24) and in acting as a catalyst for the specific hydrogenation of isonitriles and acetonitrile under mild conditions (33). This is the first example of a homogeneous catalytic hydrogenation of an isonitrile although the turnover rate based on molecules of amine per molecule of cluster is very low ($\sim$0.1/hr.) at $90^{O}C$ and 1-3 atm. Acetonitrile is also hydrogenated at $90^{O}C$ in the presence of $Ni_4[CNC(CH_3)_3]_7$ to a 5:1 mixture of $C_2H_5NH_2$ and $(CH_3)_3CNHCH_3$, again at very low reaction rates.

It is perhaps relevant to note at this stage the work of Kaesz (55) in demonstrating the stoichiometric stepwise reduction pathway for acetonitrile over a tri-iron carbonyl cluster. The following sequence has been identified (with carbonyl groups omitted for clarity):

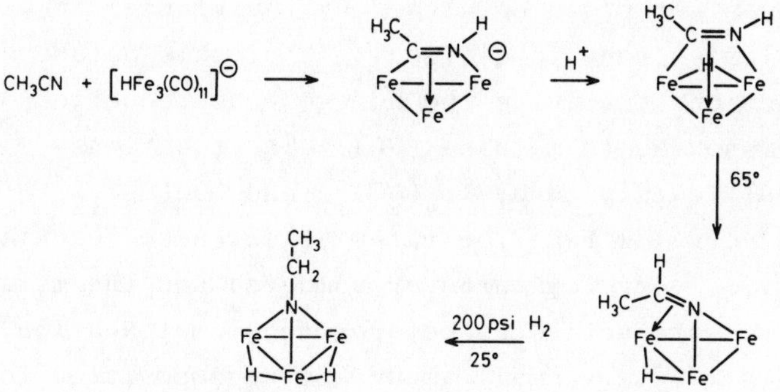

All the intermediate species have been isolated and characterised and this scheme represents a very clear example of the sequence of events which can occur on a cluster leading to the extensive reduction of the $C\equiv N$ bond of acetonitrile. This reaction sequence could be taken as evidence to support the requirement for polynuclear catalysis in the nickel system.

The apparently unique reactivity of the Ni_4L_7 cluster does not necessarily mean that catalysis by the Ni_4 entity is involved although the reaction sequence observed with the iron system would tend to suggest multinuclear catalysis.

This difficulty is a general one. First, the fact that a given product is not produced with known mononuclear catalysts, e.g., $Ni[CNC(CH_3)_3]_4$, is not necessarily evidence for catalysis by clusters, because the active catalyst may be a previously unknown mononuclear species generated under reaction conditions by fragmentation of the starting cluster. As indicated earlier this is believed to be the case in some examples. Secondly, the fact that polynuclear aggregates are the only identifiable species present again does not prove catalysis by the cluster itself. Particularly prone to difficulties of this nature are the reactions involving the homogeneous reduction of carbon monoxide to methane (30). The reactions supposedly catalysed by $Ir_4(CO)_{12}$ and $Os_3(CO)_{12}$ display very low reaction rates at extended reaction times (i.e. very low turnover numbers) and the metals themselves are active as heterogeneous methanation catalysts. A very small amount of decomposition to metallic iridium or osmium could account for the observed results.

Similar conclusions may be drawn from the Shell work (32) on the "homogeneously" catalysed formation of methane and higher (C_2-C_{30}) alkanes from CO/H_2 in the presence of $Ru_3(CO)_{12}$. The wide product distribution is more consistent with a heterogeneously catalysed process and the temperature of operation $(300^{O}C)$ must approach the limits of thermal stability of any organometallic ruthenium carbonyl species, thus leaving a question concerning catalysis of the reaction by trace amounts of metallic ruthenium rather than the supposed Ru_3 clusters. The $Ru_3(CO)_{12}$ cluster is certainly broken down to $Ru(CO)_5$ during the reaction and this has been associated with the observed catalyst deactivation.

As part of a general investigation of the reactivities of metal clusters, Muetterties and co-workers have evaluated $Ru_3(CO)_{12}$, $Os_3(CO)_{12}$ and $Ir_4(CO)_{12}$ as catalysts for a series of hydrocarbon reactions including skeletal rearrangement, dismutation, dehydrogenation, hydrogenation, isomerisation and H-D exchange between benzene and deuterium (56). None of these clusters catalyses any reaction of <u>saturated</u> hydrocarbons such as hexane or cyclohexane even in the presence of hydrogen. $Os_3(CO)_{12}$ is a catalyst for the slow H-D exchange between benzene and D_2 at 195° although no hydrogenation of benzene occurs under these conditions. $Ir_4(CO)_{12}$ is an effective catalyst for the disproportionation of 1,3-cyclohexadiene into cyclohexene and benzene at 160°. Cyclohexene also disproportionates into cyclohexane and benzene at a very slow rate in the presence of $Ir_4(CO)_{12}$. $Ru_3(CO)_{12}$ and $Os_3(CO)_{12}$ are catalysts for the isomerisation of hexenes at 70° and 140° respectively and both $Os_3(CO)_{12}$ and $Ir_4(CO)_{12}$ slowly catalyse skeletal reactions of hex-2-ene at 165° to produce small amounts of C_5 and C_7 olefins together with propane and heptane.

As Muetterties has pointed out the low activity observed with many of these systems is probably related to the fact that the majority of clusters studied are stable coordinately-saturated molecules which obey the 18-electron rule. The design and synthesis of coordinatively unsaturated clusters should lead to catalysts of higher intrinsic activity.

The reactions of various unsaturated molecules with carbon monoxide and water in the presence of basic solutions of metal cluster compounds are of potential significance. These reaction mixtures have been used to promote the hydrogenation of nitrobenzene (36) and α,β-unsaturated carbonyl compounds (23) and the hydroformylation of olefins (42,43). A plausible explanation for the course of such reactions includes the production of hydrogen <u>via</u> the water gas shift reaction followed by hydrogenation or hydroformylation. Similar systems are catalysts for the water gas shift reaction itself and, in the case of ruthenium, $[H_3Ru_4(CO)_{12}]^-$ is a major component of the active catalyst solution; it is conceivable that this anion may also be involved in the hydroformylation. Straight chain aldehydes are produced in very high selectivity (97%) with ruthenium whereas the analogous rhodium catalyst is much less selective (n:iso = 1:1) and reduces the initially formed aldehydes directly to alcohols. In the latter case $[HRh_6(CO)_{15}]^-$ is thought to be an active catalytic species.

Although the hydroformylation reaction may proceed via the intermediate generation of hydrogen through the water gas shift reaction this may not necessarily be the case with other reactions. For example, by using H_2 and D_2O in the $Fe(CO)_5$ catalysed reduction of nitrobenzene, Pettit (36) has shown that at least 80% of the reduction occurs with $CO + D_2O$ rather than H_2 acting as the reducing agent.

Of course synthesis gas (CO/H_2) is a relatively cheap raw material whereas pure carbon monoxide is expensive and it seems unlikely that processes based wholly on CO and H_2O will find wide commercial use.

Having discussed hydroformylation in the presence of CO/H_2O let us continue with an example of the more conventional hydroformylation using CO/H_2. Here, Pittman's work on catalysis of the reaction by metal clusters which are bonded by stable non-fluxional bridging groups in addition to metal-metal bonds is of significance (40,41). The cobalt clusters $Co_3(CO)_9CPh$ (I) and $Co_4(CO)_{10}(PPh)_2$ (II) have been found to hydroformylate both 1- and 2-pentenes to hexanals and 2-methylpentanal under relatively mild conditions (30-80 atm., 110-130°) but over long reaction times (20-100 hr.).

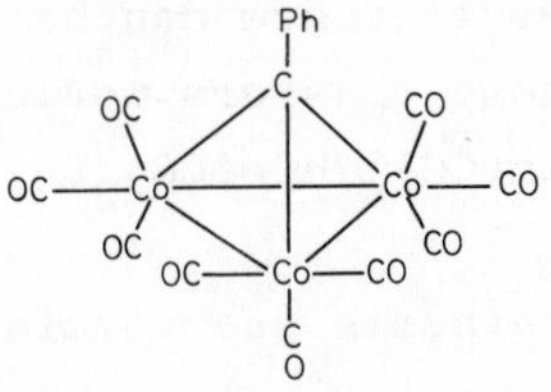

(I)

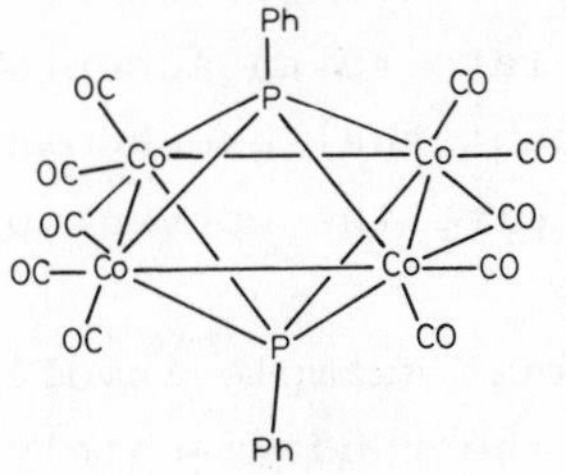

(II)

The cluster compounds are recovered unchanged in high yield at the end of the hydroformylation with no other detectable organometallic products. If fragmentation of these clusters to mononuclear catalysts occurs under reaction conditions it is difficult to visualise how they could re-form at the end of the reaction. The ratio of straight to branched chain products appears slightly lower than that observed with conventional mononuclear cobalt catalysts at corresponding temperatures although it is difficult to make direct comparisons because of different reaction pressures. The n/iso ratio decreases with increasing temperature and increases with increasing pressure in parallel with trends observed in the mononuclear system. The selectivity to the straight chain isomer may be increased by the addition of a phosphine, e.g. triphenylphosphine, to the system, from which $Co_4(CO)_8(PPh_3)_2(PPh)_2$ has been isolated after reaction.

The overall evidence appears consistent with this being a genuine case of catalysis by a cluster compound although there are no marked differences in selectivity terms over conventional mononuclear cobalt catalysts. However, if the hydroformylation were to be carried out under more forcing conditions, i.e. higher

pressures and temperatures, in order to increase the
reaction rate to an acceptable level, there can be
little doubt that breakdown of these clusters would
occur to give the conventional $HCo(CO)_4/HCo(CO)_3L$
catalysts.

A final example involving synthesis gas chemistry
concerns the rhodium-catalysed conversion of carbon
monoxide and hydrogen into ethylene glycol and
methanol (29).

$$CO + H_2 \quad \xrightarrow[\text{200--230}^{\text{o}}]{\text{1000--3000 atm}} \quad \begin{array}{c} CH_2OH \\ | \\ CH_2OH \end{array} + CH_3OH$$

Notwithstanding the very high pressure requirement and
the high concentration of rhodium this reaction is
claimed to be of direct commercial interest and is
notable for several reasons. It is one of the few
examples of the homogeneous hydrogenation of carbon
monoxide (although it should be noted that in 1953
Gresham patented a process for the production of
ethylene glycol, glycerol and their esters at 3000 atm
and 240° using cobalt acetate as a catalyst precursor
(57)). It is very selective, relative to the
heterogeneously-catalysed Fischer-Tropsch reaction,
and produces up to 75% ethylene glycol. Other
oxygenated products include propylene glycol and
glycerol at high pressures (3000 atm.) and, at lower
pressures (500-1000 atm.), methanol. The reaction
does not produce methane or other hydrocarbons. This
is a point of some significance because methane is
the preferred thermodynamic product from CO/H_2 reactions
and is always observed over heterogeneous metallic
catalysts even under conditions where a large proportion
of the products are oxygenated. The absence of methane
provides strong support for the production of ethylene

glycol _via_ a homogeneously-catalysed pathway. It may
also be significant that ethylene glycol does not appear
to be a product of the Fischer Tropsch synthesis.
Finally this system is of interest because there is at
least circumstantial evidence for the involvement of
rhodium cluster carbonyl anions.

The rhodium may be introduced in many forms,
e.g., $Rh(CO)_2acac$, $Rh_4(CO)_{12}$ and $Na_2[Rh_{12}(CO)_{30}]$, and
improved activity is observed in the presence of
promoters such as 2-hydroxypyridine or caesium acetate.
Examples of the transformations which have been
observed _in situ_ by high pressure infrared spectroscopy
(58,59), starting from $Rh(CO)_2acac$ and $[Rh_{12}(CO)_{30}]^{2-}$,
are summarised in the following reaction sequence:

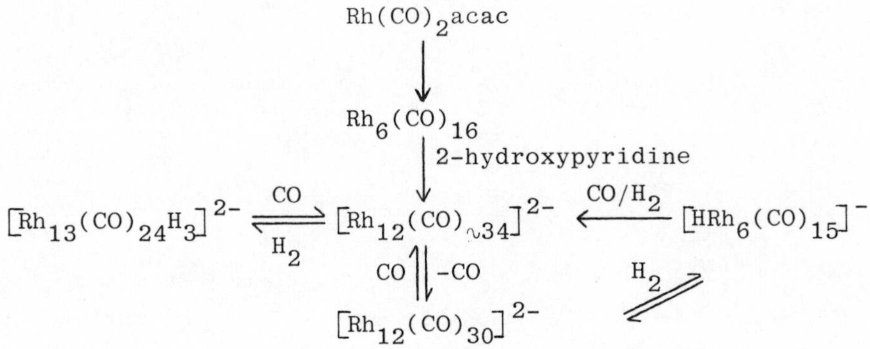

$$Rh(CO)_2acac$$
$$\downarrow$$
$$Rh_6(CO)_{16}$$
$$\downarrow \text{2-hydroxypyridine}$$

$$[Rh_{13}(CO)_{24}H_3]^{2-} \underset{H_2}{\overset{CO}{\rightleftharpoons}} [Rh_{12}(CO)_{\sim 34}]^{2-} \overset{CO/H_2}{\longleftarrow} [HRh_6(CO)_{15}]^{-}$$

$$CO \updownarrow -CO \qquad\qquad H_2 \nearrow\!\!\!\!\!\nearrow$$

$$[Rh_{12}(CO)_{30}]^{2-}$$

The predominant species present in solution at
900 atm and 200° is $[Rh_{12}(CO)_{\sim 34}]^{2-}$, the anion initially
described by Chini (60) but of still unknown structure,
together with smaller amounts of $[HRh_6(CO)_{15}]^{-}$ and
$[Rh_{13}(CO)_{24}H_3]^{2-}$, of which the last-named is only
observed under a deficiency of carbon monoxide. The
analogous iridium system $[Ir(CO)_2acac/2\text{-hydroxypyridine}]$,
which is not an active catalyst for ethylene glycol
synthesis, forms $Ir_4(CO)_{12}$ under comparable reaction
conditions and there is no evidence for the presence of

analogous polynuclear iridium carbonyl anions.

Thus the circumstantial evidence is in favour of involvement of rhodium cluster carbonyl anions in the reductive coupling of carbon monoxide to ethylene glycol.

The mechanism of ethylene glycol formation is at present unknown, but the observed product distribution is most consistent with a hydroxymethyl growth reaction, e.g.,

$$- M - CO \underset{\text{High pressure}}{\overset{CO}{\rightleftharpoons}} - M - CHO \xrightarrow{H_2} - M - CH_2OH \xrightarrow{H_2} CH_3OH$$

with $-M-CO$ having an H substituent and $-M-CHO$ having a CO substituent, then:

$$- M - CH_2OH \xrightarrow{CO} - M - COCH_2OH \xrightarrow{H_2} - M - CH(OH)CH_2OH \xrightarrow{H_2} \begin{array}{c} CH_2OH \\ | \\ CH_2OH \end{array}$$

with $\xrightarrow{CO}$ etc.

However, there is no direct evidence that glycol formation occurs by such a pathway or as to why catalysis by carbonyl clusters should be necessary. In fact Bradley (61) has recently demonstrated that, in the case of ruthenium, metal clusters are not essential for the hydrogenation of carbon monoxide under comparable reaction conditions.

$$CO + H_2 \xrightarrow[270^\circ]{1300 \text{ atm}} CH_3OH + HCO_2Me$$

Various ruthenium cluster carbonyl precursors break down to $Ru(CO)_5$ under reaction conditions. This homogeneous system gives C_1 oxygenated products with 99% selectivity and no hydrocarbons are observed (cf. (32)). Whether or not metal cluster catalysts are required for the synthesis of C_2 products such as ethylene glycol therefore remains an open question.

From the work which has been highlighted in the preceding pages it is clear that there are some strong indications of catalysis by metal clusters. In particular, by isolating individual reaction steps, Kaesz (55) has demonstrated what is probably the first plausible reaction sequence for catalysis by a cluster compound.

However, a demonstration of genuine polynuclear catalysis is still awaited. Possibly the only really conclusive proof derives from Norton's suggestion of designing a chiral cluster, using this to promote an asymmetric catalytic reaction and subsequently isolating chiral products (62). There do appear to be problems with this approach because the barriers to racemisation within the cluster framework itself appear to be low, in the chiral clusters so far studied (63,64). However, the synthesis of chiral clusters with internal non-fluxional bridging atoms such as sulphur may overcome this difficulty (65).

As an alternative approach the application of Fourier transform spectroscopic techniques (IR, NMR) to reacting systems should prove valuable in establishing the presence or otherwise of small amounts of, e.g., mononuclear species, in much higher concentrations of polynuclear aggregates.

Of the reactions i) - iii) listed on the right

hand side of Table 2 only ii), the homogeneous hydrogenation of carbon monoxide has been shown to occur in the presence of metal cluster compounds. The activation of C-C and C-H bonds in saturated hydrocarbons and the hydrogenation of molecular nitrogen remain as attractive targets for cluster chemists.

A characteristic of the "difficult" homogeneously catalysed reactions such as the hydrogenation of the triple bonds in C≡O, -C≡N etc., is that the reaction rates so far reported are, even under extreme reaction conditions, low in comparison with their heterogeneous counterparts. Therefore it seems unlikely that they will find industrial applications unless heterogeneous analogues just do not exist, as seems to be the case in the synthesis of ethylene glycol from carbon monoxide and hydrogen.

5. HETEROGENEOUS CATALYSIS

In turning to a consideration of the potential
applications of metal clusters in heterogeneous
catalysis there are two main approaches which merit
attention. First, the deposition of cluster compounds
on supports using the well-established methods for
immobilising homogeneous catalysts, and studying the
reactivity of the resultant "anchored" clusters. The
second approach concerns the deposition, either by
physical impregnation or chemical methods, of cluster
compounds onto supports followed by removal of the
ligands to produce highly dispersed supported metal
catalysts. Both methods pose severe difficulties in
terms of characterisation of the supported species.
However, before embarking on a discussion of these
problems it is appropriate to summarise some general
points from the literature relating to particle size
effects in catalysis by supported metals.

First, the specific catalytic activity (i.e.
activity per unit surface area) of a supported metal
catalyst is not necessarily independent of particle
size. Boudart (66) has classified metal catalysed
reactions as either "structure-sensitive", where
specific rates are particle size-dependent, or

"structure-insensitive", where they are independent of metal particle size.

Secondly, very small metal particles (<20 $\overset{\circ}{A}$) display "special properties", i.e., their behaviour is different from that of bulk metal. These special properties could be associated with the presence of interstitial hydrides or carbides or they may arise from the exposure of different crystal faces from those found in the bulk metal. In addition coordination and surface energy levels will differ from those found on bulk surfaces. The calculations of Burton (67) have shown that a structure having a basic icosahedral symmetry is themodynamically preferred over the normal close-packed structure for very small crystals. Thus the expected face centred cubic arrangement for a 55 atom body ($\equiv$12 $\overset{\circ}{A}$ platinum particle) is unstable with respect to the icosahedral arrangement containing triangular faces.

Similarly a 13 atom ($\sim$7 $\overset{\circ}{A}$) particle contains the icosahedral arrangement as the most stable form. The surfaces of very small particles should therefore contain only triangular 111 faces whereas the bulk metal contains both 100 and 111 faces. In terms of catalysis it may be possible to exploit these differences in affording some control over the selectivity of a given reaction.

Another important conclusion from Burton's work is that very small crystallites melt at low temperatures relative to bulk metal. This could be of considerable significance in terms of the mobility of small metal particles and their tendency to sinter under catalytic reaction conditions.

Finally, the use of small metal particles represents more efficient catalyst usage since the dispersion, D, which is defined as the fraction of

surface atoms to total atoms, approaches unity. This
aspect is of obvious importance when the use of
precious metal catalysts is considered.

With the foregoing comments in mind let us now
review the current state of the art with respect to
the preparation, characterisation and catalytic
activity of supported metal clusters.

5.1. Preparation of Supported Metal Clusters

Conventional techniques for the preparation of
supported metal catalysts are well-documented (68,69).
They fall into two principal classes, namely,
impregnation of a support with metal halide or nitrate
solutions (frequently aqueous) followed by drying and
hydrogen reduction at high temperature, or ion-exchange
followed by reduction. An example of the latter method
involves "ammoniation" of the hydroxyl groups of a
support such as silica followed by exchange of the
resultant ammonium ions for metal ions, e.g.,

$$\underset{\substack{\text{Si}\quad\text{Si} \\ \diagdown\text{O}\diagup}}{\overset{\text{OH}\quad\text{OH}}{|\quad\quad|}} \xrightarrow{\text{NH}_3} \underset{\substack{\text{Si}\quad\text{Si} \\ \diagdown\text{O}\diagup}}{\overset{\substack{\text{NH}_4^+\quad\text{NH}_4^+ \\ \text{O}^-\quad\text{O}^-}}{|\quad\quad|}} \xrightarrow{\text{M}^{2+}} \underset{\substack{\text{Si}\quad\text{Si} \\ \diagdown\text{O}\diagup}}{\overset{\substack{\text{M}^{2+} \\ \text{O}^-\quad\text{O}^-}}{|\quad\quad|}}$$

A third method, co-precipitation, is used for the
preparation of Ni/SiO_2 hydrogenation catalysts.

Production of supported metal catalysts by the
conventional impregnation techniques frequently
affords a wide range of particle size distributions
peaking in the 30-50 Å region for second row
transition metals. By carefully controlling the

experimental conditions, e.g., support pre-treatment,
temperature of reduction, etc., narrower size
distributions can be obtained. Ion-exchange methods
usually result in the production of smaller particles
with size distributions of 20-30 Å. However, neither
method permits the reproducible preparation of very
small particles (<10 Å) and hitherto the bulk of
catalyst evaluation and development of supported
metal catalysts has been carried out on metal
crystallites of diameter greater than 20 Å.

Providing that substantial aggregation can be
prevented, metal cluster compounds potentially add a
further dimension to the preparation of heterogeneous
catalysts by offering a more controlled route for the
production of supported metal catalysts of very high
dispersion (D → 1), thus approaching the situation
where all atoms are surface atoms. They can offer
the following differences/advantages over conventionally
prepared supported metal catalysts:

1. Non-aqueous methods of catalyst preparation
can be employed since metal cluster compounds are
soluble in organic solvents.

2. Organometallic clusters do not usually
contain halogen atoms and the catalysts should therefore
be halide-free. A halide-free route to supported
metal catalysts could be advantageous because the
presence of halide in conventionally-prepared
catalysts can contribute towards catalyst poisoning
and is thought to accelerate aggregation or sintering
of metal. Halide is also extremely difficult to
remove completely.

3. The high temperature reduction by hydrogen
used in conventional catalyst preparation should be
unnecessary since in the majority of metal clusters
the metal is already formally in the zero valent state.

Cluster-derived catalysts may therefore be activated
under milder reaction conditions.

4. The use of the ever-increasing number of
mixed-metal or heteronuclear cluster compounds (70)
offers the possibility of preparing bimetallics or
alloys of precisely known stoichiometry/composition in
a controlled manner. Catalysis by alloys is an area
of great interest in heterogeneous catalysis and such
materials are difficult to synthesise reproducibly.
The properties of materials produced from mixed-metal
clusters could be compared with those of the so-called
"bimetallic clusters" studied by Sinfelt and co-workers
at Exxon (7).

The two principal methods by which supported
catalysts have been prepared from metal cluster
compounds are summarised below.

The first route involves impregnation of a support
with solutions of polynuclear organometallic complexes,
M_xL_y, followed by either partial or complete removal
of L under vacuum, thermally, photochemically, or by
reaction with hydrogen, oxygen etc. This approach has
feen followed by, amongst others, Anderson (71)
particularly for the situation where M_xL_y corresponds
to polynuclear metal carbonyls. Although systematic
studies are lacking the results of this work suggest
that significant aggregation of metal atoms occurs
during ligand removal. Careful support pre-treatments
and decomposition at the lowest possible temperature
may be necessary to reproducibly obtain very small
supported metal crystallites.

An alternative approach, aimed at limiting the
extent of metal aggregation, is to chemically modify
the support by functionalisation with, for example,
hydrido, phosphino or amino groups. Chemical bond
formation between the polynuclear organometallic and

the functionalised support should produce an "anchored" cluster in the manner followed extensively in recent years for immobilising homogeneous catalysts (72), e.g.

The extent of aggregation during partial or complete removal of ligands should be strictly limited. Gates (73-75) and Evans (76) have recently described studies of this type in anchoring the polynuclear carbonyls of iridium, rhodium and osmium to functionalised supports.

A variation on these two methods of preparation comprises exchanging the cluster compound (if size constraints permit) or a precursor into the cavities of zeolites. This approach has been successfully used by Mantovani et al (77) in the formation of a zeolite entrapped rhodium cluster which is an active catalyst for the hydroformylation of olefins and dienes.

Finally, metal vapour techniques such as those of Klabunde (78), matrix isolation methods (8,79) and laser evaporation of metals (80) may provide alternative routes for the production of very small supported metal particles. However, it is very difficult to visualise any practical use for matrix

isolation methods under even the mildest catalytic
reaction conditions. The method of Klabunde involves
deposition of metal atom vapours into weakly
complexing organic media such as toluene, THF or
pentane at low temperature, followed by warming in
the presence of supports to allow the deposition of
metal crystallites within the support. Particle sizes
in the 30-80 $\overset{\circ}{A}$ range are observed with Ni supported on
γ-alumina prepared in this way. Silica- and
alumina-supported Pd, Pt and Ag, of undetermined
particle size, have also been prepared by this method.

5.2. Characterisation of Supported Metal Clusters

Having utilised the routes outlined in section
5.1 for the preparation of either supported metal
cluster compounds or very highly dispersed metals we
must now consider appropriate methods for the
characterisation of these materials. This is an area
which poses severe problems and which underlines the
shortcomings of physical techniques. The required
information frequently falls outside the range of
sensitivity of traditional surface science techniques
and beyond that obtainable by molecular structure
methods such as vibrational and NMR spectroscopy. No
one technique is satisfactory in isolation and until,
for example, ultra-high resolution electron microscopy
becomes available on a routine basis, reliance on a
combination of techniques and/or indirect evidence is
necessary. Techniques available for the characterisation of supported metal catalysts have been reviewed
(68,69,81) and those most relevant to metal cluster
systems, together with their limitations are

summarised below.

5.2.1. Chemisorption

The most extensively investigated method for determining the area and thus the crystallite size of a metal catalyst dispersed on a support is gas phase chemisorption (82). Nitrogen adsorption may be used to measure total surface area and H_2- and CO-chemisorption for the determination of available metal area. This is a very accurate technique for the characterisation of metal particles of size >20 Å. However, the interpretation of the chemisorption measurements requires assumptions concerning the stoichiometry of adsorption, such as M:H = 1:1 and M:CO = 1:1. Although this assumption is valid for larger metal crystallites it may no longer be true for particles of <10 Å. For example, the carbon monoxide to metal ratios in polynuclear metal carbonyls, e.g. $Ir_4(CO)_{12}$, $Rh_6(CO)_{16}$, are clearly very different from unity. Thus, although chemisorption measurements can provide accurate data, the results obtained at very small particle sizes do have to be treated with caution.

5.2.2. Electron Microscopy

Electron microscopy is extensively used for the characterisation of supported metal catalysts (83) and has the advantage that the metal atoms can actually be seen. Quantitative analysis of the electron micrographs can provide information on the range of metal crystallite sizes present on the support.

Modern electron microscopes can resolve very small aggregates of metal atoms (10 Å in diameter) in favourable circumstances. The observation of individual rhodium atoms in a silica-supported rhodium catalyst has been claimed by objective aperturing of a conventional transmission electron microscope (84). More recently an interpretation of electron micrographs of silica-supported ruthenium and Ru-Cu "bimetallic clusters" has claimed to distinguish two-dimensional rafts and three-dimensional clusters on the surface of silica (85). However, the technique of electron microscopy is subject to limitations which do not appear to be universally recognised and particle size distributions become increasingly subject to error as the fraction of particles with sizes below 25 Å increases (86). For practical purposes the limit of resolution is 10-15 Å dependent upon the metal content of the material and the nature of the metal. Sensitivity of the sample to the electron beam is also a feature which can cause difficulties.

5.2.3. EXAFS

The phenomenon of extended X-ray absorption fine structure (EXAFS), which has been known for many years, has, with the availability of synchrotron radiation, recently become a powerful structural technique (87). Although the major applications may ultimately lie in the determination of the coordination environment of metals in biological systems such as metalloproteins

and metalloenzymes (EXAFS is the only method
currently capable of providing such information), the
technique has considerable potential for the
characterisation of amorphous heterogeneous systems
such as supported metals and alloy catalysts.
Advantages of the technique are that it may be
selectively tuned to different elements by appropriate
choice of irradiation frequency, no special sample
preparation is required, except for low Z elements,
and samples can in principle be examined in situ
under catalytic reaction conditions. Information on
the local environments (up to 10 $\overset{o}{A}$) of a given
element is obtained and this is, of course, particularly
relevant to supported metal clusters. Disadvantages
include the fact that at present a source of
synchrotron radiation is necessary for rapid accumulation of data and that methods of analysis and interpretation of the results are sources of considerable
uncertainty.

In our work, the EXAFS spectra of highly
dispersed ruthenium (particle size <20 $\overset{o}{A}$), produced by
the impregnation of silica with toluene solutions of
$Ru_3(CO)_{12}$, followed by decarbonylation under hydrogen,
is very different from the spectrum of bulk ruthenium
metal. This behaviour has been interpreted as
indicative of a reduction in the coordination number
of ruthenium at very small particle sizes (88).

In the context of this chapter the technique has
also been used to study Ru-Cu bimetallic clusters (89)
and to investigate the structure of mononuclear
rhodium-phosphine hydrogenation catalysts immobilised
on polymeric phosphine supports (90). The implications
for the characterisation of cluster compounds bonded
to supports are obvious. When the methods of data
analysis have passed through the phase of optimising

on model compounds and are capable of handling truly
unknown systems then EXAFS should become a very
powerful technique for the characterisation of
supported metal clusters.

5.2.4. SIMS

Secondary ion mass spectrometry, a recently
developed surface analytical technique, is a powerful
tool for the study of individual monolayers on surfaces
(91). It displays high sensitivity, can detect
chemical compounds, and is claimed to cause only very
small disturbance of the surface during the analysis.
Unfortunately it requires ultra high vacuum
conditions, and is therefore rather removed from the
domain of the practising catalyst chemist. The
technique has provided significant information on, for
example, the adsorption of carbon monoxide on
conductors such as polycrystalline nickel and copper
(92), and can also be used to study insulators such
as silica and alumina. In principle the technique
should be capable of providing information on metal
cluster compound/support and metal/support interactions
from the observation of metal-support vs. metal only mass
fragments.

5.2.5. Vibrational Spectroscopy

Vibrational spectroscopy has been used quite
extensively in attempts to characterise the species
formed on initial interaction of metal cluster compounds

with supports. Several groups have reported infrared
spectra in the ν(CO) region for both physically
supported and chemically bound metal cluster carbonyls.
Examples in the first category include $Ru_3(CO)_{12}$
adsorbed on silica (93) and HY zeolite (94),
$Co_2Rh_2(CO)_{12}$ on γ-alumina (71), $Rh_6(CO)_{16}$ (95,96),
$Ir_4(CO)_{12}$ (71,97) and $\left[Pt_3(CO)_6\right]_n^{2-}$ (n = 2-5) (98) on
γ-alumina and silica.

In the context of the platinum clusters it is
relevant to note that Primet et al (99) have found
the position of the ν(NO) vibrations of nitric oxide
chemisorbed on a series of Pt/γ-Al_2O_3 samples to be
dependent upon the platinum particle size (in the
range 15-35 $\overset{o}{A}$). This has been interpreted in terms
of the collective electronic influence of the size of
the metal crystallite on the strength of the metal-
adsorbate (in this case Pt-NO) bond. This work
underlines the potential contribution of electronic
effects in catalysis by small metal particles in
addition to the geometric factors discussed earlier.

The infrared spectra of unidentified rhodium
clusters in Y-type zeolites have also been reported
(77). Spectral measurements on chemically bound
systems include the attempted identification of
products from the reactions of $Rh_6(CO)_{16}$ (74,75,100),
$Ir_4(CO)_{12}$ (73) with phosphinated polymers and
$Os_3(CO)_{11}L$ (76), $Rh_6(CO)_{16}$ (101) with chemically
modified silicas. In most cases the spectra reported
are typical of solid state (e.g. KBr) spectra,
containing broad ill-defined absorptions for which it
is extremely difficult to make convincing assignments.
In addition, it is usually impossible to measure other
vibrational modes, e.g. δ(MCO) and ν(MC) because the
commonly used inorganic oxide supports such as silica
and alumina almost totally absorb all the infrared

radiation below ca. 1250 cm^{-1}.

Of possibly greater potential are techniques which depend on light scattering rather than absorption phenomena. In particular laser Raman spectroscopy is an attractive possibility for providing structural information on supported metal clusters since metal-metal bonds are generally highly polarised and give rise to very intense metal-metal vibrations in the Raman effect (102). They are also observed within 100-250 cm^{-1} of the exciting line and therefore the traditional problem of fluorescence associated with this technique is minimised. In principle, laser Raman spectroscopy can be used to probe the changes which occur within the metal skeleton on supporting a cluster compound and on subsequent removal of ligands to produce supported metals.

Using this approach we have demonstrated that $Ru_3(CO)_{12}$ and $Os_3(CO)_{12}$ are adsorbed physically unchanged on to silica. The Raman spectra of the pure compound and supported species (containing ca. 1% metal) are virtually identical in the range 30-250 cm^{-1} from the exciting line (103). In contrast the metal-metal framework in $Rh_4(CO)_{12}$ apparently fragments on inter-action with silica and no strong signals assignable to metal-metal vibrations are observed. This conclusion is confirmed by the infrared spectrum of the supported material in which sharp $\nu(CO)$ absorptions are observed in positions corresponding only with the presence of terminal carbonyl groups $[\nu(CO)$ 2109 sh, 2090, 2031 cm$^{-1}]$. By analogy with the spectra of model compounds such as $[Rh(CO)_2OR]_2$, where R = Me, Et etc., and the reactions of these with phosphines to produce $Rh(CO)(OR)L_2$ (L = PBu^n_3, PPh_3) (104,105), we have assigned dimeric Rh(I) structures to these supported materials in which rhodium is chemically bound through Rh-O bonds to the

silica surface, i.e.,

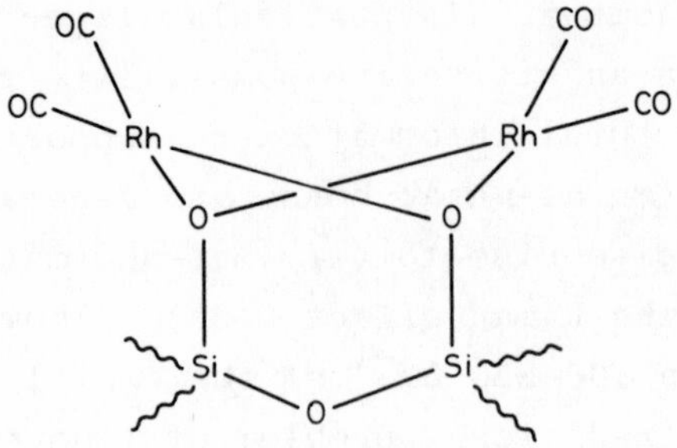

OC
OC—Rh   Rh—CO
         O   O   CO
              CO

~Si   Si~
      O

These supported rhodium dimers have been shown by infrared spectroscopy to undergo an extensive series of reversible reactions under pressures of CO, H_2, O_2 etc., which are thought to involve $Rh(O) \rightleftharpoons Rh(I)$ transformations (103).

Indirect evidence concerning aggregation or otherwise of metal atoms in supported clusters may be obtained by following spectroscopically the decarbonylation of pressed discs of supported metal carbonyls when heated under vacuum. Subsequently the discs may be pressurised with carbon monoxide to determine whether the starting spectrum can be reproduced. If so then it may be assumed that little or no aggregation of metal particles has occurred. In our experience the initial spectrum can be reproduced with Rh, where chemical bonding has taken place, but not with Ru or Os carbonyls, where simple physical adsorption has occurred.

5.2.6. NMR Spectroscopy

The development of solid state NMR techniques
should, in the future, enable direct information on
the nature of metal-support interactions in supported
cluster compounds to be obtained. A ^{13}C nmr study of
Ni(CO)$_4$ adsorbed on various supports is an indication
of the information available from the application of
"conventional" Fourier transform nmr (106) although
it has not .yet been applied successfully to the study
of polynuclear compounds.

5.2.7. Miscellaneous

Other physical methods, which are not generally
applicable, but which provide useful information in
appropriate situations include the use of ESR,
magnetic susceptibilities and Mössbauer spectroscopy.
For example Garten (107) has used Mössbauer spectroscopy
to obtain direct evidence for the formation of
"bimetallic" Fe-Pd clusters (of particle sizes $\sim$25 Å)
supported on η-Al$_2$O$_3$. The chemical state of iron in
the presence and absence of palladium after various
reduction and oxidation treatments has been determined.

5.2.8. Chemical Characterisation

In addition to the physical methods referred to
above, reactions such as buta-1,3-diene hydrogenation

and pent-1-ene isomerisation can be used as sensitive
chemical probes of the intimate environment of metal
atoms. Such reactions have recently been shown to
provide information concerning site congestion and
selectivity and in determining whether the extent of
molecular crowding at active sites influences
catalytic activity (108). These chemical methods may
provide considerable indirect information relevant to
the characterisation of very small metal particles
and should to some extent counteract the inadequacies
of physical techniques in this area.

To conclude this section therefore, the techniques
currently most useful for the characterisation of
supported metal clusters are chemisorption and electron
microscopy with EXAFS as a technique of considerable
future potential. Vibrational spectroscopy is useful
for the characterisation of supported cluster compounds.

5.3. Heterogeneous Catalysis by Metal Clusters

Having summarised methods for the preparation
and characterisation of supported metal clusters let
us now consider their catalytic properties. These
can be conveniently discussed in two sections, namely
catalysis by supported metal clusters, where partial
or, more probably, complete removal of ligands has
occurred (cf. highly dispersed supported metal
catalysts), and catalysis by supported metal cluster
compounds (cf. heterogenised homogenous catalysts).

5.3.1. Catalysis by Supported Metal Clusters

As indicated earlier in this chapter the many
reactions which occur over supported metal catalysts
may be broadly classified as either structure-
sensitive (demanding) or structure-insensitive
(facile) (66). Clearly, when considering catalysis
by small metal clusters, reactions where useful
effects are most likely to be encountered are those
which fall into the structure-sensitive class.

Much of the available evidence on particle size
effects in heterogeneous catalysis is limited to Pt
and Ni and the literature is not clear-cut. There
are many artefacts and even conflicting evidence on
the same reactions, which has been attributed to the
presence of chloride, and oxygen poisoning, etc.
However, it is possible to make some generalisations
and classes of reaction which are thought to be
structure-sensitive include those which involve the
formation and/or breaking of C-C bonds in saturated
hydrocarbons, e.g. hydrogenolysis of ethane over
Rh/SiO_2 (109) and $Ni/SiO_2-Al_2O_3$ (110), the reduction
of carbon monoxide, e.g. methanation over Pt/SiO_2 (111),
and ammonia synthesis (112). It will be recalled that
these reaction types coincide with those which are
relatively unknown in homogeneous catalysis.

Generally, the specific activity is found to
increase with decreasing particle size although in the
case of ethane hydrogenolysis over Rh/SiO_2 there is
some indication that the activity reaches a maximum
and then decreases as the dispersion approaches

unity (109). In this case the highest catalytic activity is associated with an intermediate level of dispersion.

Reactions which are thought to be structure-insensitive are principally those involving the hydrogenation of unsaturated hydrocarbons, e.g. hex-1-ene and benzene hydrogenation over Pt/SiO_2 (113) and Ni, Ir/SiO_2 (114).

Having discussed these general comments let us now turn to a consideration of the results which have been reported concerning catalysis by supported metal clusters derived from cluster compounds. The first detailed description concerns the work of Robertson and Webb (93) on the isomerisation, hydroisomerisation and hydrogenation of but-1-ene (all structure-insensitive reactions) over silica-supported ruthenium catalysts derived from $Ru_3(CO)_{12}$. A pink partially-decarbonylated material obtained by heating $Ru_3(CO)_{12}/SiO_2$ to 150^OC under vacuum, and formulated as $Ru_3(CO)_5$ chemically bonded to silica, is an active catalyst for all these reactions. In contrast, complete decarbonylation at $>200^OC$ produces a material which is inactive for isomerisation but which catalyses the hydrogenation and hydroisomerisation of but-1-ene. The latter behaviour closely resembles that of conventional supported ruthenium catalysts in butene hydrogenation (115), as might have been predicted, but the catalytic behaviour of the partially decarbonylated material is signficantly different.

An example of the application of a structure-sensitive test reaction derives from the work of Anderson and Mainwaring (116) on the preparation of a supposedly bimetallic catalyst using $Co_2Rh_2(CO)_{12}$ as catalyst precursor. This silica-supported material, of undefined constitution, which is obtained by heating $Co_2Rh_2(CO)_{12}/SiO_2$ in hydrogen and then in vacuo at 380^oC, has been tested for activity in the hydrogenolysis of methylcyclopentane at 260^oC. Selective conversion into the non-cyclic C_6 isomers 2- and 3-methylpentane occurs with very little ring expansion to higher cyclic molecules or cracking to lower hydrocarbons. This is significantly different from the behaviour observed with conventional Co and Rh catalysts. For example, the predominant reaction of methylcyclopentane and hydrogen over a rhodium film catalyst is fragmentation to methane (117).

Ichikawa (118) has used a similar structure-sensitive test reaction for the evaluation of platinum aggregates prepared by the dispersion and pyrolysis of the Chini (119) stacked platinum triangles $[Et_4N]_2[\{Pt_3(CO)_6\}_n]$ (n = 2-5) on γ-alumina. In this case the test reaction which has been used is the dehydrocyclisation and skeletal isomerisation of n-hexane at ca. 300^oC. Significant differences in product distribution are observed, depending upon the value of n in the precursor cluster, although the supported platinum catalysts have not themselves been well-characterised. For n = 3-5, the hexane dehydrogenation products methylcyclopentane and cyclohexane are obtained in >75% selectivity together with small amounts of 2- and 3-methylpentanes via skeletal isomerisation. In the case of $[Pt_6(CO)_{12}]^{2-}$ and $Pt_3(CO)_3(PPh_3)_4$ as precursor clusters, negligible skeletal isomerisation of n-hexane occurs and methyl-cyclopentane is produced in 84-93% selectivity. When neopentane is used as feed at similar temperatures skeletal isomerisation to isopentane and isobutane (1:4) is the predominant reaction for n = 4 or 5, whereas for n = 1 or 2 hydrocracking is the preferred reaction with negligible skeletal isomerisation taking place.

Pyrolysed silica-supported nickel clusters have also exhibited some interesting differences in catalytic activity in reactions of the structure-insensitive class (120). For example, materials produced from $(\eta^5-C_5H_5)_2Ni$ readily trimerise acetylene to benzene at room temperature (cf. the homogeneously catalysed reaction in the presence of $Ni_4[CNC(CH_3)_3]_7$ (51)), whereas those derived from $(\eta^5-C_5H_5)_2Ni_2(CO)_2$ and $(\eta^5-C_5H_5)_3Ni_3(CO)_2$ are inactive in the temperature range $25-120^{\circ}C$, this inactivity being ascribed to poisoning of the catalyst by the formation of stable complexes. Ethylene and benzene hydrogenation and H-D exchange occur readily at room temperature over the supported Ni_2 and Ni_3 aggregates. These materials, but not those derived from $(\eta^5-C_5H_5)_2Ni$, also show activity for the hydroformylation of ethylene at $50-100^{\circ}C$ and atmospheric pressure.

Returning to structure-sensitive reactions, Ichikawa has recently demonstrated some remarkable selectivity dependences for the reduction of carbon monoxide to ethanol, methanol or methane using pyrolysed rhodium carbonyl clusters on various oxide supports (121,122). The nuclearity of the starting cluster, e.g. $Rh_2(\eta^5-C_5H_5)_2(CO)_3$, $Rh_4(CO)_{12}$, $Rh_6(CO)_{16}$, $[Rh_7(CO)_{16}]^{3-}$ and $[Rh_{13}(CO)_{24}H_3]^{2-}$ makes only detail differences but the nature of the support clearly plays a major role in influencing the oxygenated vs. hydrocarbon product distribution at $150-250^{\circ}$ and atmospheric pressure. Ethanol, together with smaller amounts of methanol and methane, are the predominant products from the reaction of carbon monoxide and hydrogen at $>200^{\circ}C$ over rhodium clusters dispersed on amphoteric oxides such as TiO_2, ZrO_2 and La_2O_3. Catalysts prepared from Rh_2, Rh_4 and Rh_6 clusters display higher selectivities towards ethanol than those obtained from Rh_7 and Rh_{13} derivatives. The optimum catalyst, $Rh_4(CO)_{12}/La_2O_3$, gives ethanol, methanol and methane in 61, 20 and 12% selectivities respectively, at 36% CO conversion (1 atm., 224°C, 5 hr.). For comparison, a more conventional rhodium catalyst, prepared by the hydrogen reduction of $RhCl_3/La_2O_3$ at the rather high temperature (for a true comparison) of 350°C, gives only 5% conversion to methane (65%), ethanol (17%) and methanol (11%) together with other hydrocarbons under the same reaction conditions.

In contrast, when the rhodium clusters are supported on strongly basic oxides such as ZnO and MgO, methanol is obtained in >90% selectivity at 220°C with traces of methane as the only other product. Methanol is also produced, in 95-99% selectivity, from $[Et_4N]_2[Pt_3(CO)_6]_5$ supported on ZnO and MgO, although at lower conversions. $Ir_4(CO)_{12}$ supported on ZnO is less effective but also yields only methanol and methane. In further contrast, methanation of carbon monoxide is the only significant reaction observed with $Rh_4(CO)_{12}$ dispersed on acidic oxide supports such as silica and γ-alumina.

The differences in selectivity observed with pyrolysed cluster compounds may be contrasted with the thermodynamically favoured methanation reaction, which is expected and usually observed over conventional supported metal catalysts (123). Ichikawa's results may hold some promise for controlling the product distribution in the heterogeneously-catalysed Fischer Tropsch reaction.

In related work Smith et al (124) have reported the stoichiometric formation of various hydrocarbons, together with CO_2 and H_2 on heating alumina-supported polynuclear metal carbonyls (Ru, Os, Rh and Ir) under

argon (no hydrogen) to 150-250°C. Above 250°C methane
is the major product. Hydrogen, the source of which
is thought to be either adsorbed water or the surface
hydroxyl groups of the support, is produced by the
water gas shift reaction. Interestingly, when
$Rh_6(CO)_{16}$/alumina samples are heated to 200°C under
CO rather than argon the major products are ethylene
and butenes.

In the preceding work on the catalytic properties
of pyrolysed supported cluster compounds definitive
evidence for the nature of the material obtained after
ligand removal is usually lacking. Also a comparison
of the activity of such materials with that displayed
by conventional supported catalysts under similar
reaction conditions is not generally available.
Careful characterisation and systematic study of the
catalytic properties of selected materials is
therefore required. However, evidence is accumulating
that these materials, whatever their intimate
constitution, behave differently and more specifically
than bulk metals dispersed on a surface and this
serves to encourage the belief that supported metal
clusters could form the basis of a new generation of
heterogeneous catalysts.

A discussion on catalysis by supported clusters
would be incomplete without reference to the
polymetallic or bimetallic clusters discovered by
Sinfelt et al at Exxon (7). These materials are
obtained by conventional methods of catalyst
preparation, i.e., not involving metal cluster
precursors, but in some cases the production of very
small particles ($\sim$10 $\overset{o}{A}$) has been claimed. Supported
Ir-Pt combinations are used as hydrocarbon reforming
catalysts and offer advantages over the commonly used
Pt and Pt-Re catalysts (125). Evaluation of Cu-Ru and

Cu-Os bimetallics in the hydrogenolysis of ethane has shown that the high hydrogenolysis activity characteristic of Ru and Os is dramatically inhibited by the incorporation of copper (126). In contrast, the activity for dehydrogenation of cyclohexane to benzene is only marginally affected. How the structure and catalytic properties of these bimetallic clusters relates to the materials derived from cluster compounds remains an open question.

5.3.2. Catalysis by Supported Metal Cluster Compounds

Relatively few examples of catalysis by cluster compounds chemically bonded to supports have been reported although applications with mononuclear systems are common (127). Collman et al (128) described the interaction of $Rh_4(CO)_{12}$ and $Rh_6(CO)_{16}$ with poly(styrene-divinylbenzene) resins functionalised with PPh_2 groups. Decarbonylation of the resulting species in air produced materials which displayed similar activity to commercial Rh/Al_2O_3 catalysts for the hydrogenation of arenes under ambient conditions. This work has been extended by Gates (74) to include phosphinated poly(styrene-divinylbenzene) membranes, the use of which greatly facilitates characterisation of the resultant materials by transmission i.r. spectroscopy. Reaction of the membrane with $Rh_6(CO)_{16}$ at room temperature produces what is thought to be a trisubstituted $-Rh_6(CO)_{13}$ derivative. Both membranes and resins containing $Rh_6(CO)_{16}$ are active catalysts for the hydrogenation of cyclohexene, ethylene and benzene at $80^{\circ}C$ and 1 atm. The kinetics of cyclohexene hydrogenation has been used to follow the catalyst aging process from supported cluster compound to supported metal in which particle sizes of 15 and 20-25 Å are obtained after 40 and 350 hr. use respectively (75). Phosphinated polymers containing iridium, thought to be of the form $Ir_4(CO)_{11}Ph_2P(polymer)$ are also active catalysts for the hydrogenation of ethylene under ambient conditions (73).

A heterogeneous analogue of the homogeneously-catalysed hydrogenation of α,β-unsaturated carbonyl compounds, referred to earlier, has been reported (23,129). $Rh_4(CO)_{12}$ supported on cross-linked polymeric amines such as Amberlyst resins provides a more active catalyst for these transformations at $70^\circ C$ in the presence of carbon monoxide (100 atm.) and water. The higher activity is thought to be associated with the stabilisation of Rh_6 species on the resins whereas Rh_{12} aggregates, presumably $[Rh_{12}(CO)_{\sim 34}]^{2-}$, predominate in solution. Rhodium catalysts supported on similar polymeric amines, e.g. poly(N,N-dimethyl-benzylamine), or anion exchange resins have been used to hydroformylate olefins directly to alcohols in addition to the normally produced aldehydes (130). Spectroscopic evidence suggests that rhodium cluster carbonyl anions are present on the resins under operating conditions (70 atm. CO/H_2, $100^\circ C$).

The production in situ of metal clusters in zeolites is of potential significance (77). For example, a rhodium zeolite prepared by exchanging $[Rh(NH_3)_6]Cl_3$ into NaY zeolite is reacted with CO/H_2 (80 atm., $130^\circ C$) to produce a rhodium carbonyl cluster trapped in the supercages. This material is an active catalyst for the hydroformylation of olefins to n- and iso-aldehydes in the expected ratio (ca. 1:1) for homogeneous rhodium-catalysed hydroformylation. In contrast, dienes such as 1,5-hexadiene are hydroformylated to dialdehydes and monoaldehydes in 60 and 40% yields respectively, whereas monoaldehydes are usually the major products from non-conjugated dienes in the homogeneous process. Zeolites clearly offer a promising method for heterogenising homogeneous catalysts and their stability towards vigorous working conditions may provide advantages over the functionalised polymers referred to previously.

6. CONCLUSIONS

From the work which has been summarised in this chapter it is clear that the subject of catalysis by metal clusters is a rapidly expanding area of research. The question of whether this activity will result in technical innovation is difficult to answer at this stage. However, it seems probable that industrial applications will emerge, if only indirectly, within the next 10-20 years.

Results obtained from studies of the reactivity of metal cluster compounds will clearly enhance our understanding of reaction mechanisms, particularly in relation to heterogeneous catalysis. Homogeneous systems will only find industrial use if high selectivities outweigh low activities in a given reaction (or if their activities can be dramatically increased over those currently attainable) or if heterogeneous analogues just do not exist, as appears to be the case with the conversion of CO/H_2 into ethylene glycol.

Supported metal clusters, whatever their constitution, have already produced significant activity and selectivity differences over conventional supported metal catalysts, differences which encourage

the expectation that they could form the basis of a new
generation of heterogeneous catalysts. Advances seem
most likely to arise from the area of structure-
sensitive reactions, namely, activation of C-C and C-H
bonds in saturated hydrocarbons, and the hydrogenation
of carbon monoxide and nitrogen. However, these
activity/selectivity differences must be sufficiently
great to produce an economic case for industrial
concerns to transfer from well-tried and tested
bucket and shovel methods to large scale sophisticated
methods of catalyst preparation.

Finally, we can expect the further development/
refinement of physical techniques to facilitate
characterisation of systems of the types referred to
in this chapter.

598.

7. REFERENCES

1. P. Chini, G. Longoni and V. G. Albano, Adv.
    Organometal. Chem., 14, 285 (1976).
2. H. Vahrenkamp, Struct. Bonding (Berlin), 32, 1
    (1977).
3. E. L. Muetterties, Bull. Soc. Chim. Belg., 84,
    959 (1975); Science, 196, 839 (1977).
4. H. Conrad, G. Ertl, H. Knözinger, J. Küppers and
    E. E. Latta, Chem. Phys. Lett., 42, 115
    (1976).
5. R. Ugo, Catal. Rev. -Sci. Eng., 11, 225 (1975).
6. P. Wynblatt and N. A. Gjostein, Prog. Solid State
    Chem., 9, 21 (1975).
7. J. H. Sinfelt, Acc. Chem. Res., 10, 15 (1977);
    Science, 195, 641 (1977).
8. G. A. Ozin, Catal. Rev. -Sci. Eng., 16, 191 (1977).
9. A. K. Smith and J. M. Basset, J. Mol. Catal., 2,
    229 (1977).
10. J. M. Basset and R. Ugo, Aspects Homog. Catal., 3,
    137 (1977).
11. L. M. Slifkin, Sci. Prog., 60, 151 (1970);
    L. K. H. van Beek, Philips Technical Rev.,
    33, 1 (1973); J. F. Hamilton, J. Vac. Sci.
    Technol., 13, 319 (1976).

12. T. J. Coutts, Electrical Conduction in Thin Metal
         Films, Elsevier (1974); J. E. Morris and
         T. J. Coutts, Thin Solid Films, 47, 3 (1977).

13. R. J. Hodges, D. E. Webster and P. B. Wells,
         J. Chem. Soc. (A), 3230 (1971).

14. J. W. Rathke and H. M. Feder, J. Amer. Chem. Soc.,
         100, 3623 (1978).

15. J. M. Manriquez, D. R. McAlister, R. D. Sanner
         and J. E. Bercaw, J. Amer. Chem. Soc., 98,
         6733 (1976).

16. J. Chatt, A. J. Pearman and R. L. Richards,
         Nature, 253, 39 (1975).

17. G. W. Parshall, J. Mol. Catal., 4, 243 (1978).

18. J. Falbe, J. Organometal. Chem., 94, 213 (1975).

19. J. F. Roth, J. H. Craddock, A. Hershman and
         F. E. Paulik, Chem. Technol., 1, 600 (1971).

20. Chem. Eng. News, 26 April 1971, p. 30; Chem. Week,
         12 May 1971, p. 32; Eur. Chem. News,
         26 April 1974, p. 15.

21. W. S. Knowles, M. J. Sabacky and B. D. Vineyard,
         Chem. Technol., 2, 590 (1972); Homogeneous
         Catalysis II, Adv. Chem. Series 132, ed.
         D. Forster and J. F. Roth, ACS, 1974, p. 275;
         Chem. Week, 20 November 1974, p. 71.

22. G. Pregaglia, A. Andreetta, G. Ferrari and R. Ugo,
         J. Organometal. Chem., 33, 73 (1971).

23. T. Kitamura, N. Sakamoto and T. Joh, Chem. Lett.,
         379 (1973).

24. M. G. Thomas, W. R. Pretzer, B. F. Beier and
         E. L. Muetterties, J. Amer. Chem. Soc., 99,
         743 (1977).

25. D. Labroue and R. Poilblanc, J. Mol. Catal., 2,
         329 (1977).

26. M. Bianchi, F. Piacenti, G. Menchi, P. Frediani,
         U. Matteoli, C. Botteghi, S. Gladiali and

600.

        E. Benedetti, Chim. Ind. (Milan), <u>60</u>, 588
        (1978).

27. B. Heil and L. Markó, Acta Chim. Acad. Sci. Hung.,
        <u>55</u>, 107 (1968).

28. P. Frediani, U. Matteoli, M. Bianchi, F. Piacenti
        and G. Menchi, J. Organometal. Chem., <u>150</u>,
        273 (1978).

29. U. S. Patents (to Union Carbide), 3,833,634 (1974);
        3,878,214, 3,878,290, 3,878,292 (1975);
        3,944,588, 3,948,965, 3,952,039, 3,957,857,
        3,968,136 (1976).

30. M. G. Thomas, B. F. Beier and E. L. Muetterties,
        J. Amer. Chem. Soc., <u>98</u>, 1296 (1976).

31. G. C. Demitras and E. L. Muetterties, J. Amer.
        Chem. Soc., <u>99</u>, 2796 (1977).

32. Ger. Offen. (to Shell), 2,644,185 (1977);
        Chem. Abstr., <u>87</u>, 41656h (1977).

33. E. Band, W. R. Pretzer, M. G. Thomas and
        E. L. Muetterties, J. Amer. Chem. Soc., <u>99</u>,
        7380 (1977).

34. F. L'Eplattenier, P. Matthys and F. Calderazzo,
        Inorg. Chem., <u>9</u>, 342 (1970).

35. A. F. M. Iqbal, Tet. Lett., 3385 (1971).

36. K. Cann, T. Cole, W. Slegeir and R. Pettit, J. Amer.
        Chem. Soc., <u>100</u>, 3969 (1978).

37. N. S. Imyanitov and D. M. Rudkowskii, Zh. Prikl.
        Khim., <u>39</u>, 2020, 2029 (1967).

38. B. Heil and L. Markó, Chem. Ber., <u>101</u>, 2209 (1968);
        <u>102</u>, 2238 (1969).

39. G. Braca, G. Sbrana, F. Piacenti and P. Pino,
        Chim. Ind. (Milan), <u>52</u>, 1091 (1970).

40. R. C. Ryan, C. U. Pittman and J. P. O'Connor,
        J. Amer. Chem. Soc., <u>99</u>, 1986 (1977).

41. C. U. Pittman and R. C. Ryan, Chem. Technol., <u>8</u>,
        170 (1978).

42. H. Kang, C. H. Mauldin, T. Cole, W. Slegeir,
    K. Cann and R. Pettit, J. Amer. Chem. Soc.,
    99, 8323 (1977).

43. R. M. Laine, J. Amer. Chem. Soc., 100, 6451 (1978).

44. R. M. Laine, R. G. Rinker and P. C. Ford, J. Amer.
    Chem. Soc., 99, 252 (1977); P. C. Ford,
    R. G. Rinker, C. Ungermann, R. M. Laine,
    V. Landis and S. A. Moya, J. Amer. Chem.
    Soc., 100, 4595 (1978).

45. D. Bingham, B. Hudson, D. E. Webster and
    P. B. Wells, J. Chem. Soc. Dalton, 1521
    (1974).

46. A. J. Deeming and S. Hasso, J. Organometal. Chem.,
    114, 313 (1976).

47. R. P. Ferrari and G. A. Vaglio, Inorg. Chim. Acta,
    20, 141 (1976).

48. M. Castiglioni, L. Milone, D. Osella, G. A. Vaglio
    and M. Valle, Inorg. Chem., 15, 394 (1976).

49. M. Valle, D. Osella and G. A. Vaglio, Inorg. Chim.
    Acta, 20, 213 (1976).

50. P. Pino, G. Braca, G. Sbrana and A. Cuccuru,
    Chem. Ind., 1732 (1968).

51. V. W. Day, R. O. Day, J. S. Kristoff, F. J. Hirsekorn
    and E. L. Muetterties, J. Amer. Chem. Soc.,
    97, 2571 (1975).

52. G. D. Mercer, J. Shing Shu, T. B. Rauchfuss and
    D. M. Roundhill, J. Amer. Chem. Soc., 97,
    1967 (1975); G. D. Mercer, W. B. Beaulieu
    and D. M. Roundhill, J. Amer. Chem. Soc.,
    99, 6551 (1977).

53. C.-S. Chin, M. S. Sennett and L. Vaska, J. Mol.
    Catal., 4, 375 (1978).

54. E. L. Muetterties, Bull. Soc. Chim. Belg., 85,
    451 (1976).

602.

55. M. A. Andrews and H. D. Kaesz, J. Amer. Chem. Soc.,
    99, 6763 (1977).

56. K. G. Caulton, M. G. Thomas, B. A. Sosinsky and
    E. L. Muetterties, Proc. Natl. Acad. Sci.
    U.S.A., 73, 4274 (1976).

57. U.S. Patent (to du Pont), 2,636,046 (1953);
    Chem. Abstr., 48, 2087h (1954).

58. R. L. Pruett, Ann. N.Y. Acad. Sci., 295, 239
    (1977).

59. S. Rigby and R. Whyman, unpublished results.

60. P. Chini and S. Martinengo, Inorg. Chim. Acta,
    3, 299 (1969).

61. J. S. Bradley, First Internat. Symp. Homog. Catal.,
    Corpus Christi, Texas, November 1978,
    Abstract 54.

62. J. R. Norton, Preprints, Petroleum Chemistry
    Division, 172nd National ACS Meeting, 343
    (1976).

63. A. Agapiov, S. E. Pederson, L. A. Zyzyck and
    J. R. Norton, J. Chem. Soc. Chem. Comm.,
    393 (1977).

64. A. Agapiov, R. F. Jordan, L. A. Zyzyck and
    J. R. Norton, J. Organometal. Chem., 141,
    C35 (1977).

65. F. Richter and H. Vahrenkamp, Angew. Chem. Internat.
    Edit., 17, 864 (1978).

66. M. Boudart, A. Aldag, J. E. Benson, N. A. Dougharty
    and C. Girvan Harkins, J. Catal., 6, 92 (1966).

67. J. J. Burton, Catal. Rev. -Sci. Eng., 9, 209 (1974).

68. R. L. Moss, Experimental Methods in Catalytic
    Research, Vol. 2, ed. R. B. Anderson and
    P. T. Dawson, Academic Press, 1976, p.43.

69. J. R. Anderson, Structure of Metallic Catalysts,
    Academic Press, 1975.

70. See for example: J. R. Shapley, G. A. Pearson,

M. Tachikawa, G. E. Schmidt, M. R. Churchill
and F. J. Hollander, J. Amer. Chem. Soc., 99,
8064 (1977); R. Bender, P. Braunstein,
Y. Dusausoy and J. Protas, Angew. Chem.
Internat. Edit., 17, 596 (1978);
T. V. Ashworth, M. Berry, J. A. K. Howard,
M. Laguna and F. G. A. Stone, J. Chem. Soc.
Chem. Comm., 45 (1979).

71. J. R. Anderson, P. S. Elmes, R. F. Howe and
D. E. Mainwaring, J. Catal., 50, 508 (1977).

72. J. C. Bailar, Catal. Rev. -Sci. Eng., 10, 17
(1974); Z. M. Michalska and D. E. Webster,
Chem. Technol., 5, 117 (1975).

73. J. J. Rafalko, J. Lieto, B. C. Gates and
G. L. Schrader, J. Chem. Soc. Chem. Comm.,
540 (1978).

74. M. S. Jarrell and B. C. Gates, J. Catal., 54, 81
(1978).

75. M. S. Jarrell, B. C. Gates and E. D. Nicholson,
J. Amer. Chem. Soc., 100, 5727 (1978).

76. S. C. Brown and J. Evans, J. Chem. Soc. Chem.
Comm., 1063 (1978).

77. E. Mantovani, N. Palladino and A. Zanobi,
J. Mol. Catal., 3, 285 (1978).

78. K. J. Klabunde, D. Ralston, R. Zoellner,
H. Hattori and Y. Tanaka, J. Catal., 55, 213
(1978).

79. W. E. Klotzbücher and G. A. Ozin, J. Amer. Chem.
Soc., 100, 2262 (1978).

80. E. A. Koerner von Gustorf, O. Jaenicke,
O. Wolfbeis and C. R. Eady, Angew. Chem.
Internat. Edit., 14, 278 (1975).

81. T. E. Whyte, Catal. Rev. -Sci. Eng., 8, 117 (1973).

82. J. Muller, Rev. Pure Appl. Chem., 19, 151 (1969).

604.

83. C. M. Sargent and J. D. Embury, Experimental
       Methods in Catalytic Research, Vol. 2,
       ed. R. B. Anderson and P. T. Dawson,
       Academic Press, p. 140 (1976).

84. E. B. Prestridge and D. J. C. Yates, Nature, 234,
       345 (1971).

85. E. B. Prestridge, G. H. Via and J. H. Sinfelt,
       J. Catal., 50, 115 (1977).

86. P. C. Flynn, S. E. Wanke and P. S. Turner,
       J. Catal., 33, 233 (1974); L. A. Freeman,
       A. Howie and M. M. J. Treacy, J. Microscopy,
       111, 165 (1977).

87. P. Eisenberger and B. M. Kincaid, Science, 200,
       1441 (1978)

88. A. D. Cox, R. T. Murray and R. Whyman, unpublished
       results.

89. F. W. Lytle, G. H. Via and J. H. Sinfelt,
       Preprints, Petroleum Chemistry Division,
       172nd National ACS Meeting, 366 (1976).

90. J. Reed, P. Eisenberger, B.-K. Teo and
       B. M. Kincaid, J. Amer. Chem. Soc., 99, 5217
       (1977).

91. A. Benninghoven, Surf. Sci., 35, 427 (1973).

92. M. Barber, J. C. Vickerman and J. Wolstenholme,
       J. Chem. Soc. Faraday I, 72, 40 (1976).

93. J. Robertson and G. Webb, Proc. Roy. Soc. London,
       A., 341, 383 (1974).

94. P. Gallezot, G. Coudurier, M. Primet and B. Imelik,
       Molecular Sieves-II, ACS Symposium Series 40,
       ed. J. R. Katzer, p. 144 (1977).

95. G. C. Smith, T. P. Chojnacki, S. R. Dasgupta,
       K. Iwatate and K. L. Watters, Inorg. Chem.,
       14, 1419 (1975).

96. J. L. Bilhou, V. Bilhou-Bougnol, W. F. Graydon,
       J. M. Basset, A. K. Smith, G. M. Zanderighi

and R. Ugo, J. Organometal. Chem., <u>153</u>, 73
(1978).

97. R. F. Howe, J. Catal., <u>50</u>, 196 (1977).

98. M. Ichikawa, Chem. Lett., 335 (1976).

99. M. Primet, J. M. Basset, E. Garbowski and
M. V. Mathieu, J. Amer. Chem. Soc., <u>97</u>, 3655
(1975).

100. K. Iwatate, S. R. Dasgupta, R. L. Schneider,
G. C. Smith and K. L. Watters, Inorg. Chim.
Acta, <u>15</u>, 191 (1975).

101. H. Knözinger and E. Rumpf, Inorg. Chim. Acta, <u>30</u>,
51 (1978).

102. M. J. Ware, Essays in Structural Chemistry, ed.,
A. J. Downs, D. A. Long and L. A. K. Staveley,
MacMillan, p. 404 (1971).

103. R. Whyman, unpublished results.

104. A. Vizi-Orosz, G. Pályi and L. Markó, J. Organometal.
Chem., <u>57</u>, 379 (1973).

105. G. Giordano, S. Martinengo, D. Strumolo and
P. Chini, Gazz. Chim. Ital., <u>105</u>, 613 (1975).

106. E. G. Derouane, J. B. Nagy and J. C. Védrine,
J. Catal., <u>46</u>, 434 (1977).

107. R. L. Garten, J. Catal., <u>43</u>, 18 (1976).

108. D. McMunn, R. B. Moyes and P. B. Wells, J. Catal.,
<u>52</u>, 472 (1978).

109. D. J. C. Yates and J. H. Sinfelt, J. Catal., <u>8</u>,
348 (1967).

110. J. L. Carter, J. A. Cusumano and J. H. Sinfelt,
J. Phys. Chem., <u>70</u>, 2257 (1966).

111. M. A. Vannice, J. Catal., <u>40</u>, 129 (1975).

112. J. A. Dumesic, H. Topsøe, S. Khammouma and
M. Boudart, J. Catal., <u>37</u>, 503 (1975).

113. T. A. Dorling and R. L. Moss, J. Catal., <u>5</u>, 111
(1966).

606.

114. R. Van Hardeveld and F. Hartog, Adv. Catal., $\underline{22}$, 75 (1972).

115. G. C. Bond, G. Webb and P. B. Wells, Trans. Faraday Soc., $\underline{61}$, 999 (1965).

116. J. R. Anderson and D. E. Mainwaring, J. Catal., $\underline{35}$, 162 (1974).

117. J. R. Anderson and B. G. Baker, Proc. Roy. Soc. London, A, $\underline{271}$, 402 (1963).

118. M. Ichikawa, J. Chem. Soc., Chem. Comm., 11 (1976).

119. J. C. Calabrese, L. F. Dahl, P. Chini, G. Longoni and S. Martinengo, J. Amer. Chem. Soc., $\underline{96}$, 2614 (1974).

120. M. Ichikawa, J. Chem. Soc. Chem. Comm., 26 (1976).

121. M. Ichikawa, Bull. Chem. Soc. Japan, $\underline{51}$, 2268 (1978).

122. M. Ichikawa, J. Chem. Soc. Chem. Comm., 566 (1978); Bull. Chem. Soc. Japan, $\underline{51}$, 2273 (1978).

123. M. A. Vannice, Catal. Rev. -Sci. Eng., $\underline{14}$, 153 (1976).

124. A. K. Smith, A. Theolier, J. M. Basset, R. Ugo, D. Commereuc and Y. Chauvin, J. Amer. Chem. Soc., $\underline{100}$, 2590 (1978).

125. U.S. Patent (to Exxon), 3,953,368 (1976).

126. J. H. Sinfelt, J. Catal., $\underline{29}$, 308 (1973).

127. R. H. Grubbs, Chem. Technol., $\underline{7}$, 512 (1977).

128. J. P. Collman, L. S. Hegedus, M. P. Cooke, J. R. Norton, G. Dolcetti and D. N. Marquardt, J. Amer. Chem. Soc., $\underline{94}$, 1789 (1972).

129. T. Kitamura, T. Joh and N. Hagihara, Chem. Lett., 203 (1975).

130. A. T. Jurewicz, L. D. Rollmann and D. D. Whitehurst, Homogeneous Catalysis II, Adv. Chem. Series 132, ed. D. Forster and J. F. Roth, ACS, p.240 (1974).

Transition Metal Clusters
Edited by Brian F.G. Johnson
© 1980 John Wiley & Sons Ltd.

CHAPTER 9

Electrons in Transition Metal Cluster Carbonyls

R.G. Wooley

Department of Physics, Cavendish Laboratory,
University of Cambridge, Cambridge, U.K.

1. INTRODUCTION

In recent years it has become evident that chemical compounds in which metal atoms are part of a chemically bonded metal-metal network are numerous and widely distributed throughout the Periodic Table. A significant and interesting class of such compounds consists of clusters of transition metal atoms bonded to carbon monoxide with generic formula $M_m(CO)_n^{2Q-}$, $m \gtrsim 4$, $4m \gtrsim n \gtrsim 2m$, $2 \gtrsim Q \gtrsim 0$. Metal cluster carbonyls are stable, stoichiometric compounds that can be prepared in pure form and have attracted much interest recently not least because of an obvious analogy with chemical processes involving metal particles. Whereas mononuclear and/or binuclear binary carbonyls are formed with most of the elements of the transition metal block, cluster carbonyl formation appears to be confined almost exclusively to the Group VIII transition metals, with the exception of Pd.

VIII			IB
Fe	Co	Ni	Cu
Ru	Rh	(Pd)	Ag
Os	Ir	Pt	Au

   Although much is known about cluster chemistry
and the utility of cluster compounds in homogeneous
catalysis (1,2), rather little is known about their
electronic properties simply because few relevant
experiments have so far been conducted. From the
theoretical point of view a priori molecular orbital
methods, even if restricted to the valence electrons,
seem out of the question because of the enormous
computational effort involved, and hence the best
procedure we can follow is to use all the available
experimental information to guide the theoretical
framework. I shall therefore begin by reviewing
briefly experimental information from three
particular areas namely (i) the bonding systematics
of chemisorption of CO on transition metals at low
coverage (3), (ii) valence electron photoemission
experiments on cluster carbonyls and chemisorbed CO
on transition metals (4,5), and (iii) systematics of
cluster carbonyl stereochemistry and fluxionality (6,
7), (see also Chapter 7). Taken together this
experimental information suggests that the
intuitively appealing physical picture of a small
metal particle encased in a sheath of dielectric
material (a monolayer of carbon monoxide) provides a
useful framework for a discussion of the
characteristics of the electronic structure of
transition metal cluster carbonyls: such a model
need not require a close correspondence between the
chemical reactivity of metal cluster carbonyls, and
analogous transition metal-CO chemisorption systems.

2. REVIEW AND DISCUSSION OF SOME EXPERIMENTAL EVIDENCE

In recent years chemisorption of first-row diatomic molecules on transition metal surfaces has been extensively studied by a variety of techniques, and one can now fairly confidently describe the systematics of adsorbate bonding in relation to the Periodic Table. At low coverage one finds that chemisorption can take place either through dissociative adsorption (D) or by associative adsorption (A) in which the diatomic molecule largely preserves its identity as a ligand. Figure 1 summarizes the known systematics for carbon monoxide chemisorbed on transition metal surfaces at low coverage (3). It is evident that one of the characteristic energies that should be borne in mind when looking at these systematics is the energy required to rupture the molecular bond (the diatomic dissociation energy) E_b: some values are given below (8).

Molecule	E_b (e.V.)[†]
O_2	4.8
NO	6.5
N_2	9.8
CO	11.1

[†]1 e.V. = 96.49 kJ mol^{-1}

IV	V	VI	VII		VIII			IB
Ti	V	Cr	Mn	Fe	Co	Ni		Cu
D				D/A	A	A		A
Zr	Nb	Mo	Tc	Ru	Rh	Pd		Ag
	D	D		A	A	A		A
Hf	Ta	W	Re	Os	Ir	Pt		Au
	D	D/A	D/A	A	A	A		A

D = Dissociative Adsorption   CO → C* + O*

A = Associative Adsorption   CO → CO*

Figure 1

Systematics of CO chemisorption on transition metal surfaces (ref. 3).

Experimentally, one finds that the dividing line between dissociative/associative adsorption moves steadily to the right across the Periodic Table as the energy E_b decreases.

Although there is no quantitative theory of this general behaviour it seems intuitively reasonable to suppose that dissociation occurs via a charge transfer mechanism in which electrons are transferred from some one-electron orbital in the metal, $|\phi^{metal}\rangle$, with energy E_{metal}, to a suitable empty (antibonding) orbital of the perturbed CO molecule (9): since the carbonyl 2π levels are not seen in photoemission experiments involving chemisorbed CO, we conclude that these levels are pushed up above the Fermi energy of the metal and are the probable acceptor orbitals; the orbital $|2\pi\rangle$ has energy $E_{2\pi}$ in the perturbed molecule. An energy E_c, associated with such a charge transfer process mediated by an interaction potential V, has the form (frontier orbital theory)

$$E_c \simeq \frac{|<\phi^{metal}|V|2\pi>|^2}{E_{metal}-E_{2\pi}}$$ 2-1

but which characteristic energy in the metal should we choose? There are really only two possibilities: we could reasonably take E_{metal} to be the Fermi energy, E_F i.e. the energy of the highest filled one-electron orbital = - (work function + surface dipole energy), or alternatively we could choose the energy of the highest filled localized valence levels in the metal, i.e. the centre of the d-band, C_d. If we put $E_{metal} = E_F$ this implies that the orbital $|\phi^{metal}>$ from which an electron is transferred, is an extended electron state (Bloch state) at the Fermi energy made up principally of atomic s and p orbitals: on the other hand, if we choose $E_{metal} = C_d$, $|\phi^{metal}>$ must be a localized orbital of mainly atomic d-orbital character. It is evident from Figure 2 which displays E_F and C_d for the 4d transition metals (10) that it is C_d rather than E_F that shows an asymmetric change across the Period with (C_d-E_F) steadily becoming more negative: thus it appears that we find molecular chemisorption when the centre of the metal d-band has moved far enough down to eliminate a significant probability of charge transfer. This interpretation implies that $|<\phi^{Bloch}|V|2\pi>|<<|<\phi^d|V|2\pi>|$ since $|E_F-E_{2\pi}|<<|C_d-E_{2\pi}|$. From the theoretical point of view this result is expected: the carbon monoxide molecules bond to specific sites on the metal surface and the inter-action of any one molecule with the surface is localized (9). In order to form the spatially localized metal orbital required for effective over-lap with the carbonyl orbitals one has to form a

614.

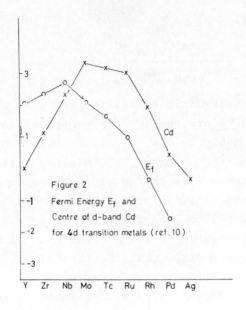

Figure 2
Fermi Energy E_f and
Centre of d-band Cd
for 4d transition metals (ref.10)

linear superposition of Bloch functions through the
whole d-band, and thus the energy of $|\phi^d\rangle$ must be C_d,
the centre of the d-band. Note particularly that the
metals that allow molecular chemisorption of CO are
also the metals that form polynuclear metal cluster
carbonyls (Group VIII) and that chemisorbed diatomic
species on the noble metals (IB) have very low
binding energies. I interpret this correlation as
evidence that the transition metal d-electrons play
an important role in cluster carbonyl molecules.

Further evidence to support this view comes from
a recent photoemission study of cluster carbonyls in
which both gas-phase and thin film measurements were
made (4). It appears that transition metal carbonyl
clusters with more than ~3 transition metal atoms
have (UPS) photoemission spectra that correspond very
well with the photoemission spectra of CO chemisorbed

on the surface of the bulk metal: such comparisons
have been made for $Ru_3(CO)_{12}$, $Ir_4(CO)_{12}$ (4) and
$Os_6(CO)_{18}$ (5). These spectra all show the same
general features: in both carbonyls and chemisorbed
CO systems ionization begins with transitions out of
the metal "d-band" which typically for clusters
extends 3-4 e.V. below the ionization threshold.
Then below the metal d-band and separated from it by
$\gtrsim 1$ e.V. is a large peak that is assigned to the
carbonyl "1π" and "5σ" levels, which are not differ-
entiated: this peak is typically 2-3 e.V. wide, and
below it is a smaller peak that is assigned to the
carbonyl "4σ" levels (c.f. Figure 7). We follow the
usual practice of identifying these transitions with
the quantum labels of the isolated subsystems in
quotation marks e.g. "d-band" for the bulk metal or
metal cluster, "5σ" for the carbon monoxide molecules.
As noted above the "2π" levels are not seen in photoemission and must therefore be pushed up in energy by
several volts with respect to free CO. The work
function ϕ of a transition metal surface with a
monolayer of carbon monoxide is increased by $\leq \frac{3}{4}$ e.V.
as compared with the clean surface for which ϕ
typically is ~ 5 e.V. (11). On the other hand, the
ionization potentials of cluster carbonyls are
several volts larger e.g. the ionization threshold
for $Os_6(CO)_{18}$ is ~ 8 e.V. (5), and this involves a
significant shift of a bulk metal property.

As might be guessed from their name, transition
metal cluster carbonyls can be visualized as
consisting of an interior polyhedron of metal atoms
with metal-metal nearest-neighbour distances
comparable to (usually a little larger) those found
in the bulk metal (6), surrounded by an approximately
spherical shell of carbonyl groups bonded to the

metal cluster through the carbon atom and pointing
out "radially" from the metal polyhedron with nearest-
neighbour carbonyl contacts somewhat shorter than the
Van der Waals distance for carbon (12). At
relatively low temperatures in solution, ^{13}C n.m.r.
spectroscopy shows that the carbonyl groups in many
cluster carbonyls are mobile over part or all of the
metal cluster: recent n.m.r. experiments involving
nuclear resonance of the metal nuclei (e.g. Pt, Rh)
have also shown fluxional behaviour in the metal
cluster at low temperatures (7). Activation free
energies for these processes are commonly $\lesssim 50$ kJ mol^{-1}
and one may conclude that the equilibrium nuclear
configuration for the ground electronic state is
readily deformed into other configurations with low
excitation energies from the (nuclear) ground state.

The comparison of cluster carbonyls with CO-
transition metal chemisorption systems strongly
suggests that the local electronic environment of a
transition metal atom in a metal cluster carbonyl
cannot be very different from that of a metal atom at
a metal surface or even in the bulk metal, and that
the major differences should be accounted for by the
differences in coordination number, and by the
presence of the dielectric layer (the carbonyls). On
the other hand fluxionality of both the metal cluster
and the carbonyl groups surely indicates that there
are a large number of possible cluster structures
with energies similar to the ground state structure
that can be excited by thermal energies: these
structures share the property that the local
coordination of a metal atom or carbonyl group does
not vary much (13). Thus a classical (localized)
bond description of the electronic structure of these
systems seems inappropriate since the metal atoms and

ligands do not sit in well defined (deep) potential
wells in the potential energy hypersurface governing
the nuclear motion, although of course the radial
("breathing") vibrations of the cluster can be
described as small-amplitude vibrations in the
conventional manner. One should not therefore expect
to be able to pick out a unique solution for the
lowest energy (equilibrium) nuclear configuration on
the basis of simple molecular orbital arguments.

This view of carbonyl clusters is rather
different from the popular approach to cluster
bonding in the chemistry literature in which one first
looks at $M(CO)_n$ (n = 2-5) fragments using semi-
empirical molecular orbital theory, and subsequently
attempts to assess what happens when the fragments
are put together (14). The main theme of this
chapter however is that the local electronic state of
a transition metal atom in a cluster carbonyl is much
closer to that found in the bulk metal than to the
isolated metal atom or isolated metal atom plus a few
ligands. I shall further argue that the carbonyl
ligands are weakly perturbed by their interaction with
the metal cluster, and that for example much of the
width of the photoemission peaks assigned to carbonyl
levels can be ascribed to the mutual interaction of
the ligands on their own when they are arranged in a
polyhedral shell of appropriate geometry. Thus a
useful way to approach the electronic structure of
these metal cluster compounds is to begin by
considering separately small clusters of metal atoms,
regarded as (possibly distorted) fragments of the bulk
metal in the sense that their local coordination can
be described in terms of close-packed spheres with
metallic or Wigner-Seitz radii (11), and polyhedral
shells of carbonyl ligands disposed about an empty

618.

spherical cavity and then finally to put the two parts
together. This programme has a qualitative part
concerned with the physical description of the
electronic interactions (e.g. the relative positions
of the metal and ligand levels, strengths of interactions, bonding mechanisms), and a quantitative part
requiring accurate one-electron molecular orbital
theory. We have not yet achieved the latter goal
although some progress has been made (15): however a
valuable qualitative insight into the electronic
structure of these cluster carbonyls can already be
described. In order to carry through such a
qualitative discussion it will be useful to begin by
reviewing some specifics of the modern one-electron
theory of the bulk transition metals.

3. ELECTRONS IN TRANSITION METALS

It is now clearly recognized that specific
electronic factors govern the principal physical
properties of the bulk transition metals such as their
equilibrium atomic volumes, crystal structures and
cohesive energies. Progress in the band theory of
metals has been such that the variation in these
quantities across the Periodic Table can now be
accurately predicted from first principles
calculations (16). Figure 3 shows the variation in
Wigner-Seitz radius ($\simeq$ metallic radius, see below)
and cohesive energy for the 4d transition metals. If
the valence d-electrons could be ignored the radius
of the atom would decrease monotonically across each
period since the valence s-orbital contracts uniformly
as nuclear charge increases. In practice we find a
distinct parabolic trend in each series with an
anomalous behaviour only in the magnetic 3d-metals.
It is now possible to perform quantitative band-theory
calculations using the local density functional method
that reproduce the curves in Figure 3 to within 1-2%
(16). Moreover if one calculates the equilibrium
properties for the 3d-metals constrained to a non-
magnetic state one indeed finds the parabolic
behaviour displayed by the heavier metals. Figure 3

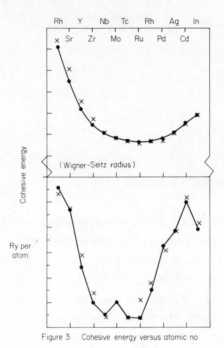

Figure 3   Cohesive energy versus atomic no

shows that a similar parabolic trend is found in the
cohesive energies of the transition metals and this is
also ascribed to the d-electron interactions - the
maximum cohesive energy coincides with a half-filled
d-band (10,17). The Wigner-Seitz radii (11), S, and
metallic radii, R, in $\overset{\circ}{A}$, for the Group VIII metals are
compared below (S>R):

Fe	Co	Ni
1.41, 1.26	1.38, 1.25	1.36, 1.25
Ru	Rh	Pd
1.48, 1.34	1.49, 1.34	1.51, 1.37
Os	Ir	Pt
1.50, 1.35	1.49, 1.36	1.52, 1.39

The metallic radii are taken from A. F. Wells,
<u>Structural Inorganic Chemistry</u>, 4th Ed[n], p.1022, 1976.

In transition metals the d-electrons remain
fairly localized and occupy a narrow band of states,
whereas the valence s,p orbitals overlap very strongly
and form a broad free-electron like band of states.
The d-sp hybridization is fairly constant across each
series, and it is therefore useful to speak of "d" and
"sp" contributions $n_d(E), n_{sp}(E)$ respectively to the
density of one-electron states $n(E)$. A schematic
representation of $n(E)$ for a transition metal is shown
in Figure 4.

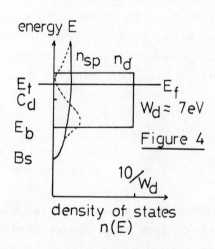

Figure 4

For <u>partially filled</u> "d-bands", the d-electron interactions are strongly attractive, whereas the "sp" electrons give rise to repulsive interactions between the atoms, because as atoms are brought together to form the solid, the "sp" electrons are squeezed more and more into neighbouring atomic cores from where they are repelled by orthogonality constraints. Compression also raises the centre of the d-band, C_d, with respect to the atomic d-level because of the correlation energy, but the overall "d"-contribution remains attractive at equilibrium. Thus as the atoms are brought together equilibrium will be achieved when the <u>repulsive "sp" force</u> (pressure) has increased sufficiently to balance the <u>attractive</u> "d"-electron force between adjacent metal atoms. The equilibrium condition can then be written in terms of the electronic partial pressures P_1 (1 = angular momentum quantum number 0, 1, 2.... or label s, p, d....) at an atomic site,

$$P_d + P_{sp} \approx 0 \qquad\qquad 3-1$$
$$P_d < 0, \; P_{sp} > 0 \text{ near equilibrium} \qquad 3-2$$
$$\text{atomic volume}$$

The bonding between transition metal atoms is thus rather different from that encountered with main group elements (sp bonded materials). Usually we think that large orbital overlap is directly related to significant net bonding for incomplete shells, but this qualitative argument is spoilt in transition metals because it fails to take into account the large energy shifts due to the core repulsion experienced by the s, p electrons in structures near equilibrium atomic volume. The cores of transition metal atoms occupy a very significant fraction, η, of the equilibrium atomic volume ($\cdot 2 \lesssim \eta \lesssim \cdot 35$) compared with $\lesssim \cdot 1$ in s, p bonded metals. As a result in transition metals the bottom of the conduction band, B_s, which corresponds to the lowest s, p bonding (valence) molecular orbital in a finite system, may be close to or even above the s-orbital energy of the free metal atom, when the solid is near its equilibrium volume. These energies for the equilibrium structures of the 4d-metals in Group VIII are given below (10).

Metal	$-(B_s - E_s^{atom})$ e.V.
Ru	0.3
Rh	1.3
Pd	2.0

Using the Virial Theorem Pettifor (18) has shown that the cohesive energy per atom, U, of a transition metal can be conveniently obtained by integrating the partial pressures P_1 as a function of atomic volume, and so U can be given an angular momentum decomposition into the different orbital contributions to the cohesion, $U = U_{sp} + U_d$. The s(p) electrons give rise to the form expected for an electron gas with Fermi wavevector k_F weakly perturbed by a potential,

$$U_s \sim N_s(B_s - \varepsilon_{xc}) + 3k_F^2/5m_s \qquad \text{3-3}$$

where $N_s = \int^{E_F} dE n_s(E)$, is the number of s-electrons, B_s = energy of bottom of conduction band, ε_{xc} is the exchange-correlation potential at the atomic sphere boundary $\simeq -3/2S$ where S = Wigner-Seitz radius (see below), and $m_s \simeq 1$ is the effective mass of the electrons. On the other hand the d-electron contribution, U_d, arises mainly from the so-called "bonding integral" μ_1,

$$\mu_1 = \int^{E_F} dE(E-C_d)n_d(E) \approx U_d \qquad 3-4$$

where C_d, the central energy of the d-band is chosen
by the requirement that $\mu_1 = 0$ for a full band
(10 d-electrons per atom). Formula (3-4) says that
μ_1, and thus the cohesive energy U_d, is calculated as
a sum over the energies E_i (measured from C_d the
centre of the d-band) of the occupied molecular
orbitals Ψ_i, weighed by the probability that a
d-electron (i.e. an electron in a d-orbital) will be
found in molecular orbital Ψ_i.

In a close-packed elemental solid such as a
transition metal it is an excellent approximation to
treat each atom as a spherical volume with radius S
chosen so that $4\pi S^3/3 = \Omega_{ws}$, the volume of the Wigner-
Seitz cell. The value of 2S is therefore larger than
the nearest neighbour separation distance 2R (by about
10%) so that the spheres overlap somewhat. In the
local density functional formalism we define a
potential about each atomic site, and use its
spherical component with the radial Schrödinger
equation to define the "atomic orbitals" for the
system: these orbitals are obviously not the same as
the atomic orbitals of the free atom, but they are
much more useful since we can use them to set up a
simple, transparent but accurate molecular orbital
theory of the solid in the LCAO framework (19-21), in
which the dependence of the molecular orbital energies
on the crystal structure and the atomic potentials is
clearly separated. The simplest form of this theory
is called the Atomic Sphere Approximation and we shall
briefly summarize some of the results of the ASA for
the d-bands of transition metals (10, 21, 22).

    In the Slater-Koster (23) form of LCAO theory
the molecular orbitals are the eigenfunctions of a
Hamiltonian matrix $\underline{\underline{H}}$ in which the diagonal elements
H_{ii} are effective orbital energies ("self-energies"),
and the off-diagonal elements, H_{ij}, are two-centre
matrix elements ("hopping integrals"): thus $\underline{\underline{H}}$ has
essentially the same structure as in the Extended
Hückel method (24) where the analogous quantities are
referred to as "VSIP's" and "resonance integrals".
The ASA leads to a particularly simple and accurate
prescription for choosing the hopping integrals for
the d-electrons. The d-orbital radial function $R_d(r)$
obtained as indicated above, with energy C_d, is
normalized to unity within the atomic sphere,

$$\int_0^S dr\, r^2 R_d^2(r) = 1 \qquad\qquad 3-5$$

If we now define the "effective mass" for the
d-electrons by (20,21),

$$m_d = \frac{2}{S^3 R_d^2(S)} \qquad\qquad 3-6$$

i.e. inversely proportional to the probability of
finding a d-electron at the boundary of the atom ($r =$
S), the Slater-Koster hopping integrals between two
transition metal atoms A and B a distance R_{AB} apart
are given in the ASA by,

$$<d_\lambda^A | H | d_\lambda^B> \equiv dd\lambda = -\frac{10 A_\lambda}{m_d S^2}\left(\frac{S}{R_{AB}}\right)^5 \qquad\qquad 3-7$$

where $\lambda = \sigma,\ \pi,\ \delta$ and,

$$A_\sigma : A_\pi : A_\delta = +6 : -4 : +1 \qquad\qquad 3.8$$

A typical value for m_d in a transition metal is ~ 5

(m = 1 for free electrons). Andersen (21) has shown
that the simple formula

$$W_d \simeq \frac{25}{m_d S^2}$$
3-9

gives the width of the d-band W_d in close-packed
transition metals with an error $\leq$ 10%. Since values
for S and W_d are published in the literature (10,11)
one can use equation 3-9 to calculate values for m_d
and so generate values for the hopping integrals ddλ:
if these are put into the Slater-Koster secular
equation one has a very simple means of studying the
variation of the d-band energies with structure in a
cluster.

General arguments show (25,26) that the r.m.s.
width of the d-band $\bar{W}_d$ is directly related to the
square root of the second moment of the d-electron
density of states μ_2, integrated over the whole band
(filled and unfilled states)

$$\mu_2 = \int dE E^2 n_d(E):$$
3-10

since the hopping integrals (3-7) vary as R^{-5} it is a
reasonable approximation to retain only nearest
neighbour interactions and one then finds that

$$\mu_2 = Z\beta^2$$
3-11

where Z = the coordination number of a metal atom
(Z = 12 in the bulk F.C.C. and h.c.p. structures, = 4
in an octahedron), and β^2 is given by (25)

$$5\beta^2 = dd\sigma^2 + 2dd\pi^2 + 2dd\delta^2$$
3-12

We then have that (26)

$$\bar{W}_d \simeq C\sqrt{\mu_2} = \frac{C'\sqrt{Z}}{R^5} \qquad\qquad 3\text{-}13$$

where the constants are unimportant: if we now assume
that the one-electron states in the d-band are
uniformly distributed, as in the rectangular density
of states shown in Figure 4, $\bar{W}_d \simeq W_d$, the actual width
of the d-band (17).

On the other hand we can easily evaluate the
bonding integral μ_1, (3-4) which determines the
cohesive energy, for this model density of states.
We have that

$$n_d(E) = \begin{cases} 10/W_d & E_b < E < E_t \\ 0 & \text{otherwise} \end{cases} \qquad 3\text{-}14$$

with $W_d = E_t - E_b$. The number of d-electrons N_d is
given by

$$N_d = \int^{E_F} dE\, n_d(E) \equiv \frac{10(E_F - E_b)}{(E_t - E_b)} \qquad 3\text{-}15$$

and since,

$$\int^{E_t} dE(E - C_d) n_d(E) = 0 \qquad\qquad 3\text{-}16$$

by definition, we must have $C_d = \tfrac{1}{2}(E_t + E_b)$. Then we
easily find

$$\int^{E_F} dE\, E n_d(E) = \frac{N_d^2 W_d}{10} \qquad\qquad 3\text{-}17$$

so that

$$\mu_1 = \frac{W_d N_d(N_d - 10)}{20} \simeq U_d \qquad\qquad 3\text{-}18$$

μ_1 obviously takes its maximum value $-5W_d/4$ when
$N_d = 5$, corresponding to all bonding orbitals filled
and all antibonding orbitals empty, and (3-18) is
consistent with the trend in cohesive energy shown in
Figure 3. Thus this simple model allows us to
represent the cohesive energy per atom in a transition
metal in the form,

$$U \simeq CN_d(10-N_d) \sqrt{Z}\left(\frac{S}{R}\right)^5 \qquad\qquad 3\text{-}19$$

by combining equations (3-13) and (3-18) where C is a
constant. Although we cannot expect the ASA to yield
a quantitative determination of the numerical factors
in (3-19) we would certainly expect this equation to
be useful for a semi-empirical treatment of cohesive
energies and in §7 we shall apply it to the metal
clusters in metal cluster carbonyls.

4. ELECTRONS IN SMALL METAL CLUSTERS

The only finite homonuclear systems involving
transition metal atoms for which quantitative
molecular orbital studies have been carried out are
the diatomic molecules formed by the first series of
transition metals (3d metals)(27,28). The latest
calculations (28) use the local spin density
functional formalism with the linear muffin-tin
orbital method (20), and involve no adjustable
parameters. The great dearth of experimental data
on transition metal diatomics means that only very
limited comparisons can be made with the theoretical
results for binding energies E_b, force constants k_e
and equilibrium internuclear separation r_e: however
it is encouraging to find that the theoretical values
of k_e and r_e satisfy Badger's rule (29)

$$(r_e - r_o)k_e^3 = C \qquad\qquad 4-1$$

with the usual accuracy (within 5% (8)) and with the
same constants r_o and C as other diatomic molecules
from the first Long Period for which spectroscopic
data is available. These calculations suggest the
following picture of bond formation. At large
separations, bonding 4s-like orbitals provide an

attractive force which, neglecting the effects of spin
and d-electrons, would give a bonding contribution
varying from ~0.5 e.V. (K_2) to ~2 e.V. (Cu_2), the
increase resulting from the contraction of the outer
parts of the 4s-orbitals. As d-orbital overlap
becomes substantial, an additional force draws the
nuclei together and increases the binding energy.
The equilibrium separation is determined by the point
where this force, together with the 4s-force which
becomes repulsive at small r_e due to core penetration,
balances the electrostatic repulsion due to the
incomplete screening of the nuclei (28). In the
ground electronic state, the equilibrium bond-lengths
of the molecules at the beginning and end of the 3d
transition series are less than the nearest-neighbour
distances in the bulk metals especially at the right-
hand end of the series (Group VIII) where

$$r_e \simeq R_{bulk} - 0.4 \text{ Å}. \qquad\qquad 4\text{-}2$$

The binding in these systems is accordingly
attributed mainly to the overlap of the d-orbitals
since at these short distances core penetration
cancels out most of the energy gain from the strong
overlap of the 4s-orbitals. In simplistic terms one
can say that relative to the bulk metal the repulsive
s(p) force can be substantially alleviated as the s-
electron density "spills out" into empty space, and
one thus expects contraction of the bond under the
influence of the attractive d-electron force, (this
argument is applied to the $[M_3(CO)_6]_n^=$ clusters in §6,
M = Pt, Ni). The magnitude of the binding energy of
the diatomic molecule is more difficult to calculate
since the spin correlation energy between low-spin and
high-spin configurations must be taken into account.

However in a cluster of N atoms this spin energy increases as N whereas the bonding energy due to orbital overlap increases more nearly as N^2 since any one atom can bind to more than one neighbour: hence in larger systems the spin correlation energy becomes unimportant and we will not need to consider it further except to caution that comparisons between metal clusters and _magnetic_ bulk metals are of doubtful value since the magnetism has a considerable effect on the equilibrium properties such as nearest-neighbour distance, cohesive energies etc. (17). In summary, we can say that for the bulk metals (16,17) and the metal diatomic molecules (28) we have a _quantitative_ molecular orbital theory which justifies these simple notions of competition between attractive d-electron bonding and the repulsive s(p) pressure due to core penetration by the strongly overlapping valence s, p orbitals. For larger metal clusters and their compounds we have only much poorer molecular orbital theories which give partially contradictory results: however it seems reasonable to expect an important bonding role for the metal d-electrons in any system containing aggregates of transition metal atoms, and this view is certainly consistent with the experimental information discussed in §2.

In general the solid state physics versions of molecular orbital theory suggest that metal clusters M_m, $m \geqslant 4$ with M-M bond-lengths ~ bulk nearest-neighbour distances, have gross electronic structures that can be related to the bulk metal electronic structures shown in Figure 4: the broad continuous s(p)-electron density of states $n_s(E)$ breaks up into a number of discrete states that overlap and hybridize with the narrow cluster "d-band". As an example, we can look at the width of the cluster "d-band", which according to equation (3-13) should scale as $W_d \propto \sqrt{Z}$ where Z is the number of nearest neighbours: in an octahedral cluster, M_6, we have Z = 4, so that compared with a bulk f.c.c. or h.c.p. structure (Z = 12) we expect

$$\frac{W_d^{\text{octahedron}}}{W_d^{\text{bulk}}} \sim \frac{1}{\sqrt{3}} \qquad\qquad 4\text{-}3$$

provided that the bond-lengths are the same. Most solid state physics methods agree with this sort of scaling of d-band widths in small metal clusters (15, 22, 30-32). By contrast conventional semi-empirical quantum chemical methods like CNDO and Extended Hückel often give much narrower "d-bands" in such clusters (33, 34), for example in an EH calculation for Co_6, Mingos (33) found $W_d^{\text{Oct}} \sim 1$ e.V. whereas scaling of the bulk d-band width for Co (11) as above, (4-3), would give $W_d^{\text{Oct}} \sim 2\frac{1}{2}$ e.V., and thus a markedly different d-electron contribution to the cohesive energy if equation (3-18) holds approximately in a cluster. Harris and Jones (28) remark that many EH results have been highly parameter dependent, and in general one would not expect the simple prescription of making the resonance integrals H_{ij} simply proportional to the orbital overlap integral S_{ij} (24) to work well in polynuclear transition metal systems where core penetration effects that destabilize s,p electron interactions are important (10, 17, 28), and "d-orbitals" are significantly less contracted than free atom d-orbitals (10, 35). Thus in the absence of evidence to the contrary we shall suppose that small metal clusters can be related to the bulk metal as indicated above, as far as gross electronic properties are concerned.

5. AGGREGATES OF CARBON MONOXIDE

Since the carbonyl groups are fairly well
separated in cluster carbonyls it is instructive to
study the intermolecular potential $\Phi(r)$ of a set of
carbon monoxide molecules distributed over the surface
of an empty spherical cavity of radius r in different
polytope arrangements. The interaction energy between
a pair of identical molecules i and j can be written
in the form,

$$V(r_{ij}) \simeq \underbrace{\frac{-\frac{3}{4}I\alpha^2}{r_{ij}}}_{\substack{\text{London} \\ \text{dispersion} \\ \text{formula}}} + \underbrace{\varepsilon \, \exp\left\{-t\left(\frac{r_{ij}-r_m}{r_m}\right)\right\}}_{\substack{\text{repulsion due to} \\ \text{overlap of charge} \\ \text{clouds}}} \qquad 5\text{-}1$$

where (numerical values refer to CO),

v_{ij} = distance between centres of molecules i
and j treated as spheres

I = first ionization potential $\simeq 13\frac{1}{2}$ e.V.

α = scalar polarizability $\simeq 2$ $\overset{o}{A}^3$

r_m = position of minimum in $V_{ij} \simeq 4$ $\overset{o}{A}$

$2\varepsilon/k_B$ = depth of minimum in V_{ij} = $65^{\circ}K$

t = measure of steepness of repulsive wall =

17

634.

Then,

$$\Phi(r) = \tfrac{1}{2} \sum_{\substack{\text{pairs} \\ i,j}} V(r_{ij}(r)) \qquad\qquad 5\text{-}2$$

Figure 5 shows typical results for 12 carbonyls distributed over a sphere of radius r (36): the Van der Waals interaction leads to a shallow minimum in the potential e.g. $\Phi_{min} \sim -0.5$ e.V. for the most stable polytope, the icosahedron with $r_{min} \simeq 3.3$ Å. The r value for which $\Phi(r_{HS}) \simeq 0$ gives the hard-sphere radius of the carbonyls: numerous $M_n(CO)_{12}$ species are known that have carbonyl arrangements close to the hard-sphere prediction (12). Since $r_{HS} < r_{min}$ we conclude that (a) the observed CO polyhedra are not

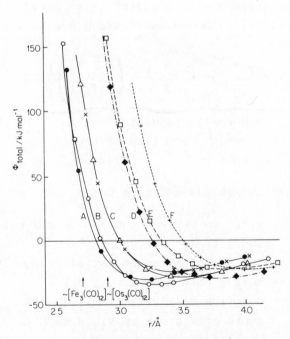

Figure 5 Φ total / kJ mole^{-1} versus r/Å

equilibrium structures and must be stabilized by metal cluster-ligand interactions and (b) the metal cluster evidently "sucks in" the carbonyl shell until the steep repulsive wall of $\Phi(r)$ is encountered. Thus the long-range behaviour of $\Phi(r)$ beyond r_{min} is irrelevant and only the dependence of $\Phi(r)$ on polytope arrangement for $r \approx r_{HS}$ is likely to be important for the ground state energy.

It is also possible to study these shells of carbon monoxide molecules using molecular orbital theory. In the local density functional formalism that underlies the chemical pseudopotential method (15, 37, 38) one cannot expect to account for the Van der Waals contribution to the energy which must be dealt with separately, and so a molecular orbital calculation gives information about the repulsive part of the intermolecular potential. As a group of carbon monoxide molecules are brought together to form the polyhedral shell found in metal cluster carbonyls, the individual orbital levels of the isolated molecule (1π, 5σ, 2π etc.) interact with one another and broaden out into "bands" of levels. Thus in a planar D_{3h} configuration for $(CO)_6$ (at distances appropriate to the clusters $M_3(CO)_6^=$, M = Ni, Pt), the 5σ-orbitals, which point in to the centre of the structure, and the π-orbitals that are perpendicular to the molecular plane (the π_z levels) do not mix with one another strongly and give "bands" of levels of width $\lesssim 1$ e.V. The in-plane π-orbitals do, however, overlap appreciably and the isolated molecule π_y levels fan out into "bands" with widths $\lesssim 2$ e.V. Nevertheless the various bands of states are clearly identifiable since they do not overlap one another and it is therefore useful to continue to label each band by the symmetry species (in $C_{\infty v}$) of the isolated molecule from which

it originates. When two of these planar units are
brought together to form $(CO)_{12}$ as in the $M_6(CO)_{12}^=$
species (M=Ni, Pt) the in-plane levels (5σ, π_y) are
relatively unaffected, but there is further interaction
between the π_z levels that overlap parallel to the
three-fold axis of the structure, and this broadens
the π_z bands to widths of $\leq$ 2 e.V. (15). An
essentially similar pattern is found in e.g. an icosahedral structure for 12 carbonyls: as one would expect
the individual bands broaden with increasing local
coordination but do not loose their identity.

Figure 6 shows a schematic density of states for
$(CO)_m$ polyhedra in structures appropriate to metal
cluster carbonyls: it is quite noticeable from these
calculations that the centres of gravity of especially
the 1π and 2π "bands" are shifted upwards, and this
is consistent with the repulsive interaction between
the carbonyl groups predicted by the semi-empirical
calculation of $\Phi(r)$. It is also consistent with the
photoemission results on cluster carbonyls (§2) which
show that the 1π levels are moved up enough in energy
to merge into the bottom of the 5σ band. Figure 6
also shows the corresponding energy levels for an
isolated transition metal atom M_1, and a small
transition metal cluster M_m(m~6) with values
appropriate to rhodium, the central element in the
block of metals that form cluster carbonyls. It
should be noted that this is a schematic representation
and that these finite systems have closely spaced
groups of discrete one-electron energy levels: hence
one can think of Figure 6 as a representation of the
density of states looked at under low resolution. The
upward shift of the metal levels in the cluster has
been calculated by extrapolation from the results
for the bulk metal (10). This concludes the discussion

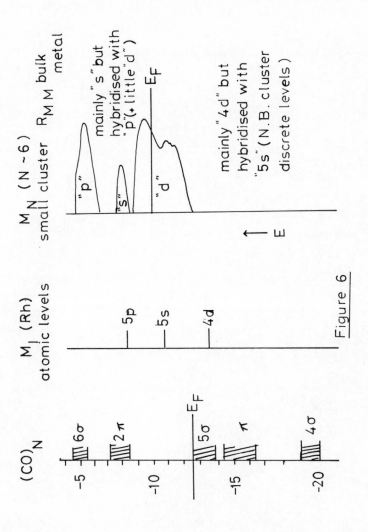

Figure 6

of the metal cluster, M_m, and carbonyl shell, $(CO)_n$, fragments of metal cluster carbonyls, and we must now consider how to put the pieces together.

6. ELECTRONS IN METAL CLUSTER CARBONYLS

Transition metal cluster carbonyls form solids in which the basic packing units are clusters of close-packed metal atoms surrounded by a "sheath" of carbon monoxide molecules. The metal cluster can be enclosed in a sphere of radius A with $3\text{ Å} \lesssim A \lesssim 5\text{ Å}$ for many polynuclear cluster carbonyls. There is a general theorem concerning the effects of the bounding surface on the density of states in an enclosure, namely the "Black-Body Radiation Theorem" of Laue. To quote Kittel (39), the theorem states that the particle density per unit energy range (essentially our density of states $n(E)$) is approximately independent of the form of the boundary at distances from the boundary greater than a characteristic particle wavelength at the energy concerned. To apply the theorem we imagine that initially we have clusters of close-packed metal atoms embedded in the bulk metal and a matrix composed of polyhedral carbonyl shells, and we then transfer the metal clusters into the spherical carbonyl shell cavities and enquire how $n(E)$ might be modified at energies near E_F (see Figures 4 and 6). This requires an estimate of the characteristic wavelengths of the metal electrons at energies close to E_F. The s(p) electrons in the bulk metal form a nearly free electron gas, and so if we have N_s sp electrons in an atomic cell of volume Ω_{ws}, the Fermi wavevector at the Fermi energy for s electrons, k_F^S is given by (40),

$$k_F^S = \left(\frac{3\pi^2 N_s}{\Omega_{ws}}\right)^{1/3} = \frac{1}{S}\left(\frac{9\pi N_s}{4}\right)^{1/3} \qquad 6\text{-}1$$

in terms of the Wigner-Seitz radius S. The Fermi wavelength is then given by,

$$\lambda_F = 2\pi/k_F \qquad 6\text{-}2$$

and since $N_s \lesssim 1$, $S \lesssim 3\ \overset{\circ}{A}$, λ_F^S for the Group VIII metals is $\sim 5\ \overset{\circ}{A}$. Hence $\lambda_F^S \gtrsim A$, the carbonyl cavity radius, and we can always expect metal cluster carbonyls to show significant changes in the s-electron density of states in relation to the bulk metal or a bare metal cluster: these changes simply relect the formation of bonding and antibonding orbitals between the metal s-orbitals and suitable carbonyl orbitals (instead of metal-atom orbitals) due to overlap, and possible concomitant changes in the s-d hybridization (recall Figure 4). On the other hand the metal d-electrons are much more localized and since $(N_d/N_s)^{1/3} \sim 2.3$ we have $\lambda_F^d \lesssim 2\ \overset{\circ}{A} < A$: according to the Laue theorem we would thus expect the d-electron density of states in the cluster carbonyl to be related to the density of states in a metal cluster of the same structure.

In general one finds that nearest-neighbour metal-metal distances are somewhat larger than in the bulk metal (6), and this can be interpreted as reflecting the changed environment of the metal cluster s-electrons in going from the bulk metal to the cluster carbonyl. It seems physically reasonable that the presence of the dielectric (insulating) monolayer should increase the confinement of the s-electrons and so increase the repulsive s-electron force between adjacent metal atoms. Since there is no obvious mechanism for enhancing the attractive d-electron force, the equilibrium structures of the carbonyls usually have longer metal-metal bond lengths (c.f. equations 3-1, 3-1). A striking exception is found in the "2-dimensional" cluster carbonyls $M_3(CO)_6^=$ (M = Ni, Pt) and oligomers of this planar fragment: in this case the s-electrons are not confined by carbonyls in the z-direction, perpendicular to the molecular plane, and so just as in the homonuclear metal diatomic molecules (§4), some of the repulsive s(p) force between metal atoms can be relieved by the s-electron density spilling out into the vacuum above and below the molecular plane. As a result the M-M bonds in the $M_3(CO)_6^=$ units are significantly shorter than in the bulk metals, and this remains true in the oligomers $[M_3(CO)_6]_n^=$, n>1 in which interfragment M-M bonds are significantly larger than in the bulk metals (6).

We thus expect that these ideas about competition between repulsive s-electron forces and attractive d-electron forces between metal atoms will apply generally in transition metal cluster compounds, and that the discussion of bonding in metal cluster carbonyls attempted in this chapter should provide a useful starting point for discussion of the bonding in other polynuclear transition metal cluster compounds (halides, chalcogenides etc.). We have just seen that the s-electrons of the metal cluster are much more sensitive to the cluster environment (the ligands) than are the cluster d-electrons which act as the "glue" holding the metal atoms together, so that the cluster geometry should alter in response to changes in the s-electron pressure, P_s, (equation (3-1)) induced by variation in the ligands. On the other hand, when the environment is kept fixed (or nearly so) structures should be governed by the d-electron forces, which, in both cases, provide the principal contribution to the cohesive (binding) energy of the metal cluster.

Apart from these general arguments, overlap and energy considerations suggest that the important orbital interactions when the fragments (M_m, $(CO)_n$) are brought together will involve metal cluster s- and d- orbitals and carbonyl 5σ- and 2π-orbitals (See Figure 6). The bonding mode envisaged here is the conventional synergic bonding mechanism in which the carbonyl 5σ-orbitals act as electron donors while electron density is pushed off the metal cluster through hybridization between the metal d_π-orbitals and the carbonyl "2π" band of levels. Given that the bands of energy levels of the carbonyl shell are fixed, the efficacy of this bonding mode depends strongly on the energies of the metal cluster levels, and the number of metal valence electrons i.e. the position of E_{HOMO} in the metal cluster (see Figure 6). The position of E_{HOMO} in the incompletely filled cluster "d-band", and the position of the metal cluster "s" and "d" bands between E_{HOMO} and E_{LUMO} for the carbonyl shell shown in Figure 6 for Rh, is ideal for cluster carbonyl formation. In the bulk Group IB metals, $N_d = 10$ and the d-electron interactions between metal atoms are weakly repulsive: the cohesion in the metal is due to the overlap of s- and p-orbitals which leads to an attractive force because the metal cores are much smaller. Hybridization between the carbonyl 5σ-orbitals and the metal s-orbitals will weaken cluster bonding, while the metal d-orbitals are no longer useful because C_d is now so low that favourable interaction with the carbonyl "2π" levels is not possible,

and cluster carbonyls are not found with any of the
post transition series metals. On the other hand as
one moves from Group VIII to the earlier transition
metals the energy levels for the metal cluster move
up markedly as may be seen from Figures 2 and 6. In
this case the cluster s-orbital levels become well
removed from the carbonyl 5σ-levels and cannot mix
significantly, while the occupied part of the cluster
"d-band" may move up to, or above, the empty carbonyl
"2π" levels so making a dissociative charge transfer
mechanism likely (c.f. my discussion of the systematics
of CO chemisorption in §2). Even within the Group VIII
metals there are important variations in the electronic
characteristics of the metal fragment, and these must
affect the chemical properties of the cluster carbonyls.
The lack of binary cluster carbonyls of Pd depends
ultimately on these factors: the d-band width of bulk
Pd is much narrower than W_d for the other 4d and 5d
metals of Group VIII (11), and E_F for Pd is more than
1 e.V. lower than for Pt (43), so that Palladium
appears to be rather more like Gold than a Group VIII
metal in relation to the carbonyl energy levels.
Experimental evidence for this view can be seen in the
existence of phosphine substituted carbonyl clusters
like $Pd_3(CO)_3(PPh_3)_3$ and the occurrence of several
gold phosphine clusters e.g. $[Au_9(PPh_3)_8]^{3+}$ although
neither metal has any binary carbonyls. We have
$E_{5\sigma} < E_d < E_s < E_{2\pi}$ in the non-interacting fragments
and this order is thought to be preserved in the metal
cluster carbonyl: thus the photoemission assignments
discussed earlier (§2) are based on this ordering (4,
5). It is useful to describe some electron counting
arguments, bearing in mind that hybridization will
have to be taken into account if we require true
values for N_s, N_d etc. as defined in §3.

The generic cluster carbonyl has the formula

$$M_m(CO)_n^{2Q-} \qquad 0 \leqslant Q \leqslant 2, \; m \geqslant 4 \qquad\qquad 6\text{-}3$$

Metal M has N_M valence electrons with $8 < N_M < 10$ for Group VIII metals. Each carbonyl 5σ ("lone-pair") orbital has 2 electrons so that the total number of valence electrons N_{val} is given by,

$$N_{val} = mN_M + 2n + 2Q \qquad\qquad 6\text{-}4$$

which is normally an even integer. There are $n(5\sigma)$ orbitals, $5m$ d-orbitals and m s-orbitals available (I neglect the carbonyl 2π-levels as these formally remain empty), and the total number of valence orbitals is $N_{orb} = 6m+n$. Thus

$$\Delta \equiv 2N_{orb} - N_{val} = (12-N_M)m - 2Q > 0 \qquad\qquad 6\text{-}5$$

so we always have a partially filled "band".

A common assumption in the chemistry literature has been that the cluster "d-band" is full, and can therefore be ignored (33, 41). Let us make this assumption and see what it implies. A full "d-band" contains $10m$ electrons. Let $(10-N_M) = n_M$ an integer $\geqslant 0$. If we discard the "d-band" we are left with $2n+2Q-mn_M$ electrons requiring $(n+Q-\tfrac{1}{2}mn_M)$ orbitals, and we have a total of $(n+m)$ available orbitals. We can therefore distinguish several possible orbital occupation schemes depending on the magnitudes of Q and mn_M:

(A) $Q>\tfrac{1}{2}mn_M$:   (n-k) carbonyl 5σ-orbitals and $[(Q-\tfrac{1}{2}mn_M+k] \leqslant m$ metal s-orbitals filled $k \geqslant 0$.

(B) $Q=\tfrac{1}{2}mn_M$:   (n-k) carbonyl 5σ-orbitals and $k \leqslant m$ metal s-orbitals filled, $k \geqslant 0$

(C) $Q<\frac{1}{2}mn_M$: 5σ-orbitals incompletely filled
         (necessarily).

The calculation can be illustrated by the following
typical examples, taken from the hexanuclear cluster
carbonyls; N_{val} = 86, m = 6:

$Pt_6(CO)_{12}^{2-}$, n_M = 0, Q = 1   (A)

$Ir_6(CO)_{16}$, n_M = 1, Q = 0   (C)

$Os_6(CO)_{18}^{2-}$, n_M = 2, Q = 1   (C)

This electron counting argument suggests that a
full cluster "d-band" is plausible for $Q>\frac{1}{2}mn_M$ (in
practice this means M = Ni, Pt) but otherwise the
cluster "d-band" can only be filled if some of the 5σ-
orbitals are shifted $\gtrsim$ 4 e.V. in energy (i.e. above
the "d-band"), and the rest are not, and this seems
highly unlikely. It is more natural to suppose that
the n 5σ-orbitals remain below the top of the "d-band"
and are filled with 2n-electrons. The remaining
$(nN_M + 2Q)$ electrons fill $\frac{1}{2}mn_M$ + Q of the available
6m "metal" orbitals. Hence for $N_M<10$ we necessarily
have a partially filled cluster "d-band": it should be
noted that even when the cluster "d-band" is formally
full as with cluster carbonyls of Ni and Pt, the
number of d-electrons, N_d, that enters the cohesive
energy formula, (3-18) is <10, and so the d-electrons
are still important (15). Thus an accurate
representation of the electronic structure of metal
cluster carbonyls near E_F (energy of HOMO) will
certainly involve a careful calculation of the metal
d-electron interactions as a first step, since, as we
now discuss, there is experimental evidence requiring
the existence of a band-gap above E_{HOMO} which usually
lies in the "d-band".

The neutral cluster carbonyls form molecular crystals and it seems reasonable to suppose that they will have a high electrical resistivity: $Os_6(CO)_{18}$ for example has been shown to be a wide band-gap semi-conductor in the solid-state (42). The interactions between clusters can be expected to be rather weak and so the energy bands of the crystal should be closely related to the energy levels of a single cluster molecule. Hence we can conclude that $E_{LUMO}-E_{HOMO}$ for carbonyl clusters are likely to be $\gtrsim$ 1 e.V., a conclusion that is not inconsistent with the very dark colours of many cluster species. Although it is true that the nuclear framework will deform so as to maximize $E_{LUMO}-E_{HOMO}$, by virtue of a generalized Jahn-Teller argument, this process only continues while a lowering of the total energy results, and it is not easy to estimate a priori what gap will result from any given structure: one might interpret fluxionality as indicating the existence of many structures with comparable band gaps. Structural variation, however, is not the only mechanism that can lead to a large value for $E_{LUMO}-E_{HOMO}$. In the third row of transition metals the spin-orbit coupling effects are large: in the bulk metals it is known that the spin-orbit coupling parameter for the d-electrons, averaged over the d-band, is comparable to the atomic value (~1 e.V.) and markedly reduces the density of states at the Fermi energy $n(E_F)$ (43). One may expect similar behaviour in the cluster carbonyls of the heavy (5d) transition metals although spin-orbit effects are likely to be relatively unimportant in 3d and 4d metal clusters. Cluster carbonyls with an even number of electrons are invariably diamagnetic, and since the exchange integral for d-electrons in transition metals is

typically $\frac{3}{4}$-1 e.V. (44) this again suggests
$E_{LUMO}-E_{HOMO} \gtrsim 1$ e.V.

At this point it seems appropriate to recall the
situation in the one-electron theory of the bulk
transition metals (§3): there we found that molecular
orbital theory could give excellent agreement with the
experimental values for equilibrium atomic volumes,
cohesive energies, crystal structure variations etc.,
(16) and could be interpreted in a simple way. The
simple physical interpretation of these properties in.
terms of the different orbital contributions (U_d vs U_s
equations 3-3, 3-4 etc.) is only possible, however,
because full self-consistent calculations have been
performed for each metal, so that one has an accurate
picture of the energy bands in the metals, which can
then be parameterized in terms of simple physical
quantities like B_s, C_d, N_d, N_s, m_s and so on (17).
Such quantitative molecular orbital calculations are
not yet possible for cluster carbonyls, and so we will
conclude this discussion with a more qualitative
account of their energy level structure as inferred
from recent experimental studies (42) and molecular
orbital calculations (15).

We recall that the conventional interpretation of
photoemission spectra for CO chemisorbed on transition
metal surfaces, and metal cluster carbonyls suggests
that the main effect of the interaction between the
cluster and the carbonyl shell is for hybridization
to cause shifts in energy and broadening of the
individual bands of energy levels of the fragments
which preserve their identity. It is reasonable to
suppose that the same is true for the carbonyl "2π"
and "6σ" bands which are moved up in energy with
respect to their position in the carbonyl shell
fragment. Thus a plausible guess for the density of

states for a metal cluster carbonyl can be obtained by
simply superposing the densities of states for the
metal cluster and carbonyl shell, Figure 6, having
first increased the gap between the occupied and
unoccupied states in the carbonyl shell. Such a
picture is broadly in agreement with chemical pseudopotential calculations recently performed on
$[M_3(CO)_6]_n^{2-}$, n = 1, 2, M = Ni, Pt (15). Spectroscopic
investigations of the excited electronic states of
transition metal cluster carbonyls have been very
limited although a study of thin films and single
crystals of $Os_6(CO)_{18}$ has recently been started (42),
and the preliminary findings can be summarized here.

Electrically, both crystals and powder samples
behave as excellent insulators with resistivities
$>>10^{12}$ ohm cm: this corresponds well to the ready
solubility in organic solvents like acetone, and to
the weak interactions between clusters suggested by
the crystal structure of $Os_6(CO)_{18}$ (45). Transmission
infra-red spectroscopy using powder samples pressed in
KBr discs reveals an absorption feature at ~2,000 cm^{-1}
that corresponds closely to the feature found in
solutions and identified with the carbonyl stretching
frequencies. The optical work (transmission and
reflection) suggests that $Os_6(CO)_{18}$ is a semi
conductor with a band gap $\gtrsim$ 1 e.V. and this is in
agreement with the resistivity measurements. Single
crystal reflectivity measurements reveal electronic
transitions at ~2.5 e.V. (±0.2 e.V.), ~4.2 e.V. (±0.2
e.V.) and ~5.2 e.V. The lowest energy feature
corresponds quite well with the first feature in the
solution visible absorption spectrum at ~2.7 e.V.,
although additional broad features at ~3.4 e.V. (±0.4
e.V.) are also seen in the solution spectrum. Electron
energy loss measurements show a strong loss feature at

~8.5 e.V. consistent with a set of strong electronic transitions at ~10 e.V. (±1 e.V.): the main plasmon loss peak is centred at about 20 e.V. and is definitely overlaid by fine structure due to high energy electronic transitions in the range 20-24 e.V. (42).

Gas-phase photoemission studies of $Os_6(CO)_{18}$ show that the ionization threshold is at ~8 e.V., and the cluster "d-band" which shows some fine structure extends down to about 12 e.V. The carbonyl "$5\sigma + 1\pi$" feature peaks at ~$14\frac{1}{2}$ e.V. and extends over ~3 e.V. (5). Although the "4σ" feature was not recorded, experience with other carbonyls and with chemisorbed CO on transition metals suggests that it should lie at ~18 e.V. (4). Thus taking all these results together we can construct a schematic density of states diagram for the cluster $Os_6(CO)_{18}$, Figure 7. It is largely in agreement with an unpublished chemical pseudopotential calculation using the observed crystal structure (46), although our molecular orbital theory does not give the correct gap in the d-band for $E_{LUMO}-E_{HOMO}$: as suggested above it is very likely that explicit inclusion of spin-orbit coupling will be necessary to obtain $E_{LUMO}-E_{HOMO}$ correctly. The visible and near u.v. transitions can be assigned as charge transfer transitions from the metal cluster "d-band" into the empty carbonyl "2π" levels, whereas the intense transitions at ~10 e.V. seen in electron energy loss measurements are presumably due to transitions from the metal cluster into the empty "6σ" orbitals. It seems reasonable to suppose that Figure 7 is a faithful representation of the density of states of metal cluster carbonyls in general, although of course quantitative details like $E_{HOMO}-E_{LUMO}$ and spectral transition energies will require

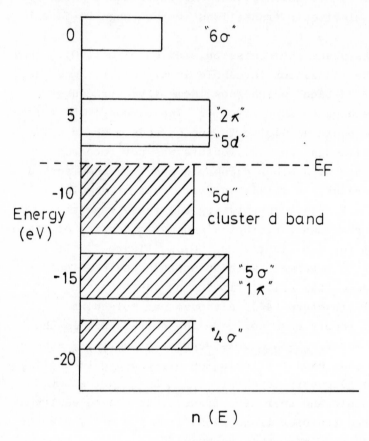

n(E)

Figure 7

accurate computations in individual cases. It is
evident that more experimental studies along the lines
described above for $Os_6(CO)_{18}$ would be highly desirable.

7. BOND STRENGTHS IN METAL CLUSTER CARBONYLS

It has recently been suggested that a realistic
set of bond energies for metal carbonyl clusters can
be obtained by assuming that the metal-metal bonds in
clusters are like those in the metals and that, like
covalent bonds in general, they have energies that
decrease as their length increases (47). This point
of view is closely related to the main theme of this
chapter, although I shall argue that the coordination
number of the metal atom is also an important quantity
and obtain results somewhat different from ref. 47.
We argued in §3 that the cohesive energy in the bulk
metals is dominated by the d-electron contribution
and has the form,

$$U \simeq \frac{A \sqrt{Z}}{R^5} \qquad \qquad 7\text{-}1$$

where R is the nearest-neighbour distance, and Z is
the number of nearest neighbours (coordination number)
c.f. equation 3-19. The constant A is determined by
the number of d-electrons (defined as $\int^{E_F} dE\ n_d(E)$), and
the details of the atomic potential at each site (e.g.
the value of the effective mass m_d, equation 3-6,
which determines the magnitude of the resonance

integrals). For the sake of argument we now propose
that these factors are relatively unchanged in
passing to the metal cluster carbonyl and that the
main change in bond-energy arises from the reduction
in coordination number and the change in nearest-
neighbour distance. The assumption that the constant
A is approximately the same in the bulk metal and
metal cluster carbonyl allows us to write the
contribution from each metal atom to the cohesive
energy of the cluster as,

$$U^{cluster} \simeq \left(\frac{R^{bulk}}{R^{cluster}}\right)^5 \left(\frac{Z^{cluster}}{Z^{bulk}}\right)^{\frac{1}{2}} U^{bulk} \qquad 7-2$$

an equation which is similar to that proposed by
Housecroft et al (47) except that we have an explicit
square root dependence on the coordination number Z.
This is important because of the large change in
coordination number in passing from the bulk metal to
the cluster carbonyl: however, if we use equation
(7-2) to study the variation in $U^{cluster}$ between
cluster carbonyls of the same metal the trends will
probably be similar to those previously reported
because the changes in coordination number are much
smaller (47) (48). If $\Delta H_{disrupt.}$ is the enthalpy of
disruption for the metal cluster carbonyl i.e. the
enthalpy for the process

$$M_m(CO)_n \rightarrow mM(g) + nCO(g) \qquad 7-3$$

and D(M-CO), the metal-carbonyl bond-energy we have,

$$nD(M-CO) = \Delta H_{disrupt} + mU^{cluster} \qquad 7-4$$

after converting $U^{cluster}$ into kJ mol^{-1}. It is more
conventional in the chemistry literature to work with
bond energies rather than the cohesive energy per

atom, and so we define the metal-metal bond-energy b as the cohesive energy per atom divided by half the number of nearest neighbours,

$$b = \frac{2U}{Z} \qquad \text{7-5}$$

If we combine equation 7-2 and 7-5 we can write,

$$b^{cluster} = \left(\frac{Z^{bulk}}{Z^{cluster}}\right)^{\frac{1}{2}} \left(\frac{R^{bulk}}{R^{cluster}}\right)^5 b^{bulk} \qquad \text{7-6}$$

This estimate of $b^{cluster}$ is probably a little too large since a better calculation would result from using a $(Z^{cluster})$effective $> Z^{cluster}$ (but $< Z^{bulk}$ of course), to take account of the fact that each metal atom has some carbonyl ligands with which the d-electrons interact. Put another way, the exponent of $\frac{1}{2}$ in equation 7-6 is probably too large.

As an example of the use of these equations we can look at the metal rhodium and its cluster carbonyls $Rh_4(CO)_{12}$ and $Rh_6(CO)_{16}$ using the data in ref. 47. Bulk rhodium metal has the f.c.c. structure so that $Z^{bulk} = 12$, and $R^{bulk} = 2.69$ Å: in the cluster $Rh_4(CO)_{12}$ we have $Z = 3$, $R = 2.73$ Å, and in $Rh_6(CO)_{16}$ we have $Z = 4$, $R = 2.78$ Å. Finally ΔH disrupt for $Rh_4(CO)_{12} = 2,648$ kJmol^{-1} and for $Rh_6(CO)_{16} = 3,874$ kJ mol^{-1}. The cohesive energy in bulk rhodium is -5.78 e.V./atom. Using these values and equation (7-2) we calculate

$$U^{oct} = 0.49 \ U^{bulk}, \quad U^{tet} = 0.46 \ U^{bulk} \qquad \text{7-7}$$

the reduction occurring mainly because of the large change in Z. Both values when used with equation 7-4 lead to $D(M-CO) \simeq 140$ kJ mol^{-1} which is somewhat less than the values obtained previously (namely 166 kJ mol^{-1} (48), 182 kJ mol^{-1} (47)). On the other hand

using equation (7-6) we obtain $b^{oct} = 136$ kJ mol^{-1},
and $b^{tet} = 173$ kJ mol^{-1}, considerably larger than the
bond energy in the bulk metal, 93 kJ mol^{-1} (47).
Since the d-electron interactions between pairs of
transition metal atoms are attractive, the d-electron
density tends to pile up between the atoms and so if
the coordination number Z is reduced in passing to the
cluster from the bulk one would expect more d-electron
density per pair of atoms to lie in the bonding regions
and hence a larger value for $b^{cluster}$. This effect
will, of course, be eventually outweighed if $R^{cluster}$
increases significantly: in practice metal-metal bond-
lengths in metal cluster carbonyls are close to the
bulk metal values and so one would generally expect
$b^{cluster} > b^{bulk}$.

One last comment should be made about an energy
formula like equation (7-2) which we were lead to in
§3 and have used above. It is generally true that the
second moment of the density of one-electron states,
μ_2, equation 3-10, is proportional to the product of
Z, the number of nearest-neighbours, and a quantity
like β^2, equation 3-12, which is related to the square
of the resonance integrals between adjacent atoms.
If the energy obtained by filling the one-electron
levels is the dominant contribution to the bonding we
will thus have a cohesive energy $\alpha \sqrt{\mu_2} \; \alpha \sqrt{Z} \beta$,
irrespective of the kind of orbitals (s, p, d) that
are interacting to produce the one-electron energy
levels. In this situation we can expect structures
in which Z is maximized (other things being equal!),
and hence the widespread occurrence of polyhedral
structures based on triangulated faces (49, 50) can be
understood in terms of increasing coordination
increasing the cohesive energy compared with more open
structures, independently of any particular bonding
picture based on s, p or d orbitals.

656.

It is a pleasure to acknowledge many stimulating discussions with Professors V. Heine, F.R.S. and J. Lewis, F.R.S., and Drs. D. W. Bullett, B. F. G. Johnson and D. G. Pettifor. I thank Dr. A. D. Yoffe for providing the preliminary results on $Os_6(CO)_{18}$ prior to publication. Financial support under the SRC Advanced Fellowship programme is acknowledged.

8. REFERENCES

1. J. H. Basset and R. Ugo, in R. Ugo, Ed. "Aspects
      of Homogeneous Catalysis" Vol. 3. D. Reidel
      Publishing Co., 1977, p.138.

2. E. Muetterties, Science, $\underline{194}$, 1150 (1976), $\underline{196}$,
      839 (1977).

3. G. Brodén, T. N. Rhodin, C. Brucker, R. Benbow
      and Z. Hurych, Surf. Sci., $\underline{59}$, 593 (1976).

4. E. W. Plummer, W. R. Salaneck and J. S. Miller,
      Phys. Rev., $\underline{B18}$, 1673 (1978).

5. J. Green, E. A. Seddon and D. M. P. Mingos,
      J. Chem. Soc. Chem. Comm., 94 (1979).

6. P. Chini, G. Longoni and V. G. Albano, Adv.
      Organomet. Chem., $\underline{14}$, 285 (1976).

7. J. B. Heaton, private communication (1978).

8. G. Herzberg, Spectra of Diatomic Molecules, 2nd ed.,
      Van Nostrand Reinhold Co. 1960.

9. R. Haydock and A. J. Wilson, Surf. Sci., $\underline{82}$,
      425 (1979).

10. D. G. Pettifor, J. Phys-Metal Physics, $\underline{F7}$, 613
      (1977).

11. R. M. Nieminen and C. H. Hodges, J. Phys-Metal
      Physics, $\underline{F6}$, 573 (1976).

12. B. F. G. Johnson, J. Chem. Soc. Chem. Comm. 211
      (1976).

658.

13. B. F. G. Johnson and R. E. Benfield, J. Chem.
        Soc. Dalton, 1554 (1978).

14. M. Elian and R. Hoffmann, Inorg. Chem., 14, 1058
        (1975).

15. K. W. Chang and R. G. Woolley, J. Phys.-Solid
        State Physics, C12, 2745 (1979).

16. V. L. Moruzzi, A. R. Williams and J. F. Janak,
        Phys. Rev., B15, 2854 (1977).

17. D. G. Pettifor, CALPHAD, 1, 305 (1977).

18. D. G. Pettifor, Comm. on Physics, 1, 141 (1976).

19. O. K. Andersen, Solid State Commun., 13, 133
        (1973).

20. O. K. Andersen and R. G. Woolley, Mol. Phys., 26,
        905 (1973).

21. O. K. Andersen, Phys. Rev., B12, 3060 (1975).

22. O. K. Andersen, W. Klose and H. Nohl, Phys. Rev.,
        B17, 1209 (1978).

23. J. C. Slater and G. F. Koster, Phys. Rev., 94,
        1498 (1954).

24. R. Hoffmann, J. Chem. Phys., 39, 1397 (1963).

25. F. Cyrot-Lackmann and F. Ducastelle, Phys. Rev.,
        B4, 2406 (1971).

26. F. Ducastelle and F. Cyrot-Lackmann, J. Phys.
        Chem. Solids, 31, 1295 (1970).

27. A. B. Anderson, J. Chem. Phys., 66, 5108 (1977).

28. J. Harris and R. O. Jones, J. Chem. Phys., 70,
        830 (1979).

29. R. M. Badger, J. Chem. Phys., 2, 128 (1934).

30. R. P. Messmer, S. K. Knudson, K. H. Johnson,
        J. B. Diamond and C. Y. Yang, Phys. Rev.,
        B13, 1396 (1976).

31. R. P. Messmer, D. R. Salahub, K. H. Johnson and
        C. Y. Yang, Chem. Phys. Letters, 51, 84
        (1977).

32. G. Seifert, E. Mrosan and H. Müller, Phys. Stat.
        Solidi, 89, 553 (1978).

33. D. M. P. Mingos, J. Chem. Soc., A, 133 (1974).

34. R. C. Baetzold, J. du Physique, C2-175 (1977).

35. J. M. Gallagher and R. Haydock, Phil. Mag., 35, 845 (1977).

36. P. B. Hitchcock, R. Mason and M. Textor, J. Chem. Soc. Chem. Comm., 1047 (1976).

37. P. W. Andersen, Phys. Rev., 181, 25 (1969).

38. D. W. Bullett, J. Phys.-Solid State Physics, C8, 2695, 2707 (1975).

39. C. Kittel, Quantum Theory of Solids, Wiley, New York 1963.

40. N. W. Ashcroft and N. D. Mermin, Solid State Physics, Holt Reinhart Winston, 1976, Ch.2.

41. D. M. P. Mingos, J. Chem. Soc. Dalton, 1163 (1976).

42. A. D. Yoffe, private communication (1979).

43. O. K. Andersen, Phys. Rev., B2, 883 (1970).

44. O. Gunnarsson, J. Phys.-Metal Physics, F6, 587 (1976).

45. R. Mason, K. M. Thomas and D. M. P. Mingos, J. Am. Chem. Soc., 95, 3802 (1973).

46. D. W. Bullett, E. P. O'Reilly and R. G. Woolley, unpublished calculations.

47. C. E. Housecroft, K. Wade and B. C. Smith, J. Chem. Soc. Chem. Comm., 765 (1978).

48. J. A. Connor, Topics Current Chem., 71, 71 (1977).

49. R. B. King, J. Am. Chem. Soc., 94, 95 (1972).

50. F. C. Frank and J. S. Kasper, Acta Cryst., 11, 184 (1958), 12, 483 (1959).

Index

668

CHAPTER 3

CHAPTER 4

CHAPTER 5

CHAPTER 6

CHAPTER 7

Compounds

CHAPTER 8

CHAPTER 9